T0323402

Core Logic

Core Logic

Neil Tennant

OXFORD
UNIVERSITY PRESS

OXFORD
UNIVERSITY PRESS

Great Clarendon Street, Oxford, OX2 6DP,
United Kingdom

Oxford University Press is a department of the University of Oxford.
It furthers the University's objective of excellence in research, scholarship,
and education by publishing worldwide. Oxford is a registered trade mark of
Oxford University Press in the UK and in certain other countries

Published in the United States of America by Oxford University Press
198 Madison Avenue, New York, NY 10016, United States of America

British Library Cataloguing in Publication Data

Data available

Library of Congress Control Number: 2017944010

ISBN 978–0–19–877789–2

Printed and bound by
CPI Group (UK) Ltd, Croydon, CR0 4YY

For Timothy Smiley

Preface

core, *n*. The central or innermost part, the 'heart' of anything.

logic, *n*. The branch of philosophy that treats of the forms of thinking in general, and more especially of inference and of scientific method. (Prof. J. Cook Wilson.) Also, since the work of Gottlob Frege (1848–1925), a formal system using symbolic techniques and mathematical methods to establish truth-values in the physical sciences, in language, and in philosophical argument.

Oxford English Dictionary

Acknowledgments

This book arises from a desire to bring together, between two covers, and in a more systematic way, new material along with old material that would otherwise have had to remain scattered over different venues and dates of publication. The system of Core Logic needs a definitive statement, explanation, and justification in one up-to-date work, and I hope to provide that here. I am grateful to the editor, Peter Momtchiloff, and two anonymous readers for sage advice on how best to reach my intended readership. If I fail to do so, it will be my fault.

For permission to reuse some of the material from my previous publications, I thank Oxford University Press itself, as well as the editors and publishers of *The Review of Symbolic Logic*, *Philosophia Mathematica*, *Notre Dame Journal of Formal Logic*, and *Synthese*.

My work on different aspects of Core Logic has benefited from comments from various scholarly audiences over many years, in Britain, the United States, Europe, and Australasia. I acknowledge with gratitude all the constructive feedback that has helped my work along, after presentations spanning a considerable period.

I am indebted to many other scholars for their private comments, criticism, and encouragement ever since I began developing the ideas set out in this book. Some of them have passed away, as indicated with [†]. I should like to thank Jimmy Altham, Nuel Belnap, John Burgess, Jon Cogburn, Julian Cole, Roy Cook, Roy Dyckhoff, Salvatore Florio, Curtis Franks, Torkel Franzen[†], Harvey Friedman, Dov Gabbay, Nathan Kellen, Seppo Keronen, Robert Kraut, Jonathan Lear, David McCarty, Michael MacRobbie, Robert Meyer[†], Peter Milne, Pierluigi Miraglia, Julien Murzi, Alex Oliver, Adam Podlaskowski, John Pollock[†], Dag Prawitz, Graham Priest, Stephen Read, Patrick Reeder, Alan Robinson[†], Gemma Robles, Gil Sagi, Joseph Salerno, Kevin Scharp, Richard Schienes, Peter Schroeder-Heister, George Schumm, Stewart Shapiro, Wilfried Sieg, John Slaney, Jack Smart[†], Timothy Smiley, Peter Smith, Matthew Souba, Florian Steinberger, Tadeusz Szubka, Joseph Vidal-Rosset, Heinrich Wansing, Alan Weir, and Crispin Wright.

My abiding debt on this topic is to the teaching and writings of Timothy Smiley—the first 'perfectionist' in the sense of Burgess [2005]. His classic paper 'Entailment and Deducibility' (Smiley [1959]) posed the challenge to which this work is a sustained response:

> ... the whole point of logic as an instrument, and the way in which it brings us new knowledge, lies in the contrast between the transitivity of 'entails' and the non-transitivity of 'obviously entails', and all this is lost if transitivity cannot be relied on. Of course if there is an effective way of predicting when transitivity will hold then most of the objection vanishes; ... but I do not even begin to see how the thing might be done in, say, predicate logic ... (p. 242)

I can only hope that Tim would be pleased by the system of Core Logic and its transitivity properties.

There is another person to whom I would like to express my gratitude, but I do not recall her name. She was a student in a large lecture audience in the mid-1970s, in the Appleton Tower at the University of Edinburgh. I was teaching an introductory Logic course that induced considerable anxiety among students who needed to pass it in order to fulfill certain degree requirements. Back then I was using cyclostyled lecture notes, *samizdat* copies of which would pass among students after the initial public offering at 25 pence, and whose material eventually found its way into my book *Natural Logic*. At that time I was content to adopt the usual forms of the rules of natural deduction, including the Absurdity Rule (*Ex Falso Quodlibet*). This student inquired what the consequences would be if one were to drop the Absurdity Rule altogether. Not having given the matter any thought before, I said I did not know, but would look into it. Almost forty years later, here now is my considered reply to this earnest inquirer. I am glad she was not like so many students today, who just wish to learn the set material and maximize their grade in an unquestioning way. I imagine she still lives somewhere near shops that sell shortcake, and different brands of tea ... a place and time so far, far away now for me. Unknown Student, I salute you; and I hope that you might one day recognize yourself from this *memoir*, should you ever be browsing among remaindered volumes in Half-Price Books.

Last, but not least, I want to thank my wife Janel Hall and our sons Henry and Hugo. They keep me grounded by reminding me—without any dithyrambic outbursts—that life is greater than logic.

Contents

Chapter 1

Introduction and Overview

Abstract

As this, the opening chapter, has the word 'Overview' in its title, it may strike the reader as rather odd to provide a further overview of it here; but it is important to let the reader know at the beginning of each chapter what it sets out to achieve. The first chapter is no exception.

§1.1 makes clear that that this is a foundational work, written not just for philosophers of logic, but for logicians and foundationalists generally. The work is concerned to revise the very foundations of modern logic, at its Fregean inception, dealing with the formal first-order language of mathematics. It revisits Gentzen's proof theory in order to build *relevance* into proofs, while leaving intact all the logical power one is entitled to expect of a deductive logic for mathematics and for scientific method generally.

§1.2 stresses the primacy of inference and proof in the logician's scheme of things. It acknowledges the role and value of model theory, or formal semantics, while reserving pride of place for, and primary focus on, our *methods of proof*. It also suggests an important reconstrual of what the completeness of a system of proof with respect to a formal semantics really amounts to.

§1.3 summarizes the main topics in the debate over logical reform—changes to 'local' rules of inference governing logical operators, and/or 'global', or structural, rules of inference supposedly governing deducibility itself. Both constructivization and relevantization (of the *deducibility* relation) loom large in this regard. We discuss how proof systems are constituted by particular choices of rules of inference. We also raise the important issue of the *reflexive stability* of any argument for a particular choice of logic as the 'right' logic.

§1.4 turns to the question of pluralism v. absolutism in choice of logic. It suggests that the informal notion of valid argument is stable and robust enough for us to be able to 'get it right' with our formal system of proof—be this a system for constructive reasoning, or one for non-constructive reasoning.

§1.5 describes the main 'conventional' systems that are now well known from the literature in proof theory, and that will obviously serve as markers in the logical landscape within which the Core systems of logic have to be located. These are the system **C** of Classical Logic, the system **I** of Intuitionistic Logic, and the system **M** of Minimal Logic.

§1.6 describes an important remaining system, the system **R** of Relevance Logic, with its 'relevantized' object-linguistic *conditional connective*. We illustrate core proofs at work by actually deriving all the (many) axioms of **R** within Core Logic or Classical Core Logic, as appropriate.

1.1 The aim of this book

I am setting out to present a definitive, detailed, and up-to-date account of the beautiful, elegant, and powerful system of Core Logic.

Wherever we engage in deductive reasoning—in mathematics or in the empirical testing of scientific theories—we in effect employ Core Logic or its classicized extension. Logicians, foundationalists, mathematicians, and philosophers have yet to recognize clearly that these Core systems *suffice* for those deductive explorations. Moreover, these systems do so while keeping one's reasoning absolutely *relevant*—that is, while sustaining a proper 'connection of subject matter' between the premises and the conclusion of any *proof*. In this work we shall repeatedly stress relevantizing the *deducibility relation*, rather than trying to relevantize the object-linguistic conditional *connective*.

The book describes all the theoretically interesting aspects of Core Logic—philosophical, metamathematical, proof-theoretic, methodological, computational, and revision-theoretic. Here for the first time they are all examined together in a single work. A unified and all-encompassing treatment of Core Logic is called for because its many and various aspects have thus far been dealt with in relative isolation from each other, within different sub-specialist realms of the literature (philosophy of science; philosophical logic; history of logic; proof theory; automated deduction; belief-revision theory; truthmaker theory). Another reason why a unified and more holistic coverage is called for is that no single work has yet drawn together all the different aspects to show how they are mutually illuminating and how they fruitfully interconnect.

My target audience is logicians and philosophers, from advanced undergraduate students upwards. But because of the way their respective disciplines will be contributing to the overall mix the book might appeal also to at least some readers from diverse but neighboring areas such as metamathematics and foundations of mathematics, proof theory, automated deduction, formal epistemology, and the theory of belief-revision. Priority is being given, however, to explaining all our considerations in a manner accessible to non-specialists. The more difficult or extensive formal details (proofs of metatheorems, etc.) at the appropriate level of rigor for technical experts can be found in the author's peer-reviewed journal publications, to which references will be made as necessary. The present work is intended to minimize the risk of having the non-technical reader get bogged down by formal material in the middle of a more informal discussion of philosophical or methodological issues. In those places where technicalities thicken as the dialectic deepens, I shall try to confine them to footnotes, or to briefer subsections that the non-technical reader may skim or skip. I shall, however, present some new formal material in the main text when doing so is justified by the following three considerations: first, it is not overly difficult; second, it really does help in order to clarify the philosophical points and arguments being developed; and third, the formal treatment in question is innovative, and goes beyond what has already appeared in the journal literature.

Providing a unified and all-encompassing account of Core Logic, with *all* its metalogical properties uncontroversially delineated by metatheorems, affords at last a perspective from which to deal effectively with various published criticisms of it (and of its classicized extension). Replies appear here in print for the first time to objections from Dirk Hartmann, from JC Beall and Greg Restall, from John Burgess, and from Arnon Avron and Harvey Friedman. These concern such issues as intersubstitutivity of interdeducibles, transitivity of deduction, intuition and reflective equilibrium, methodological adequacy, burden of proof (on the would-be logical reformer), and the (alleged) indispensability of the rule EFQ (*Ex Falso Quodlibet*) for the formalization of mathematical reasoning. I have purposely postponed publishing my replies to Hartmann, Beall and Restall, and Burgess, in order to put them into a book that describes the properties of Core Logic with the necessary clarity and thoroughness. My reply to Avron and Friedman, although given at some length in various postings on the moderated email list fom@cs.nyu.edu, is also included here in a more summary—and I hope more dispositive—form.

As the reader will already have gleaned from the mention of these few names of living logicians, and can suspect from anticipated mention of other contemporary writers in subsequent pages, this work would appear to rest upon—even if only by way of deviant departure from—what the logical tradition has slowly accumulated and afforded until quite recently. But while engagement with the more current tradition will indeed prove to be useful in locating the main new ideas in this work and providing a broader theoretical framework, I hope the reader will forgive me for offering the following expository conceit, for it will serve to focus attention on the fixed purpose that I have in mind.

I honestly believe that a book like this, with such new ideas as it might contain, could, in principle, have been written by some suitable immediate intellectual successor to the three giants Kurt Gödel, Gerhard Gentzen, and Alfred Tarski of the late 1930s. For all the motivating ingredients, and all the basic methods of solution, had been invented by then. My book can in some ways be thought of as an outcome of prolonged imaginary discussions with those three great figures *back then*. The discussion with Gentzen would have been of ways he might have perfected his proof-theoretic breakthroughs, by tweaking his logical rules and dropping his structural ones. It would also have involved urging him to seek a *Dilution-Elimination* Theorem to match his famous *Hauptsatz*, the Cut-Elimination Theorem (for the sequent calculus). (See §2.3.8 for further development, and §6.17 for the recommended results that Gentzen could have obtained.) The discussion with Tarski would have been about how to perfect his (then) new theory of truth, by reifying truthmakers, so that truth *à la* Tarski would consist in the existence of a truthmaker. (See §8.2.) And the discussion with Gödel would have been about how to rework his reasoning about the unprovability of consistency by employing a logic still complete in his sense, but in which not every inconsistent theory in a given language is the whole language itself. At the forefront of their concerns, as practicing mathematicians and formal logicians, would have been the overriding question of how best *faithfully* to formalize the deductive reasoning carried out by mathematicians. With the 1936 Lewis Paradox (A, $\neg A$; *therefore* B; much more on which anon— see §1.3.2) to focus their minds on the issue of relevance, I would like to think that all the ideas in this work could have been made both vivid and acceptable to these three great figures in modern logic.

1.2 Inference and proof as primary in logic

Logic is *primarily* the study of inference and proof. Some contemporary logicians might balk at this description. They might point to developments over the last century in model theory, or formal semantics, and say that *that*, now, is really 'what logic is about'. They might even complain that proof theory is now a very small and specialized area of logical investigation, somehow not meriting any continued serious study. They might indeed; and today the discipline is all the more impoverished for that. Model theory, or formal semantics, was born rather recently, only well into the 1930s. Ironically, the birth of model theory coincided with the full flowering of proof theory, in what can now be seen, with the benefit of several decades' hindsight, to have been a near-perfect presentation of a formal canon of deductive inference for the languages of mathematics. I am referring, of course, to the pioneering contribution of the aforementioned Gentzen [1934, 1935], who created from whole cloth both the systems of natural deduction, and the closely related sequent calculi.

Logic has *always* drawn its main sustenance from deductive reasoning. Deductions, or proofs, take one from their premises to their conclusions. The whole point of a proof is to convince a rational thinker that *if* its premises are true, *then* its conclusion is true also. To be sure, the various 'ways' in which a sentence might be true are studied in formal semantics (or *should* be, if properly pursued). But for all that, investigations in formal semantics really only *complement* the primary focus of logic, which is on *proofs*—studying every aspect of their structure.

From the modern point of view, we can acknowledge the importance of formal semantics, and of the soundness and completeness results that can be established for a given proof system with respect to a particular formal semantics for the language involved. What we want to know, first and foremost, is how to deduce conclusions from collections of premises that 'really do' logically imply them (in terms of a formal semantics that respects the meanings of logical expressions). When we succeed in providing a proof (in a sound system), we are done. The proof bears witness to the validity of the argument in question; the premises of the argument can be seen to have the conclusion as a logical consequence. But, if there is *no* proof to be had, there is something else we can furnish, in order to witness the *absence* of any proof: namely, we can describe or construct a *counterexample* to the argument. This will be a complex, perhaps infinite, mathematical object defined by the formal semantics (a 'model', or an 'interpretation'). It 'makes the premises true' while 'making the conclusion false'. The completeness theorem for a system of proof with respect to a formal semantics assures us that any non-provable argument can be given a counterexample in this sense.

Note, however, that a completeness theorem can hold even if the formal semantics is highly artificial, providing no philosophical illumination of matters to do with truth, meaning, assertability, probability, accuracy, reliability, warrant, or any remotely connected normative notion involved in explaining what we are aiming at in our deductive reasoning, and how we can be understood as properly achieving those aims. The literature abounds with varieties of formal semantics that are, as it were, merely combinatorial in their ability to diagnose unprovability or non-deducibility in a particular system of proof. We genuinely prize, by contrast, those varieties of formal

semantics that provide, in the useful word of Susan Haack, some explanatory
or interpretative 'patter' on how the language actually works (Haack [1978],
at p. 190).

The foregoing is very much a post-Gödelian way to appreciate the service
rendered by formal semantics. It is also a way based on a conception of
completeness that has become conventional to the point of being set in
stone. It is expressed formally (in the metalanguage) as follows:

$$\forall\Delta\,\forall\varphi\,(\Delta\models\varphi \;\Rightarrow\; \Delta\vdash\varphi).$$

Here Δ ranges over sets of sentences, and φ over sentences; $\models$ is the orthodox
notion of classical logical consequence due to Tarski; and $\vdash$ is the relation
of deducibility in any of a number of proof systems, from Frege through
Hilbert to Gentzen, that embody irrelevance by incorporating the first Lewis
Paradox. Note also that $\Delta\vdash\varphi$ abbreviates the claim that there is a proof
of φ from premises in Δ. Such a proof does not have to use *all* the available
premises in Δ. When we have a particular proof system $\mathcal{S}$ in mind, we shall
frequently write $\Delta\vdash_{\mathcal{S}}\varphi$ for deducibility *in that system*.

It should not escape notice that we may have infinite Δ in instances of the
assumption $\Delta\models\varphi$ in the foregoing conventional statement of completeness.
So, since proofs themselves are finite, that statement of completeness can
be rendered more accurately as follows:

$$\forall\Delta\,\forall\varphi\,(\Delta\models\varphi \;\Rightarrow\; \exists\text{ finite }\Delta'\subseteq\Delta\text{ s.t. }\Delta'\vdash\varphi),$$

or, equivalently:

$$\forall\Delta\,\forall\varphi\,(\Delta\models\varphi \;\Rightarrow\; \exists\text{ finite }\Delta'\subseteq\Delta\text{ s.t. the sequent }\Delta':\varphi\text{ has a proof}).$$

And this in turn is to say the following:

> Every sequent in the extension of $\models$ has a subsequent in the
> extension of $\vdash_{\mathcal{S}}$

—where, note, the conventional notion of subsequent that is in use is un-
naturally asymmetrical. For the conventional notion of subsequent (the one
that is involved in the conventional construal of completeness) has it that a
sequent $\Delta':\varphi$ can count as a subsequent of $\Delta:\varphi$ only by having $\Delta'\subseteq\Delta$.
On this unnatural reading *any 'subsequent' of $\Delta':\varphi$ still has to have the
very same conclusion φ*. A more natural and symmetrical approach would
be to count $\Delta':\bot$ *also* (note the conclusion $\bot$!) as a legitimate subsequent
of $\Delta:\varphi$ when $\Delta'\subseteq\Delta$. This is because:

1. $\Delta : \varphi$ is really short for $\Delta : \{\varphi\}$;

2. $\Delta' : \perp$ is short for $\Delta' : \emptyset$; and

3. $\emptyset \subseteq \{\varphi\}$.

So the completeness of a proof system S with respect to even the orthodox logical consequence relation $\models$ would amount to no more than this:

> Every sequent in the extension of $\models$ has a subsequent in the extension of $\vdash_S$.

Thus one may continue, if one wishes, to construe $\models$ in the conventional way. Yet *even with that concession*, for our proof system S we can take the system $\mathbb{C}^+$ of Classical Core Logic—its proofs are *relevant*, and it does not contain the Lewis Paradox.

When Aristotle began the study of logic in his *Prior Analytics*, proofs of syllogisms were his main topic. When Frege launched modern logic with his *Begriffsschrift* in 1879, it was again with unrelenting focus on the nature of fully formalized, logically watertight passages of deductive reasoning— once again, on *proofs*. They look rather rebarbative by today's standards, to be sure; but they were completely rigorous, effectively checkable finite objects serving a crucial epistemic purpose: that of *demonstrating* preservation of truth from premises to conclusions. The difference from Aristotle at this more mature stage was that the Fregean proofs involved sentences of impressive expressive power compared to those that Aristotle had treated. At long last logicians could master the intricacies of the logical expressions 'some' and 'all', whose iterative interplay had been applied to such great effect in nineteenth-century real analysis. The pioneering contributions of both Augustin-Louis Cauchy and, shortly after him, Karl Weierstraß, made rigorous sense of the notions of limit, convergence, sums of infinite series, continuity, uniform continuity, derivatives and integrals. For the first time in the history of logic, logicians had supplied a formal language in which, arguably, every recognizably mathematical statement could be rigorously expressed (particularly after the development of set theory). And they had a formal system of proof in which, arguably, every convincing 'informal' mathematical proof could be fully formalized.

Logic had achieved its apotheosis as a formal science. Its primary and enduring *raison d'être* can now be transformed into the following exigent but tractable challenge: provide natural, homologous and completely rigorous formalizations of 'informal' but suasive proofs and disproofs. These arise, in the main, in mathematics and in scientific theorizing. And the formal proofs that 'regiment' them in an optimally structure-respecting and structure-revealing way will be the sorts of formal objects to be constructed and checked in a *computational* science of deductive reasoning. This is the area of formal study in which we harness the power of electronic computers to make short work of proof *discovery* and proof *verification*. Proofs are *finite* objects, so computer programs can handle them. Models, by contrast, are *infinite* structures in the main, so computer programs *cannot* handle them. Thus another irony is that the birth of model theory coincided, more or less, with the birth of the theory of computable and recursive functions, leading eventually to computational logic. And the theory of recursive functions furnished Gödel's argument for the incompleteness of any consistent, sufficiently strong, axiomatizable theory of the natural numbers.

The most obvious things for logic to study—actual proofs and disproofs— are available in abundance. They are owed to our current experts or their intellectual forebears. These proofs and disproofs have of course been written up in a more *informal* way than would satisfy the formal logician. They provide, however, a great deal of raw material for the logician both to treat as data and eventually to vindicate (by formalizing them as proofs whose correctness, down to the last primitive detail, can be mechanically checked). There are in addition infinitely many other *potential* proofs and disproofs— 'potential' in the sense that they exist as finitary formal objects, but have not yet been tokened by any rational thinker. They await discovery by contemporary experts or their intellectual successors.

A formal logic is at heart a system of proof. It has to answer to deductive practice, but without simply recapitulating it in more painstaking fashion. A system of proof has to provide for *all* the moves of deductive reasoning that we might encounter in our mathematical and scientific investigations; and it has to be able to formalize every possible informal proof that would carry conviction among the experts.[1] As a reflective and perfecting discipline, logic involves both *modeling* and *idealization*; and, as we all know, some models and idealizations reveal themselves to be way better than others. The problem of *choice of logic* for an established discourse is that of finding the system that furnishes the best possible model and idealization of expert deductive reasoning within that discourse and that can serve it as a set of

---

[1] This *includes* proofs making important use of diagrams. *Either* one has to find a way to codify such proofs entirely symbolically, using formal sentences only; *or* one has to develop a so-called heterogeneous system of diagrammatic reasoning, in which diagrams themselves feature *as such* within formal proofs in a precise and rigorous way, making the correctness of any heterogeneous proof an effectively decidable matter. See Barwise and Etchemendy [1996].

norms, or canon.[2]

Logic seeks both to *atomicize* and to *anatomize* the structure of deductions.

By *atomicizing* we mean *exhibiting the simplest parts that cannot be split any further into subparts.* The logician must identify the atomic steps of deduction—the steps that are *absolutely basic* and that cannot themselves be derived by interpolating yet more steps between their starting points and their endpoints.

By *anatomizing* we mean *revealing the overall configuration, identifying the parts within the whole and their roles therein.* The logician must characterize the ideal ways in which deductive steps can be arranged so as to effect the transition from given starting points to deductively 'distant' endpoints. The starting points of any proof are its *premises* and its endpoint is the *conclusion.*

A completely convincing arrangement of steps from premises to conclusion is called a *proof* (or, if the conclusion is absurdity, a *disproof*). Premises and conclusions are *sentences* in a formal language. When a proof is given, one can *read off from it* both its premises and its conclusion. The proof *shows* beyond any doubt that *if* its premises are true, *then* its conclusion is true also. That is, the proof shows that the transition *from* those premises *to* that conclusion is *valid.* Should the conclusion be absurdity, the proof shows that the premises *cannot all be true.*

1.3   The debate over logical reform

There is much dispute over which system of logic is the right one—indeed, over whether there could even be such a thing as *the* right logical system rather than a spectrum of logical systems variously suited for different applications in different areas or within different discourses.

Note that the talk here is about *systems* of logic. Having characterized logic as 'the study of inference and proof', one can hardly be understood as suggesting that there might be more than one such study. But there is a plethora of logical *systems* generated by this study. Having made this terminological matter clear, we shall now proceed in the usual careless fashion, and talk of the debate over which logic is the right logic (rather than more

---

[2] There are, of course, formal logics that have been invented without there having been any established discourse for which they might serve as a canon. The logic of the quantifier 'there exist uncountably many x such that ...' is an example.

carefully, of the debate over which system of logic is the right system of logic).[3]

Absolutists about logic regard talk about *the* right logic as justified; *pluralists* have their principled doubts. Among the participants in the absolutism v. pluralism debate are those we can call the *quietists*. They are willing to accept the full canon **C** of Classical Logic. Their opponents— intuitionists and relevantists prominent among them—argue that certain rules of Classical Logic lack validity and have no right to be in the canon. These opponents of quietism—let us call them the *reformists*—are opposed to the quietists in various different ways. And while some reformists—this writer included[4]—are absolutists, yet others are pluralists. Frege was a quietist absolutist. Brouwer was a reformist absolutist. JC Beall and Greg Restall are reformists because they are pluralists. One cannot be a quietist pluralist.

1.3.1   Debate over rules governing logical operators

Intuitionists, on one hand, draw inspiration for their critique of Classical Logic from what are best understood as their requirements of constructivity in mathematical proof. According to the constructivist's notion of existence a mathematical existence claim of the form 'there is a natural number n such that $F(n)$' requires its asserter to be able to provide a justifying instance—a constructively determinable number t for which one can prove (constructively!) that $F(t)$:[5]

$$\frac{F(t)}{\exists n F(n)}$$

This means that one may *not* use the 'back-door', or indirect, reasoning that would be available to a classical mathematician, whereby in order to derive

---

[3] I am grateful to an anonymous referee for making me aware of the need for this clarification.

[4] To be both more precise and more honest: I am a *bifurcated* absolutist reformist. I offer Core Logic to the constructivist and Classical Core Logic to the classicist. If push comes to shove, I side with constructivists; but I can tell a good story about what the classicist is really up to. See §2.4 for further details.

[5] As McCarty [2005] puts it at p. 358:

> The identity badge of intuitionism as a branch of constructive mathematics is the insistence that every rule underwriting an existential statement about numbers $\exists n P(n)$ must provide, if implicitly, an appropriate numerical term t and the knowledge that $P(t)$ holds.

the conclusion that there is a natural number n such that $F(n)$ it would be sufficient simply to assume that *no* natural number has the property F and then (classically!) derive an absurdity (in symbols: $\bot$) from that assumption:

$$\frac{\quad\quad\quad}{\neg\exists n F(n)}{}^{(i)}$$

$$\vdots$$

$$\frac{\bot}{\exists n F(n)}{}^{(i)}$$

Thus the rule of *Classical Reductio ad Absurdum (CR)* ends up being rejected by intuitionists:

$$\frac{\quad\quad}{\neg\varphi}{}^{(i)}$$

$$\vdots$$

$$\frac{\bot}{\varphi}{}^{(i)}$$

as does any rule equivalent to it *modulo* the set of rules that the intuitionist can eventually motivate or justify in a more direct fashion. Among these intuitionistic equivalents of (*CR*) is the *Law of Excluded Middle (LEM)*:[6]

$$\frac{\quad\quad\quad}{\varphi \vee \neg\varphi}$$

through whose rejection Intuitionistic Logic is perhaps better known.

1.3.2 Debate over EFQ and relevance

Intuitionists reject the Law of Excluded Middle and other equivalent strictly classical rules of inference; but they *retain* the rule *Ex Falso Quodlibet* (EFQ), also known as the Absurdity Rule:

$$\frac{\bot}{\varphi}$$

This allows one to infer *any* conclusion one wishes as soon as one has derived an absurdity.

---

[6] The inference stroke over $\varphi \vee \neg\varphi$, with nothing above it, indicates that $\varphi \vee \neg\varphi$ is taken (by the Classical logician) to be assertible outright, without any need for proof. $\varphi \vee \neg\varphi$ is a proof of itself from nothing.

This residual rule within Intuitionistic Logic **I** is anathema to relevantists, since it affords an easy proof of the infamous 'paradox of irrelevance' mentioned above, known as *Lewis's First Paradox*:

$$A, \neg A : B$$

Its intuitionistic (hence also classical) proof is

$$\frac{\dfrac{A \quad \neg A}{\bot}}{B}$$

Note how this paradox does not explicitly involve the object-linguistic conditional →. It arises in the form of an intuitively unacceptable *proof*, which is responsible for irrelevance infecting the associated relation ⊢ of *deducibility*.

Relevantists refuse to accept Lewis's First Paradox on the grounds that there need not be any connection in meaning between the sentence A in its premises and its conclusion B. Relevantists regard such a 'lack of relevance' between the premises and conclusions of certain classically (and intuitionistically) approved inferential transitions as compromising their claim to inclusion in the deductive canon. One need not spill any ink on the controversial question whether the inferential transition from $\neg A, A$ to (arbitrary) B is *genuinely* valid. One need only respect the strong relevantist intuition that such moves, surely, have no place at all in any proof system aspiring to codify the deductive reasoning in which mathematicans actually engage.

Many relevantists are still otherwise *classical* in their orientation in endorsing (as relevantly valid) such inferences as *Double Negation Elimination* (*DNE*), another intuitionistic equivalent of *CR* (and of *LEM*):

$$\frac{\neg\neg\varphi}{\varphi}$$

The intuition that deductive reasoning should be relevant is a very strong one. It is well known to teachers of logic, who find that their beginning students balk at EFQ, and are often highly resistant to attempts to 'brainwash them into the tradition'. Here significant work needs to be done to vindicate the good sense behind the acolyte's intuition of relevance. How does one explain the 'genuineness' of connection between premises and conclusion of any suasive mathematical proof? What is it, over and above the rather thin notion of validity-by-virtue-of-truth-preservation, that the relevantist is demanding? The first stab at an answer is: some kind of *connection via sharing of expressions* between the premises and the conclusion of any genuinely valid argument. One is intuitively convinced that such a connection is mediated by, or made manifest in, the proofs that we provide to 'move us in thought' from the premises to the conclusion. One never crosses any *thematic gulf* when reasoning deductively; one never moves from premises involving certain extra-logical expressions to a conclusion involving none of them (but involving quite different ones instead). The basic intuition is that such connections are always present in mathematical reasoning. Mathematicians are *never* forced to resort to the 'irrelevantist excesses' available within both Intuitionistic and Classical Logic. The challenge of explicating such a connection between the premises and the conclusion of a proof, and showing that Classical Core Logic (hence also Core Logic) *always* respects the need for it, is met in Chapter 10.

It has been alleged that 'the case for relevance isn't easy to make'. The best way to answer this is to lay one's methodological cards on the table.

> The case is straightforward. We wish to avoid the positive and negative forms of the First Lewis Paradox: 'A, $\neg A$, so B'; and 'A, $\neg A$, so $\neg B$'. Nothing could be simpler by way of motivation. Avoid *those*, and everything falls into place, as we learn from the Relevance Metatheorem 13 for Classical Core Logic that is explained in Chapter 10. We are not being diverted by captious considerations about 'acceptability in English' (or any other natural language). We are simply seeking the best possible regimentation of *mathematical* reasoning, which is (i) committed to avoiding contradictions, and (ii) tenaciously relevant (i.e. mathematical axioms are used to prove mathematical theorems by wholly relevant means).

It has also been suggested that 'the reader might think that the standard arguments for relevance require that $\to$ be relevant, so that the Anderson–Belnap system **R** (see §1.6), which relevantizes $\to$, is after all preferable to $\mathbb{C}^+$, which relevantizes $\vdash$.' The answer to *this* surmise is as follows.

> The reader is invited to attend carefully to the metatheorems that will be established about $\mathbb{C}^+$ in this work, which establish its adequacy for classical mathematical reasoning; and then to inquire whether there have been any metatheorems to remotely similar effect concerning **R**. (Actually, there have been *negative* results in this regard—see Friedman and Meyer [1992].) We might also point out the further irony that the Relevance

Metatheorem for Classical Core Logic that is explained in Chapter 10 holds for **R**, because every sequent provable in **R** is provable in $\mathbb{C}^+$. But not a single variable-sharing result proved for **R** or any of its subsystems by workers who focused on $\to$ has come close in strength to our Relevance Metatheorem.

1.3.3 The two main lines of reform of Classical Logic

Both of the main reformist camps—intuitionist and relevantist—have variously produced philosophical, methodological, meaning-theoretic and intuitive considerations in support of their respective recommendations for restricting Classical Logic. Those recommendations can involve eschewing or restricting the range of application of various rules among those already mentioned, including the structural ones.

Classical Logic **C** is of course the orthodox system that is best known, most widely taught and used, and most deeply entrenched. So one is virtually obliged to take it as a reference point when explaining what Core Logic is, by contrast. Core Logic lies at the intersection of what might be thought of as two orthogonal lines of logical reform of Classical Logic: *constructivization* and *relevantization*.[7] Conceived this way, Core Logic $\mathbb{C}$ might appear to be a mere *residue* of sacrificial reforms, beginning with Classical Logic **C**. Core Logic would be thought of as what is left over when one eschews certain principles of classical reasoning (both those that are not constructive, and those that embody irrelevancies). But this impression, unsupplemented by any other perspective on how Core Logic might be the canon of choice, would be mistaken.

To the classical mathematician who insists on being able to use strictly classical methods of deductive reasoning, one can offer the *classicized extension* of Core Logic, the system $\mathbb{C}^+$ mentioned above. The important point is that each of these Core systems (constructive or classicized) ensures *relevance* of premises to conclusions of proofs in the same principled way. Moreover, for the reader as yet unacquainted with the systems of Core Logic

---

[7] For those concerned to track matters of provenance, I should mention that Core Logic is the system that I formerly called IR (Intuitionistic Relevant Logic), but now with what Tennant [1987a] called the *weak* conditional (i.e. the Fregean conditional—for which, in particular, the following two inferences hold: B, therefore $A \to B$; and $\neg A$, therefore $A \to B$). The new name 'Core Logic' was introduced in the two-part study Tennant [2014b] and Tennant [2014c]. The rules of Core Logic were stated in full in both Tennant [1992] and Tennant [1994b]—but there using the label IR. Also, the classical extension of the system IR—formerly called CR, as in Tennant [1984], but furnished since then with the weak conditional—is now called Classical Core Logic.

and Classical Core Logic, it should be emphasized at the outset that the *way* these systems achieve relevance of premises to conclusions of their proofs is *utterly different* from the ways explored so far in the 'relevantist tradition' of Anderson and Belnap. To put it in a nutshell: the Core systems relevantize at 'the level of the turnstile', rather than via the object-linguistic conditional connective.[8] (Classical) Core Logic does not achieve its status as a relevance logic *simply* by being paraconsistent (i.e. simply by avoiding explosion). Rather, it achieves that status by having rules of inference so carefully crafted that a remarkable 'relation of relevance' $\mathcal{R}(\Delta, \varphi)$ holds between the premise set Δ and the conclusion φ of any (classical) core proof. This relation of relevance goes well beyond anything previously conceived by way of a 'variable-sharing' requirement.[9]

1.3.4 Structural rules in the sequent calculi

Thus far we have mentioned only rules that deal explicitly with occurrences of logical constants or operators. There is another species of rules, however, called *structural rules*. These are stated without any mention of logical operators. In order to state them we need the notion of a *finite sequent*. A finite sequent is of the form $\Delta : \varphi$, where the *antecedent* Δ is a finite *set* (not a *sequence*) of premises, and the conclusion φ is a single sentence. (When context makes clear that only finite sequents are under discussion, we suppress the adjective 'finite'.) As a special case we allow for $\perp$ (absurdity) to feature as the conclusion of a sequent. If one insists on defining a sequent as having a *set* on the right of its colon (as its *succedent*), then the succedent is 'at most a singleton'. That is, either it is of the form $\{\varphi\}$ or it is $\emptyset$, the empty set. The latter alternative is identified with $\perp$. We also write 'Δ, Γ' for '$\Delta \cup \Gamma$'; 'φ' for '$\{\varphi\}$'; hence also 'Δ, φ' for '$\Delta \cup \{\varphi\}$'.

There are three important structural rules in the continuing orthodox tradition that dates all the way back to Gentzen and that can be stated using sequents of the form just described:

---

[8] It is rather unfortunate that Priest [2002], in an explicit attempt to 'clarify the relationship betweeen relevant logics and paraconsistent logic' wrote (at p. 289) that '[t]he motivating concern of relevant logic is ... to avoid paradoxes of the *conditional*.' (Emphasis added.) Core Logic and Classical Core Logic are *both* paraconsistent *and* relevant, but with relevance vested in the deducibility relation $\vdash$ rather than in the conditional $\rightarrow$.

[9] 'Variable sharing' is the term of art employed for systems of *propositional* logic, for which 'variable' means 'propositional atom'. But the 'variable' sharing described by the relation $\mathcal{R}$ for Classical Core Logic concerns not just propositional atoms, but also extralogical predicates in the first-order case. See Chapter 10 for details.

Reflexivity	$\varphi : \varphi$	

Thinning $\quad\dfrac{\Delta : \varphi}{\Delta, \Gamma : \varphi} \qquad \dfrac{\Delta : \emptyset}{\Delta : \varphi}$

Cut $\quad\dfrac{\Delta : \varphi \quad \Gamma, \varphi : \psi}{\Delta, \Gamma : \psi}$

As thus stated, these are rules *within* some system S of (sequent) proof, finding application in the construction of (sequent) proofs. They can also be stated, however, in terms of *deducibilities within the system S*, without being envisaged as applicable rules in their own right:

Reflexivity $\quad \varphi \vdash_S \varphi$

Thinning $\quad\dfrac{\Delta \vdash_S \varphi}{\Delta, \Gamma \vdash_S \varphi} \qquad \dfrac{\Delta \vdash_S \emptyset}{\Delta \vdash_S \varphi}$

Cut $\quad\dfrac{\Delta \vdash_S \varphi \quad \Gamma, \varphi \vdash_S \psi}{\Delta, \Gamma \vdash_S \psi}$

One could then inquire whether these structural rules are *admissible for S*, even if not *derivable in S* itself.[10] That is to say, are the *meta*-logical inferences above involving the deducibility-turnstile $\vdash_S$ derivable in the *meta*-logic, even if the *object* logic (the system S in question) cannot furnish proofs of the corresponding rules of Reflexivity, Thinning and Cut that are expressed by means of the colons? One of our main results is that the rule of Cut, in a much more welcome form, is admissible both for Core Logic and for Classical Core Logic—even though not derivable.

1.3.5   On inductive definition of proof-in-system-S

An important word of explanation is in order, at this juncture, of how a proof theorist *defines* the notion of proof-in-system-S, and how such proofs are then appealed to in order to define the notion deducibility-in-system-S

---

[10] The distinction between admissible and derivable rules is due to Hiż [1959].

(i.e. the relation $\vdash_S$ employed above).[11] Let us consider for the time being systems S of natural deduction. (What we say subsequently can easily be tweaked in minor details so as to apply just as well to sequent calculi.) The notion of proof-in-system-S is a *ternary* one, which can be formally expressed (in the metalanguage) as follows:

$$\mathcal{P}_S(\Pi, \Delta, \varphi).$$

This means that Δ is the exact set of premises actually used, and remaining undischarged, in the proof Π of the conclusion φ. The proof Π is constructed in accordance with the rules of the system S. As a special case, one can have $\bot$ as φ. Then Π would be more felicitously described as a *dis*proof of Δ.

We treat *axioms* as 'zero-premise' rules. The Law of Excluded Middle is, in fact, the only axiom we ever consider adding to any stock of otherwise ordinary rules that call for at least one subproof in their applications to form new proofs.

Whatever the system S might be, it is always to be characterized as a *set of rules*. The definition of the notion $\mathcal{P}_S(\Pi, \Delta, \varphi)$ is always an *inductive* definition. Every such definition has a simple *Basis Clause*:

$$\mathcal{P}_S(\varphi, \varphi, \{\varphi\})$$

(i.e. every sentence is a proof of itself as conclusion from itself as sole premise); and the familiar *Closure Clause*:

> If $\mathcal{P}_S(\Pi, \Delta, \varphi)$ holds, then this can be shown by appeal to the Basis Clause, and to the inductive clauses for S-proof formation that respectively correspond to the rules of the system S.

The proof system S is then essentially *identified* by providing the remaining clauses in the inductive definition of $\mathcal{P}_S(\Pi, \Delta, \varphi)$. These are the *inductive clauses* themselves, which correspond to those rules of inference 'of S' that we usually state in the more familiar graphic form. *That* set of rules *constitutes* the system S in question.

1.3.6 Reflexive stability

There is also the important question of the *reflexive stability* of one's whole undertaking in arguing for a particular choice of logic as the 'right' logic.

---

[11] This was explained in Tennant [1978].

It is the question whether the logic commended to the reader is one within which the commender has been able to justify his commendation. It is a well-known general phenomenon that one can argue for a particular feature to be present in an object-logic only by using that very same feature within one's metalogic. Thus when Tarski proved that every sentence of the object language is true or false (on his famous definition of truth), he used the Law of Excluded Middle in the metalanguage. (Indeed, he had to!) So we have to proceed here by avoiding any use of EFQ at all in the metalanguage, and slowly develop our whole theory about Core Logic for the object language without drawing attention to the fact that one is never using EFQ. Then, at the end of it all, one can point out that one has never needed it. This indeed is the approach in this book. All our metareasoning is formalizable in Core Logic, since it is constructive. And if one were to be pressed to provide a Henkin-style completeness proof for $\mathbb{C}^+$, it would, happily, be formalizable in meta-$\mathbb{C}^+$.

1.4 On pluralism about logic, and the explication of deductive validity

Some thinkers react to the debate among absolutists over the correct choice of logic by taking a more detached, but still informed, position on the sidelines. They are inclined to think that different logics might be employed for different theoretical purposes—that there is no single correct logic *tout court.*

In explaining the appeal of pluralism about logics, Stewart Shapiro writes (Shapiro [2014], at pp. 1–2)

> ... there are different, mutually incompatible, but equally legitimate ways to sharpen or further articulate the notion(s) of *logical consequence and validity.* (Emphasis added.)

But imagine if one were to say instead the following, which lies near at hand:

> ... there are different, mutually incompatible, but equally legitimate ways to sharpen or further articulate the notion of *computable function* (from natural numbers to natural numbers).

Hardly any foundationalist would accept this analog of Shapiro's claim. For there is widespread acceptance of the Turing–Church Thesis: The computable functions (on the natural numbers) are exactly those that can be computed by a Turing machine. And the question now arises: why should the notion of *valid argument* be any less robust and stable than the notion of *computable function on the natural numbers*?

We need an explication of logical consequence and validity that is as definitive as the *provably coextensive explications* that were provided in the 1930s for the notion of computable function by such thinkers as Turing, Gödel, Church, Kleene, and Post. The metatheorems that demonstrated that any two of their different formal explications are coextensive convinced foundationalists that computability on the natural numbers is a stable and robust notion; they provide convincing informal evidence for the truth of the Turing–Church Thesis. What I am maintaining by way of analogy is this: We need to find *different lines of thought* that all *converge on the same stable and robust logical system*, as an explication of deductive validity.[12]

We must be prepared to take the first well-known approximations thus far—Classical Logic **C**, Intuitionistic Logic **I**, Minimal Logic **M**, the relevance logic **R** of Anderson and Belnap—as *mere guides* that have brought us to the *general vicinity* of the correct system. (We hesitate to speak of 'Relevance Logic **R**', since that would seem to carry an untoward connotation of exclusivity of **R** *qua* relevance logic. We also hasten to reassure the reader that all of the systems just mentioned will be described in detail in §1.5 and §1.6.)

Let us turn now to the task of specifying exactly each of these systems, as presented within the recent tradition.

1.5   The familiar systems M, I, and C

In §1.2 we saw some examples of rules of inference stated in the graphic form familiar from the Gentzen–Prawitz treatment of natural deduction.

Here we shall provide all such rules for the systems **M**, **I**, and **C**. But in our statement of them we shall supply an important new notational ingredient not featuring explicitly in the Gentzen–Prawitz treatment. This new ingredient takes the form of annotating the discharge-strokes that are placed above assumptions to be discharged by an application of the rule in question. A *diamond* ◇ will indicate that the subproof in question *need not* have the sentence in question among its undischarged assumptions; but, if it does, then all as-yet-undischarged occurrences of that assumption are discharged

---

[12] Note how Etchemendy [1990], too, draws a parallel between our attempt to understand the notion of logical consequence and our attempt to understand the notion of computable function. He proposes (p. 6) that there should be a logicians' analog of Church's Thesis, which he calls *Hilbert's Thesis*, to the effect that 'all and only logically valid arguments of a given language are provable within a given deductive system.'

by applying the rule. We may summarize by saying: *a diamond means that vacuous discharge of assumptions is allowed*. A *box* □, by contrast, will indicate that vacuous discharge is *prohibited*: there really must be, in the subordinate proof in question, an undischarged assumption of the indicated form in order for the rule to find application at all. But of course when the rule is applied all as-yet-undischarged occurrences of that assumption are discharged.

We shall state alongside each natural-deduction rule the corresponding rule of the *sequent* calculus. To each Introduction rule there corresponds the 'Right' rule of the sequent calculus; while to each Elimination rule there corresponds the 'Left' rule. In any Elimination rule (@E) the sentence immediately above the inference stroke with the logical operator @ dominant is called the *major premise for the elimination* (abbreviated as 'MPE').

In the graphic statement of natural-deduction rules, descending dots indicate the possible presence of non-trivial proof work. They are absent from the treatment in Prawitz [1965], but faithful to his demonstrated intent. We have a reason for supplying them here: one of the important changes that will be proposed by the Core logician is that in certain important cases, namely above MPEs, *there should be no descending dots*. The reason why will emerge in due course.

In the sequent formulations of **M**, **I**, and **C** one has the three structural rules mentioned in §1.2: Reflexivity, Thinning and Cut.

1.5.1   The system M of Minimal Logic

Minimal Logic is determined by the following Introduction and Elimination rules (left column for natural deduction; right column for sequent calculus).

$$
(\neg\text{I})\qquad
\begin{array}{c}
\diamond\!\!-\!\!(i)\\
\varphi\\
\vdots\\
\dfrac{\bot}{\neg\varphi}\,(i)
\end{array}
\qquad\qquad
(\neg R)\qquad
\dfrac{\Delta:}{\Delta\setminus\{\varphi\}:\neg\varphi}
$$

$$
(\neg\text{E})\qquad
\dfrac{\begin{array}{cc}\vdots & \vdots\\ \neg\varphi & \varphi\end{array}}{\bot}
\qquad\qquad
(\neg L)\qquad
\dfrac{\Delta:\varphi}{\neg\varphi,\Delta:}
$$

$$(\wedge I) \quad \frac{\begin{matrix} \vdots & \vdots \\ \varphi & \psi \end{matrix}}{\varphi \wedge \psi} \qquad\qquad (\wedge R) \quad \frac{\Delta : \varphi \qquad \Gamma : \psi}{\Delta, \Gamma : \varphi \wedge \psi}$$

$$(\wedge E) \quad \frac{\begin{matrix} \vdots \\ \varphi \wedge \psi \end{matrix}}{\varphi} \quad \frac{\begin{matrix} \vdots \\ \varphi \wedge \psi \end{matrix}}{\psi} \qquad\qquad (\wedge L) \quad \frac{\Delta : \theta}{\varphi \wedge \psi, \Delta \setminus \{\varphi, \psi\} : \theta}$$
$$\text{where } \Delta \cap \{\varphi, \psi\} \neq \emptyset$$

$$(\vee I) \quad \frac{\begin{matrix} \vdots \\ \varphi \end{matrix}}{\varphi \vee \psi} \quad \frac{\begin{matrix} \vdots \\ \psi \end{matrix}}{\varphi \vee \psi} \qquad\qquad (\vee R) \quad \frac{\Delta : \varphi}{\Delta : \varphi \vee \psi} \quad \frac{\Delta : \psi}{\Delta : \varphi \vee \psi}$$

$$(\vee E) \quad \frac{\varphi \vee \psi \quad \begin{matrix} \square\!-\!(i) \\ \varphi \\ \vdots \\ \theta \end{matrix} \quad \begin{matrix} \square\!-\!(i) \\ \psi \\ \vdots \\ \theta \end{matrix}}{\theta}\!-\!(i) \qquad\qquad (\vee L) \quad \frac{\varphi, \Delta : \theta \qquad \psi, \Gamma : \theta}{\varphi \vee \psi, \Delta, \Gamma : \theta}$$

$$(\rightarrow I) \quad \frac{\begin{matrix} \diamond\!-\!(i) \\ \varphi \\ \vdots \\ \psi \end{matrix}}{\varphi \rightarrow \psi}\!-\!(i) \qquad\qquad (\rightarrow R) \quad \frac{\Delta : \psi}{\Delta \setminus \{\varphi\} : \varphi \rightarrow \psi}$$

$$(\rightarrow E) \quad \frac{\begin{matrix} \vdots \\ \varphi \rightarrow \psi \end{matrix} \quad \begin{matrix} \vdots \\ \varphi \end{matrix}}{\psi} \qquad\qquad (\rightarrow L) \quad \frac{\Delta : \varphi \qquad \psi, \Gamma : \theta}{\varphi \rightarrow \psi, \Delta, \Gamma : \theta}$$

$$(\exists I) \quad \frac{\begin{matrix} \vdots \\ \varphi_t^x \end{matrix}}{\exists x \varphi} \qquad\qquad (\exists R) \quad \frac{\Delta : \varphi_t^x}{\Delta : \exists x \varphi}$$

$$(\exists E) \qquad \overbrace{@ \dots \varphi_a^x \dots @}^{\square\!-\!(i)}$$

$$\begin{array}{ccc} & \vdots & \vdots \\ & \exists x \varphi \,^{@} & \psi \,^{@} \\ \cline{2-3} & & \multicolumn{1}{c}{-(i)} \\ & \psi & \end{array}$$

$$(\exists L) \qquad \dfrac{\varphi_a^x, \Delta : \psi}{\exists x \varphi, \Delta : \psi} \ @$$

$$(\forall I) \qquad \begin{array}{c} @ \\ \vdots \\ \varphi \\ \hline \forall x \varphi_x^a \end{array}$$

$$(\forall R) \qquad \dfrac{\Delta : \varphi}{\Delta : \forall x \varphi_x^a} \ @$$

$$(\forall E) \qquad \begin{array}{c} \vdots \\ \forall x \varphi \\ \hline \varphi_t^x \end{array}$$

$$(\forall L) \qquad \dfrac{\varphi_{t_1}^x, \ \dots, \ \varphi_{t_n}^x, \Delta : \theta}{\forall x \varphi, \Delta : \theta}$$

The notation $@$ indicates that the parameter a is not allowed to occur in the indicated places—in the implicitly indicated undischarged assumptions, or in the explicitly displayed sentence or sequent to which it is appended. These parametric prohibitions *never vary* across the different proof systems. Indeed, the quantifier rules are the *most stable* of all. They are *never* revised or modified in any way that might affect the resulting field of the deducibility relation. Those variations are attributable *only* to the logical behavior of the *connectives*, as determined by *their* rules and by the presence or absence of certain structural rules.

1.5.2   The system I of Intuitionistic Logic

Intuitionistic Logic is determined by the rules of Minimal Logic, plus the rule *Ex Falso Quodlibet* (also sometimes called *Ex Contradictione Quodlibet*):

$$(EFQ) \qquad \begin{array}{c} \vdots \\ \bot \\ \hline \varphi \end{array}$$

$$(T_R) \qquad \dfrac{\Delta :}{\Delta : \varphi} \qquad \begin{array}{l} \text{'Thinning} \\ \text{on the Right'} \end{array}$$

Note that both EFQ and Thinning on the Right are 'operatorless' rules. *They are both stated without any explicit mention of any particular logical operator.*

1.5.3 The system C of Classical Logic

Classical Logic is determined by the rules of Intuitionistic Logic, plus *either* the Rule of Dilemma:

$$
\begin{array}{cc}
\square\!\!-\!\!(i) & \square\!\!-\!\!(i) \\
\varphi & \neg\varphi \\
\vdots & \vdots \\
\psi & \psi \\
\end{array}
$$

(Dil) $\qquad\qquad$ (Dil) $\dfrac{\varphi, \Delta : \psi \qquad \neg\varphi, \Gamma : \psi}{\Delta, \Gamma : \psi}$

$$
\dfrac{}{\psi}\ (i)
$$

or the rule of Classical Reductio (note the *vacuous* discharge it involves):

$$
\begin{array}{c}
\diamond\!\!-\!\!(i) \\
\neg\varphi \\
\vdots \\
\bot\ (i) \\
\varphi
\end{array}
$$

(CR) $\qquad\qquad$ (CR) $\dfrac{\Delta :}{\Delta \setminus \{\neg\varphi\} : \varphi}$

or the rule of Double Negation Elimination:

$$
\begin{array}{c}
\vdots \\
\neg\neg\varphi \\
\varphi
\end{array}
$$

(DNE) $\qquad\qquad$ (DNE) $\dfrac{\Delta : \neg\neg\varphi}{\Delta : \varphi}$

or the Law of Excluded Middle:

(LEM) $\dfrac{}{\varphi \vee \neg\varphi}$ $\qquad\qquad$ (LEM) $\dfrac{\varphi\vee\neg\varphi, \Delta : \psi}{\Delta : \psi}$

Because (CR) allows vacuous discharge of assumptions, such applications of it are in effect applications of EFQ (in natural deduction) or of Thinning on the Right (in the sequent calculus).

One can also eschew the foregoing sequent-rule versions of (Dil), (CR), (DNE), and (LEM), and adapt the sequent calculus to the needs of the Classical logician by exploiting Gentzen's trick: allow for *more than one* sentence to occur on the right of the colon in sequents (i.e. in the so-called *succedents* of sequents). That affords the following quick proofs of (DNE) and (LEM), using only Left and Right rules adapted to the presence of multiple succedents:

$$\frac{\dfrac{A:A}{\emptyset:A,\neg A}}{\neg\neg A:A} \qquad\qquad \frac{\dfrac{A:A}{\emptyset:A,\neg A}}{\emptyset:A\vee\neg A}$$

To summarize this section: We have just seen that with the standard trio **C**, **I**, and **M**, the exact identification of each system is not by means of axioms, but rather by means of rules of inference. We stated those for natural deduction in the Gentzen–Prawitz format. We also characerized the sequent calculi for the three systems.

It will turn out that the Core logician, although favoring an approach based on rules of inference, has some crucial bones to pick within the standard stock of rules of inference involved in this characterization of the systems **C**, **I**, and **M** that we owe to Gentzen and Prawitz. Details will emerge in due course.

1.6 The system R of Relevance Logic

We turn now to the remaining well-known system, the Anderson–Belnap system **R** of relevance logic. Unlike the systems **C**, **I**, and **M**, the system **R** is specified by means of quite a long list of axioms. In §1.6.1 we shall provide a standard axiomatization of **R**. Then in §1.6.2 we shall *derive* each of these axioms by means of core proofs (and, in the case of the one strictly classical axiom, by means of a classical core proof). That will serve to give the reader a foretaste of core proofs, which are constructed in accordance with rules of inference specifically governing the logical operators.

1.6.1 Axioms of R

Here is a standard Hilbert-style axiomatization of the propositional part of the Anderson–Belnap system **R**.[13]

---

[13] See, for example, Mares [2012]. Unlike Mares, we treat the disjunction sign ∨ as a primitive connective. We also replace his one biconditional axiom with the two separate conditionals in each direction.

$A \rightarrow A$

$(A \wedge B) \rightarrow A$

$(A \wedge B) \rightarrow B$

$A \rightarrow (A \vee B)$

$B \rightarrow (A \vee B)$

$(A \rightarrow B) \rightarrow ((B \rightarrow C) \rightarrow (A \rightarrow C))$

$A \rightarrow ((A \rightarrow B) \rightarrow B)$

$(A \rightarrow (A \rightarrow B)) \rightarrow (A \rightarrow B)$

$((A \rightarrow B) \wedge (A \rightarrow C)) \rightarrow (A \rightarrow (B \wedge C))$

$((A \vee B) \rightarrow C) \rightarrow ((A \rightarrow C) \wedge (B \rightarrow C))$

$((A \rightarrow C) \wedge (B \rightarrow C)) \rightarrow ((A \vee B) \rightarrow C)$

$(A \wedge (B \vee C)) \rightarrow ((A \wedge B) \vee (A \wedge C))$

$(A \rightarrow \neg B) \rightarrow (B \rightarrow \neg A)$

$\neg\neg A \rightarrow A$

The only *rules* employed in **R** are a rule of adjunction (from A, B infer $A \wedge B$) and a rule of detachment (from A, $A \rightarrow B$, infer B). The 'real power' of the system is concentrated in the axioms, which are designed to make the object-linguistic conditional connective '$\rightarrow$' express a tigher, more 'relevant' connection between antecedent and consequent than is expressed by the corresponding connective of Classical Logic or Intuitionistic Logic (or Minimal Logic, for that matter).

1.6.2   Core proofs of the axioms of R

Now we provide core proofs of the foregoing **R**-axioms. A precise statement of the rules of inference for Core Logic is yet to be given. (See §4.2.) But the reader should have no trouble following these proofs.

They are of course proofs of *theorems*: their conclusions depend on the empty set of assumptions. They are also *natural deductions*, rather than sequent proofs. We first present trivial proofs of the five simplest **R**-axioms, and then move on to proofs of the more complex ones. The reader unfamiliar with the 'parallelized' forms of $\wedge$-Elimination and $\rightarrow$-Elimination will find it a pleasant diversion to identify both their degenerate and their non-degenerate applications across the proofs that follow. (The parallelized forms are employed in the Core systems.)

$$\cfrac{\cfrac{}{A}(1)}{A \to A}(1)$$

$$\cfrac{\cfrac{(1)\rule{1cm}{0.4pt}}{A \land B}\quad \cfrac{\rule{1cm}{0.4pt}(2)}{A}(2)}{(A \land B) \to A}(1)$$

$$\cfrac{\cfrac{(1)\rule{1cm}{0.4pt}}{A \land B}\quad \cfrac{\rule{1cm}{0.4pt}(2)}{B}(2)}{(A \land B) \to B}(1)$$

$$\cfrac{\cfrac{\cfrac{\rule{1cm}{0.4pt}(1)}{A}}{A \lor B}(1)}{A \to (A \lor B)}\qquad \cfrac{\cfrac{\cfrac{\rule{1cm}{0.4pt}(1)}{B}}{A \lor B}(1)}{B \to (A \lor B)}$$

$$\cfrac{\cfrac{(1)\rule{1cm}{0.4pt}}{A \to B}\quad \cfrac{\cfrac{(3)\rule{0.4cm}{0.4pt}}{A}\quad \cfrac{\cfrac{(2)\rule{1cm}{0.4pt}}{B \to C}\quad \cfrac{\rule{0.4cm}{0.4pt}(4)}{B}}{C}(4)\quad \cfrac{\rule{0.4cm}{0.4pt}(5)}{C}}{}(5)}{\cfrac{\cfrac{\cfrac{C}{A \to C}(3)}{(B \to C) \to (A \to C)}(2)}{(A \to B) \to ((B \to C) \to (A \to C))}(1)}$$

$$\cfrac{\cfrac{\cfrac{(2)\rule{1cm}{0.4pt}}{A \to B}\quad \cfrac{\cfrac{(1)\rule{0.4cm}{0.4pt}}{A}\quad \cfrac{\rule{0.4cm}{0.4pt}(3)}{B}}{B}(3)}{\cfrac{\cfrac{B}{(A \to B) \to B}(2)}{A \to ((A \to B) \to B)}(1)}}{}$$

$$\cfrac{\cfrac{(1)\rule{2cm}{0.4pt}}{A \to (A \to B)}\quad \cfrac{\cfrac{(2)\rule{0.4cm}{0.4pt}}{A}\quad \cfrac{\cfrac{(3)\rule{1cm}{0.4pt}}{A \to B}\quad \cfrac{\cfrac{(2)\rule{0.4cm}{0.4pt}}{A}\quad \cfrac{\rule{0.4cm}{0.4pt}(4)}{B}}{B}(4)}{B}(3)}{}}{\cfrac{\cfrac{B}{A \to B}(2)}{(A \to (A \to B)) \to (A \to B)}(1)}$$

$$\cfrac{\cfrac{\cfrac{\cfrac{\cfrac{\overset{(3)}{\rule{1.5cm}{0.4pt}}}{A \to B} \quad \overset{(2)}{\rule{0.5cm}{0.4pt}}{A}}{B}^{(4)} \quad \cfrac{\overset{(3)}{\rule{1.5cm}{0.4pt}}}{A \to C} \quad \overset{(2)}{\rule{0.5cm}{0.4pt}}{A} \quad \overset{(5)}{\rule{0.5cm}{0.4pt}}{C}}{C}^{(5)}}{B \wedge C}^{(3)}}{\cfrac{B \wedge C}{A \to (B \wedge C)}^{(2)}}}{((A \to B) \wedge (A \to C)) \to (A \to (B \wedge C))}^{(1)}$$

with $\cfrac{}{(A \to B) \wedge (A \to C)}^{(1)}$

$$\cfrac{\cfrac{\cfrac{\cfrac{\overset{(1)}{\rule{2.5cm}{0.4pt}}}{(A \vee B) \to C} \quad \cfrac{\overset{(2)}{\rule{0.4cm}{0.4pt}}{A}}{A \vee B} \quad \overset{(1)}{\rule{0.4cm}{0.4pt}}{C}}{C}^{(4)}}{A \to C}^{(2)} \quad \cfrac{\cfrac{\overset{(1)}{\rule{2.5cm}{0.4pt}}}{(A \vee B) \to C} \quad \cfrac{\overset{(3)}{\rule{0.4cm}{0.4pt}}{B}}{A \vee B} \quad \overset{(5)}{\rule{0.4cm}{0.4pt}}{C}}{C}^{(5)}}{\cfrac{C}{B \to C}^{(3)}}{(A \to C) \wedge (B \to C)}$$

$$\cfrac{(A \to C) \wedge (B \to C)}{((A \vee B) \to C) \to ((A \to C) \wedge (B \to C))}^{(1)}$$

$$\cfrac{\cfrac{\cfrac{\cfrac{\overset{(2)}{\rule{1.5cm}{0.4pt}}}{A \vee B} \quad \cfrac{\overset{(3)}{\rule{1.5cm}{0.4pt}}{A \to C} \quad \overset{(4)}{\rule{0.5cm}{0.4pt}}{A} \quad \overset{(5)}{\rule{0.5cm}{0.4pt}}{C}}{C}^{(5)} \quad \cfrac{\overset{(3)}{\rule{1.5cm}{0.4pt}}{B \to C} \quad \overset{(4)}{\rule{0.5cm}{0.4pt}}{B} \quad \overset{(6)}{\rule{0.5cm}{0.4pt}}{C}}{C}^{(6)}}{C}^{(4)}}{C}^{(3)}}{\cfrac{C}{(A \vee B) \to C}^{(2)}}{((A \to C) \wedge (B \to C)) \to ((A \vee B) \to C)}^{(1)}$$

with $\cfrac{}{(A \to C) \wedge (B \to C)}^{(1)}$

$$
\cfrac{
  \cfrac{
    \cfrac{\;}{B \vee C}^{(2)}
    \quad
    \cfrac{
      \cfrac{\dfrac{\;}{A}^{(2)} \quad \dfrac{\;}{B}^{(3)}}{A \wedge B}
    }{(A \wedge B) \vee (A \wedge C)}
    \quad
    \cfrac{
      \cfrac{\dfrac{\;}{A}^{(2)} \quad \dfrac{\;}{C}^{(3)}}{A \wedge C}
    }{(A \wedge B) \vee (A \wedge C)}^{(3)}
  }{(A \wedge B) \vee (A \wedge C)}^{(2)}
}{
  \cfrac{
    \cfrac{\;}{A \wedge (B \vee C)}^{(1)}
    \quad
    (A \wedge B) \vee (A \wedge C)
  }{(A \wedge B) \vee (A \wedge C)}^{(1)}
}
$$

$$
(A \wedge (B \vee C)) \to ((A \wedge B) \vee (A \wedge C))
$$

$$
\cfrac{
  \cfrac{
    \cfrac{
      \cfrac{\;}{A \to \neg B}^{(1)} \quad \cfrac{\;}{A}^{(3)}
      \quad
      \cfrac{\dfrac{\;}{\neg B}^{(4)} \quad \dfrac{\;}{B}^{(2)}}{\bot}^{(4)}
    }{\bot}^{(3)}
  }{\cfrac{\neg A}{B \to \neg A}^{(2)}}^{(2)}
}{(A \to \neg B) \to (B \to \neg A)}^{(1)}
$$

All the **R**-axioms just proved are Core, hence Intuitionistic, theorems. As it happens, they are also theorems of Minimal Logic. For we proved each **R**-axiom without availing ourselves of those parts of ¬-Introduction and/or ∨-Elimination that are available in Core Logic but not in Minimal Logic.

The one remaining **R**-axiom on the standard list is the strictly classical ¬¬$A \to A$, whose proof is as follows in the classical extension of Core Logic by the rule of Classical Reductio:

$$
\cfrac{
  \cfrac{
    \cfrac{\dfrac{\;}{\neg\neg A}^{(1)} \quad \dfrac{\;}{\neg A}^{(2)}}{\bot}^{(2)}
  }{A}^{(1)}
}{\neg\neg A \to A}
$$

Chapter 2

The Road to Core Logic

Abstract

This chapter situates Core Logic and Classical Core Logic within a wider logical landscape, stressing their main distinguishing features.

§2.1 returns to the theme of pluralism, and explains how Core Logic $\mathbb{C}$ lies at the intersection of two orthogonal lines of reform of Classical Logic **C**—*constructivization* and *relevantization* (of the relation of *deducibility*). It stresses also how Classical Core Logic $\mathbb{C}^+$ arises from Classical Logic **C** by the relevantizing alone, in exactly the same way that Core Logic $\mathbb{C}$ arises from Intuitionistic Logic **I**.

§2.2 makes clear by means of a diagram the interrelationships among all the systems mentioned—**C**, **I**, **M**, **R**, $\mathbb{C}$, and $\mathbb{C}^+$.

§2.3 explains the genesis of Core Logic and adumbrates the main features of its more carefully formulated rules of inference. It contrasts the Core systems' methodologically motivated eschewal of 'absolutely unrestricted' transitivity of deduction with the failures of transitivity in some other contemporary systems. It heralds the case for the claim that Core Logic arises as a smooth generalization of the proto-logic involved in working out the truth values of sentences under particular interpretations. It alerts the reader also to the case to be made for the complete methodological adequacy of Core Logic $\mathbb{C}$ for *constructive* deductive reasoning, and of Classical Core Logic $\mathbb{C}^+$ for *non*-constructive deductive reasoning.

Finally, §2.4 briefly explains how Core Logic deserves the word 'Core', referring the reader to considerations essayed upon elsewhere that reveal Core Logic to be both fully employed, and sufficient, as the metalogic involved in any process of rational belief revision. It ends by speculating on two possible explanations—semantic and methodological—of how Core Logic $\mathbb{C}$ might have been bloated to Classical Logic **C**.

2.1   Constructivizing and relevantizing

We can now pick up the threads of our earlier discussion of pluralism in §1.4, which were set to one side in order to specify the various extant systems already jostling for the reader's attention.

There is one important concession to be made to the pluralist. Insofar as mathematicians divide into constructivist and non-constructivist camps as *expert deductive reasoners*, our aim might have to be two-pronged: find the right logic for the constructivists (call it $\mathbb{C}$); and find the right logic for the non-constructivists (call it $\mathbb{C}^+$). Those two logics should be differentiated *only* by whatever added ingredient(s) yield(s) non-constructivism from constructivism.

The right logic $\mathbb{C}$ for constructivist reasoning must bear to the right logic $\mathbb{C}^+$ for non-constructivist reasoning the same sort of relation (represented by the solid horizontal arrows below) that Intuitionistic Logic $\mathbf{I}$ bears to Classical Logic $\mathbf{C}$.

Likewise, the right logic $\mathbb{C}$ for constructivist reasoning should bear to Intuitionistic Logic $\mathbf{I}$ the same sort of relation (represented by the dashed vertical arrows below) that the right logic $\mathbb{C}^+$ for non-constructivist reasoning bears to Classical Logic $\mathbf{C}$.

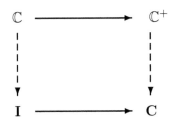

Figure 2.1: From Classical Logic to Core Logic

This work will argue that Core Logic is at the node $\mathbb{C}$ in the diagram above, and Classical Core Logic is at the node $\mathbb{C}^+$. These are the right logics for constructive and non-constructive deductive reasoning, respectively. So we use $\mathbb{C}$ and $\mathbb{C}^+$ as their names.

Note that no matter how judiciously one tries to classicize the system $\mathbf{M}$ of Minimal Logic, so as to obtain a system $\mathbf{M}^+$, say, it will turn out that $\mathbf{M}^+$ is the (non-relevant) system $\mathbf{C}$ of full Classical Logic. This is because of the following. Presumably $\mathbf{M}^+$, like $\mathbf{M}$, allows the structural rule of Cut, even if only as an admissible rule:

$$\text{Cut} \quad \frac{\Delta \vdash_{\mathbf{M}^+} \varphi \qquad \Gamma, \varphi \vdash_{\mathbf{M}^+} \psi}{\Delta, \Gamma \vdash_{\mathbf{M}^+} \psi}$$

For $\mathbf{M}$ itself we have $A \vee B, \neg A \vdash_{\mathbf{M}} \neg\neg B$; and for any classicized extension $\mathbf{M}^+$ of $\mathbf{M}$ we shall have $\neg\neg B \vdash_{\mathbf{M}^+} B$. Hence (by Cut) $\mathbf{M}^+$ will prove Disjunctive Syllogism: $A \vee B, \neg A \vdash_{\mathbf{M}^+} B$. In $\mathbf{M}^+$ we also, of course, have $A \vdash_{\mathbf{M}^+} A \vee B$. Hence (by Cut again) $\mathbf{M}^+$ will prove the First Lewis Paradox: $A, \neg A \vdash_{\mathbf{M}^+} B$. So $\mathbf{M}$ cannot occupy the top left-hand vertex of the foregoing diagram.

Despite their disagreements, all participants in the debate over logical reform have an eye to the methodological requirements of mathematics and natural science. Two central concerns have been:

> *Does one's logic afford all the mathematical theorems that are needed for applications in science?*

and

> *Does one's logic enable one to carry out the most rigorous possible tests of a scientific theory?*

We shall be providing affirmative answers to these questions for the system that is the topic of this book: Core Logic.

In order for a logical system to be able faithfully to regiment the structure of informal deductive reasoning it is essential that it formalize correctly the use that one can make of 'assumptions for the sake of argument'. Such assumptions are made, as we have seen, in order subsequently to apply such rules as (constructive) *Reductio ad Absurdum* ($\neg$I), Conditional Proof ($\rightarrow$I), Proof by Cases ($\vee$E), Classical Reductio (CR), and Classical Dilemma (Dil). In mathematical reasoning, especially, one's inferential moves are disciplined by the need to keep one's assumptions *relevant* to the conclusion towards which one is heading.

Conventional logical systems (such as Minimal Logic, Intuitionistic Logic and Classical Logic) are indifferent to this need and guilty of allowing moves that sin against canons of relevance. This is because these systems frame rules involving 'discharge of assumptions' that can be honored in the breach. (These were the rules involving diamonds in §1.5.) It turns out on closer analysis that it is absolutely crucial to lay down very precise constraints on when one can apply a discharge rule, and on what assumptions certain subordinate conclusions must have been shown to rest. One seeks, that is, to minimize the possibility of 'vacuous discharge of assumptions'. Recondite though this may sound, it is a very important precondition for the *relevance* of premises to conclusions in natural deduction.[1] We are not suggesting here that inattention to the issue of vacuous discharge of assumptions leads to *alethic* fallacies—that is, transitions from premises that may be true to a conclusion that may, in the same circumstances, be false. Rather, such inattention opens the door to fallacies of *relevance*—transitions from premises about one particular subject matter to conclusions having absolutely nothing to do with it.

Every argument provable in Core Logic is provable in Intuitionistic Logic, but not vice versa. So Core Logic may be thought of as a proper subsystem of Intuitionistic Logic. One can, however, as already indicated, *classicize* Core Logic while still preserving its virtues as a canon of *relevance* (of premises to conclusions). The classicized system may be thought of as a proper subsystem of Classical Logic. It does not, however, contain Intuitionistic Logic as a subsystem, since it is free of the irrelevancies that afflict the latter.

2.2 Core Logic in relation to familiar systems of logic

We have already stressed that Core Logic results from Classical Logic **C** by following two lines of reform: *constructivization* and *relevantization*. If one constructivizes Classical Logic, the result is Intuitionistic Logic **I**. The remaining question is then how best to relevantize Intuitionistic Logic. Historically, the system **M** of Minimal Logic (due to Johansson [1936]) has been regarded as the natural contender for the title of 'relevantized Intuitionistic Logic'.

M's claim to that title might seem to be underwritten by the very natural-looking nesting of sets of inference rules in the canonical presentation of the three standard systems **C**, **I** and **M** in Prawitz [1965], which has been reprised in §§1.5.1–1.5.3. There, the system **M** consisted of just the introduction and elimination rules for the logical operators, *in their original formulation* due to Gentzen [1934, 1935]. By adding the Absurdity Rule to **M** one obtains the system **I**. By then adding one of the usual strictly classical rules for negation (Law of Excluded Middle, Rule of Dilemma, Clas-

---

[1] The 'no vacuous discharge' requirement was first formulated in Tennant [1979], and has been applied further in Tennant [1980], [1984], [1987a], [1987b], [1989], [1992], [1994a], [1994b], [1996a], [1997], [1999], [2002], and [2005].

sical *Reductio ad Absurdum*, or Double Negation Elimination) one obtains the system **C**.

Taking the sequence in reverse: if one starts with the rules for **C** and gives up whatever strictly classical rule(s) of negation one has chosen, one is left with the rules that determine **I**. If one then gives up the Absurdity Rule, one is left with the rules that determine **M**. It seems all very natural and 'layered'.[2]

2.2.1 On a question of unique determination of subsystems

(This subsection may be skimmed or skipped by the non-technical reader.)

An anonymous referee suggested that one needs to dispose of the following possible objection: upon dropping from **C** its strictly classical rule(s) of negation *one cannot be sure that one gets* **I** *rather than any one of the uncountably infinitely many 'intermediate' logics* (i.e. logics between **I** and **C**).

One can rebut this potential objection as follows. The unique determination of **I** from **C** (upon dropping classical rules of negation), and the subsequent one of **M** from **I** (upon dropping EFQ) are secured by our observations in §1.3.5 about how the notion of $\mathcal{S}$-proof is fixed by the choice of rules for $\mathcal{S}$.

In addition to this disposive rebuttal, however, there are also the following considerations speaking against having to worry about this objection. Chagrov and Zakharyashchev [1991], at p. 200, say that Maksimova [1977] 'noted' that Intuitionistic Logic is the unique intermediate logic having both the Disjunction Property and the Craig Interpolation Property. They must have had her Theorem 3 in mind, because she never noted this explicitly herself. Maksimova's Theorem 3 says that there are exactly seven superintuitionistic (intermediate) logics satisfying Craig. Each of the six alternatives to **I** is generated by a disjunctive axiom whose disjuncts are obviously not theorems. So, **I** is the *unique intermediate logic* that has both the Disjunction Property and the Craig Interpolation Property.[3]

Moreover, Maksimova was investigating only systems of logical *theorems*. She did not consider the deducibility relation in general. Appealing as we have here to Maksimova's Theorem 3 to dispose of the potential objection framed above is, strictly speaking, overkill. The objection is already blocked by the consideration that natural-deduction rule-based systems such **M**, **I**, and **C** are uniquely determined *by their primitive rules*. So: when one drops

---

[2] Let us, however, be mindful of what Shrek told Donkey: *Ogres have layers!*

[3] These considerations also dispose of the objection raised by Roy Cook [2014], at pp. 245 *ff*.

the strictly classical negation rules from **C**, one is left with the rules of the system **I**; and when in turn one drops EFQ from **I**, one is left with the rules of the system **M**.

2.2.2   On the matter of relevantization

Our contention is that the second step, of relevantization, has not been properly carried out. The relevantization of Intuitionistic Logic should not be taken to be **M**. This is because matters have been skewed by an unsatisfactory choice of formulation of the *Ur*-rules of introduction and elimination. Ironically, Gentzen, the inventor of natural deduction as we now know it, did not get these rules quite right. He should have addressed the problem of how best to formulate the introduction and elimination rules under the constraint that the behavior of the logical operators was *all* one wished thereby to capture. The project of rule-formulation needs to be carried out not only under the constraint that one is not to commit oneself to any nonconstructive inferential moves, but also under the constraint that one is not to commit oneself to any fallacies of relevance.

   This is the path that we have followed instead. We have discovered, on traversing it, that the high road to constructivization and relevantization has a terminus other than **M**. The correct terminus is Core Logic ℂ.

2.2.3   The logical landscape of systems

The Venn–Euler diagram in Figure 2.2 shows how the system ℂ of Core Logic, and its classical extension $\mathbb{C}^+$, sit in relation to the well-known systems of Classical Logic **C**, Intuitionistic Logic **I**, Minimal Logic **M**, and the relevant system **R** of Anderson and Belnap. In this diagram, the boxes represent the respective systems' sets of provable sequents (i.e. arguments of the form $\varphi_1, \ldots, \varphi_n : \psi$).[4] The innermost white zone contains, of course, all of the axioms of **R** listed in §1.6.1 except for the last one, which is strictly classical. (We have not written them into the diagram, for want of space. They would take the form of sequents $\emptyset : \varphi$, if they were written in.)

   It is worth pointing out that the thin white zones are empty:

---

[4] The example $B : A \to B$ is due to Kevin Gibbons, answering negatively a question posed by Annie Yang: 'Is $\mathbb{C} \cap \mathbf{M}$ contained in **R**?'.

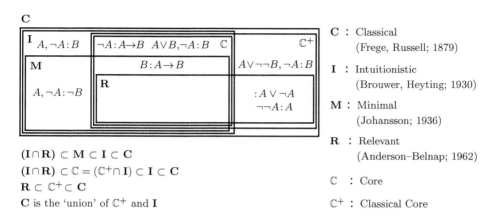

C : Classical
     (Frege, Russell; 1879)

I : Intuitionistic
     (Brouwer, Heyting; 1930)

M : Minimal
     (Johansson; 1936)

R : Relevant
     (Anderson–Belnap; 1962)

$\mathbb{C}$: Core

$\mathbb{C}^+$: Classical Core

$(\mathbf{I} \cap \mathbf{R}) \subset \mathbf{M} \subset \mathbf{I} \subset \mathbf{C}$

$(\mathbf{I} \cap \mathbf{R}) \subset \mathbb{C} = (\mathbb{C}^+ \cap \mathbf{I}) \subset \mathbf{I} \subset \mathbf{C}$

$\mathbf{R} \subset \mathbb{C}^+ \subset \mathbf{C}$

C is the 'union' of $\mathbb{C}^+$ and **I**

Figure 2.2: Logical system inclusions

2.3 What *is* Core Logic?—a Cook's tour

On the treatment to be provided here, Core Logic, like all the other systems already mentioned, proves arguments, or sequents, of the form $\Delta : \varphi$, where the *antecedent* Δ is a finite *set* (not: *sequence*) of premises, and the conclusion φ is a single sentence. As a special case we allow for $\perp$ (absurdity) to feature as the conclusion of a sequent. If one insists on defining a sequent as having a *set* on the right of its colon (as its *succedent*), then the succedent is 'at most a singleton'. That is, either it is of the form $\{\varphi\}$ or it is $\emptyset$, the empty set. The latter alternative is identified with $\perp$.

Two questions on any reader's mind will be: Exactly *which* sequents does Core Logic prove? And are there enough of them, for scientific purposes? To answer these questions, we need to attend to the rules of inference that Core Logic furnishes, and the proofs that we can construct by means of them. That provides an accessibly exact answer to the first question, and an affirmative answer to the second.

2.3.1 Comments on the rules of Core Logic

Core Logic consists of just those primitive rules of inference that directly govern the logical operators, one at a time. As we shall presently see, it matters not whether Core Logic is presented as a system of *natural deduction* or as a *sequent calculus*. Its *Introduction* rules (in natural deduction) really *are* the *Right* logical rules of the sequent calculus, for introducing logical operators on the right of the colon; and its *Elimination* rules (in natural

deduction) really *are* the *Left* logical rules of the sequent calculus, for introducing logical operators on the left of the colon. This vastly simplifies the
theoretical treatment of both natural deduction and sequent calculus, which
in the Gentzenian tradition (focused on the three standard systems **M**, **I**,
and **C**) have had a much more (and unnecessarily) complicated relationship
with one another. Core proofs in the sequent setting are isomorphic to core
proofs in the natural deduction setting, in a limpid sense that will be explained in Chapter 5. One can therefore speak of 'discharge of assumptions'
in a suitably *invariant* sense: assumptions are discharged by rule applications within sequent proofs in the same sense as they are discharged by the
respective rule applications in the corresponding natural deductions.

Core Logic therefore captures exactly the *analytically valid* inferences
from finite sets of premises to (single) conclusions. It has no need for
Gentzen's structural rules of Permutation and Contraction, which pertain
only to those formulations of sequent calculi whose sequents $\Delta : \varphi$ have *sequences* Δ of premises as their antecedents.[5] More importantly, the sequent
calculus for Core Logic does not contain—as rules that can be applied in
the construction of proofs—Gentzen's structural rules of Thinning and Cut.
These, as we have seen, purport to characterize the relation of deducibility,
but without mentioning any particular logical operators.

2.3.2 Comments on discharge rules in Core Logic

Note that with all the systems under consideration here, discharge of assumptions is *thorough*, or *total*, not *partial*. That is, one discharges *every*
undischarged occurrence of the assumption in question when one applies
a rule that discharges it. This is because we are tracking only the *set* of
undischarged assumptions as we build up a proof. Thus the sequents $\Delta : \varphi$
that are thereby proved involve (finite) *sets* Δ of sentences as their antecedents, not *sequences*. There are, of course, so-called resource-sensitive
logics, such as Linear Logic, that deal with *sequences*, rather than sets, as
the antecedents of their sequents. We merely note here the existence of such
systems, and assert, by way of justification of our total neglect of them,
that a premise is no truer or more powerful, logically, for being *repeated*,
or *repeatedly used*, in the course of a proof. Moreover, *how often* a premise
is 'repeated' can depend on quite arbitrary features of the formatting of
proofs. So, for example, the use of the serial form of ($\wedge$E) will make it
seem as though its major premise is being used more frequently than will

---

[5] Bear in mind that sequences can contain repetitions of some of their members.

be the case if one uses the parallelized form of that rule instead. (For more on serial v. parallelized forms of certain elimination rules, see §4.3.) We take this opportunity to remind the reader of the fundamental and overarching Fregean aim: to provide a logical system that will faithfully formalize deductive reasoning *as it takes place in mathematics*. And any mathematician knows full well that resorting to but a single use of a hypothesis makes whatever theorem is thereby derived *depend on it*. And such dependency is neither diminished nor increased by either rarity or frequency of such resort.

2.3.3 The admissibility of cut for Core Logic

That the sequent calculus for Core Logic does not actually contain Cut as an applicable rule for the construction of proofs does *not* imply that Cut is *inadmissible* for Core Logic. Indeed, we shall show that a *very strong* version of Cut is *admissible* for Core Logic $\mathbb{C}$:

$$\text{Cut} \quad \frac{\Delta \vdash_{\mathbb{C}} \varphi \qquad\qquad \Gamma, \varphi \vdash_{\mathbb{C}} \psi}{\text{for some } \Omega \subseteq (\Delta, \Gamma),\ \Psi \subseteq \{\psi\} :\ \Omega \vdash_{\mathbb{C}} \Psi}$$

And the same holds with $\mathbb{C}^+$ in place of $\mathbb{C}$.

Core Logic is a canon of constructive and relevant reasoning, contained 'deep within' the system of Intuitionistic (hence also of Classical) Logic. Its only structural rule in the sequent setting is

$$\text{Reflexivity} \qquad \frac{\overline{\qquad\qquad}}{\varphi : \varphi}$$

Proofs, after all, have to get started!

In the natural deduction setting, the propositional part of Core Logic consists of certain *very carefully crafted* introduction and elimination rules for the connectives. These rules effortlessly generate the connectives' familiar truth tables, on an 'inferentialist' reading (row by row, left to right). This is not to concede that the rules themselves are *beholden* to the truth tables, as some sort of independent standard for the determination of the meanings of the logical connectives. Rather, matters are the other way round. Mathematicians were reasoning correctly in accordance with rules governing their logical operators long before the invention of truth tables. Philosophical priority matches temporal priority here. It should be stressed, however, that the forms of rules that the Core logician proposes (as codifications of what the mathematicians' basic moves of inference really amount to) are subtly, and fruitfully, different from those that the proof-theoretic tradition derives from the work of Gentzen [1934, 1935] and Prawitz [1965].

2.3.4 Special features of Core Logic that ensure relevance

The following special features of the natural-deduction system for Core Logic
are designed both to ensure relevance of premises to conclusions and (in the
propositional case) to maximize efficiency in both automated proof *search*
and proof presentation.

(i) Major premises of eliminations *stand proud*. This means they have no
proof work above them.

(ii) The rules of Conditional Proof and of Proof by Cases are 'liberalized'
in crucial and entirely natural ways so as to obviate the need for any
applications of the rule *Ex Falso Quodlibet*, which is the *bête noire* of
the relevantist.

(iii) In the statement of 'discharge rules' careful attention is paid to the
much-neglected distinction between *obligatory* and *permitted* occurrences of 'assumptions for the sake of argument'.

(iv) All proofs are in normal form (because of (i)).

(v) There is no rule of *Ex Falso Quodlibet*.

These innovations within the Gentzen–Prawitz tradition were inspired
by, and in Tennant [1992] were developed to supply, the needs of the *computational* logician, whose aim is to devise efficient proof-search algorithms
for deductive problems of the form 'Is there a proof in the system S of
the conclusion Q from the premises $P_1, \ldots, P_n$?' (at least for decidable
systems S, typically at the propositional level). Hence the emphasis on
'search' above.

What sort of transitivity 'failure', then, is to be encountered among sentences in *this* language, the *Fregean* one, and embraced within Core Logic?
(I use scare quotes here to put the reader on notice that I shall be arguing
vigorously that it is no real *failure* at all, but a positive boon for the student of deducibility.) Because all proofs in Core Logic are in normal form,
an understandable worry arises: might there fail to be 'enough transitivity'
of deduction for scientific purposes?

2.3.5 Core Logic provides enough transitivity of deduction for all scientific purposes

The answer to the question just posed is reassuring: *there is transitivity
enough for all scientific purposes*. And let us stress once again: we are

talking here about transitivity of the *syntactic* or *proof-theoretic* notion of *deducibility*, not of any possibly corresponding model-theoretic or formal-semantical notion of *logical consequence* (conventionally represented by the double turnstile $\models$). Certain relations of the latter kind will receive some attention in §6.1.

The reassuring answer about transitivity takes the form of a disposi-tive metatheorem about natural deductions (equivalently: sequent-calculus proofs) in Core Logic, to the following effect: there is an effective method [,] that transforms any two core proofs

$$
\begin{array}{cc}
\Delta & A, \Gamma \\
\Pi & \Sigma \\
A & \psi
\end{array}
$$

(where $A \notin \Gamma$ and Γ may be empty) into a core proof $[\Pi, \Sigma]$ *of* ψ *or of* $\bot$ from (some subset of) $\Delta \cup \Gamma$. This is the sense in which a strong version of Cut is admissible (if one thinks in terms of a sequent system).

We shall be explaining this result in considerable detail, and drawing out all its comforting metamathematical and methodological implications. This metatheorem was first proved in Tennant [2012a] for Core Logic, and generalized in Tennant [2015b] to Classical Core Logic. In the former paper, the proof transformations involved were set out in painstaking detail. In the latter paper, a more succinct notation for proofs was devised, and the proof of the earlier result for $\mathbb{C}$ was in effect reprised on the way to the result for the classical system $\mathbb{C}^{+}$.

In this work, we take matters one step further. In Chapter 6 we show how not only the two Core systems (the one constructive, the other classical) admit of 'cut elimination' in this way, but also how the standard trio of Minimal Logic, Intuitionistic Logic, and Classical Logic admit the same. We need only ensure that in the natural deduction formulations of these systems, elimination rules are in parallelized form and can only ever be applied with their major premises 'standing proud' with no proof work above them. On the sequent-calculus side, what corresponds to this is requiring that proofs contain only single-conclusion sequents and no cuts. Let us call the combination of these corresponding requirements on ND and SC formulations of a system the *intrinsic normality* requirement concerning that system. Our 'one size fits all' method of proving Cut-Admissibility for systems meeting the intrinsic normality requirement will be presented in Chapter 6. As the reader will see, the signal advantage of the Core systems emerges from this uniform treatment of them alongside the standard trio **M**, **I**, and **C**. Like **M**, the Core systems eschew EFQ; but, unlike these three systems, the Core systems employ carefully reformulated rules governing the logical operators. These rules properly ensure the relevance of any proof's premises to its conclusion.

With the three standard systems, the 'reduct' proof $[\Pi, \Sigma]$ furnished by the foregoing metatheorem can be guaranteed only to prove a sequent whose *antecedent* (i.e. premise set) is included in $\Delta \cup \Gamma$. The *conclusion* of the sequent established by the reduct remains the same: namely, the conclusion ψ of the proof Σ. No boon there; contradiction could lurk within $\Delta \cup \Gamma$ without the standard logician becoming aware of it.

With the former (core) systems, by contrast, the 'reduct' proof $[\Pi, \Sigma]$ can establish a sequent whose conclusion might *not* be the conclusion ψ of the proof Σ—in which case *it will instead be* $\bot$ (absurdity). Since $\bot$ is the *strongest possible* conclusion, this represents *epistemic gain*—of a kind *the opportunity to enjoy which is forfeited by the standard trio*, for want of appropriately formulated logical rules for the logical connectives. This is the boon to be had with Core Logic, which we were alluding to above: lurking contradictions in $\Delta \cup \Gamma$ can be flushed out by the Cut-elimination process, revealing that the deductive advance that would supposedly be secured by 'unrestricted' Gentzenian transitivity is illusory—it is actually generated illicitly, through an unsuspected collusion between Π and Σ, whose respective premise sets might be individually consistent, but which turn out to beget inconsistency with their unholy union.

2.3.6   Philosophical arguments for Core Logic

A philosophical case for Core Logic was set out in Tennant [1987b] and Tennant [1997]. Core Logic was argued to be the correct logic, on the basis of Dummettian considerations of manifestability of grasp of meaning. This argument will no be repeated in this book. Rather, we shall be describing or supplying *three new kinds* of philosophical consideration that are quite different from the earlier, Dummettian ones. The new ones are as follows.

First, Core Logic (at first order) is the result of a smooth and natural generalization, to finite sets of *complex* sentences as premises, of the inferential methods that we employ in order to determine the truth values of complex sentences from the 'atomic information' in a model. (We use 'model' in the standard Tarski–Kemeny sense: a model consists of a domain of individuals along with extensions therein for names, function signs, and primitive predicates.) Basically, *the proto-logic involved in handling truth-table computations and evaluations of quantifications via their instances is but a small step of smooth extrapolation away from the full deductive system of first-order Core Logic.*[6] The proto-logic in question is called the Logic of Evaluation, and will accordingly be denoted as $\mathbb{E}$.

Truthmakers are the same *kind* of mathematical object as proofs: namely, trees of nodes, labeled with sentences[7] (in the case of natural deduction) or with sequents (in the case of sequent calculi). The main difference is that with truthmaker-trees there can be 'infinitely wide' branchings. This can happen when the domain of the model M with respect to which truth values are being determined is infinite. For then the M-relative verification of any universal claim $\forall x \varphi$ requires exhaustive inspection of all possible individual instances $\varphi(\alpha)$ of the predicate in question (α being drawn from the domain of M). So too does the M-relative falsification of any existential $\exists x \varphi$.

Even though truthmakers are so akin to core proofs, they nevertheless offer a novel and illuminating way to re-conceptualize logical consequence and in particular to show that core proofs are sound (details will emerge in Chapter 8). This re-conceptualization involves showing that if one is given, for any model M, M-relative truthmakers for the premises of a core proof, then those truthmakers and that proof can be 'operated on' according to definite procedures so as to produce an M-relative truthmaker for the proof's conclusion. This method will be illustrated, and shown to be adequate in general unto all core proofs and models, in Chapter 9. The required procedures are direct adaptations of those for establishing cut-admissibility in Chapter 6. So one achieves a remarkable 'conceptual congruence' between matters semantic and matters syntactic—one that provides a novel vindication for an inferentialist approach to formal semantics itself.

The assurance of *relevance*, within $\mathbb{E}$, between (the evaluation of) a compound and (the evaluations of) its constituents is transferred directly to the wider and more general deductive setting in which conclusions may be drawn from premises of any complexity (and not just from literals—i.e. from atoms and negations of atoms). Such inferential drawing is thereby guaranteed to take place along lines of genuinely relevant connection between premises and conclusions. This is a major point of departure from the Anderson–Belnap tradition of trying to capture relevance by altering the deductive behavior of the conditional connective '$\rightarrow$'. The Core logician captures it, by contrast, as already stressed above, at the 'level of the turnstile'.

The second new kind of philosophical consideration in support of Core Logic is this: it can be argued that Core Logic is the correct *logic of*

---

[6] This line of approach to Core Logic arose from the relatively recent development of a general logical theory of truthmakers and falsitymakers. Tennant [2010] provides a more informal account; Tennant [forthcoming b] gives a fully formal one.

[7] More accurately: *saturated formulae*. See §3.4.

conceptual constitution. That is, constitutive logical interrelationships among concepts are to be exhibited strictly within the confines of Core Logic. Conceptual interconnections should not trade on any logical irrelevancies such as are supplied by Intuitionistic and by Classical Logic. Unfortunately, there is no space in the present work to develop this argument; it will have to await a planned sequel using Core Logic to provide what I call *natural logicist* foundations for various branches of mathematics. The reader will, however, have a brief foretaste of it in §11.4, when we frame new Introduction and Elimination rules for the abstractive set term-forming operator $\{x \mid \ldots x \ldots\}$.

The completely formal derivations, in Tennant [1987b], of the Peano axioms for successor arithmetic from deeper logicist principles of inference governing the primitive notions 'the number of Φs' and 'n is a natural number' were given entirely within Core Logic. This is in keeping with Core Logic being the 'logic of conceptual constitution'. The same holds for the further derivations that can be given of the Peano axioms for the operations of addition and multiplication, from the logicist principles furnished in Tennant [2009].

The foregoing is not a provincial result. The adequacy of first-order Core Logic for (all of) *first-order intuitionistic mathematics* was established as a metatheorem in Tennant [1994b]. This claim of adequacy is invariant across the oft-remarked distinction between Brouwerian intuitionism and Bishop-style constructivism.[8] My claim about the adequacy of first-order Core Logic for all of first-order 'intuitionistic' mathematics is best interpreted as focusing on the underlying logic of whatever deductive reasoning is employed. The two main cases are covered.

> Case 1: One is being a Bishop-style constructivist, proving theorems, using Intuitionistic Logic, from axioms with which the classical mathematician agrees. (Examples: doing Heyting Arithmetic, which uses the Dedekind–Peano axioms, but employs only Intuitionistic Logic; doing intuitionistic Zermelo–Fraenkel set theory, which is based on axioms accepted by the classicist, but likewise employs only Intuitionistic Logic. See Friedman [1973].)

> Case 2: One is proceeding like a Brouwerian intuitionist, adopting axioms that might contradict classical theorems, or employing unusual definitions of terms with established usage (such as 'function on the reals'), and proving results contradicting classi-

---

[8] Thanks to Nathan Kellen for seeking this clarification (personal correspondence).

cal theorems—but doing so while employing only Intuitionistic Logic.

In both cases, Core Logic suffices for the formalization of the first-order reasoning involved.

The adequacy of the classicized version of Core Logic for (all of) *classical* mathematics is established in a similar fashion. And this raises the prospect of refashioning the foundations of mathematics (both intuitionistic and classical) along the natural logicist lines just mentioned, and described in somewhat more detail in Tennant [2014c].

The third and final new philosophical consideration is this: consonant with Core Logic's being the correct logic of conceptual constitution (the canon for 'unpacking concepts') is a closely related thesis concerning the logic needed in order to uncover logico-semantic paradoxes.[9] The claim is that these paradoxes are never strictly classical. The kind of conceptual trouble that such a paradox reveals will afflict the intuitionist (and relevantist) just as seriously as it does the classicist. Therefore, attempted solutions to these paradoxes, if they are to be genuine solutions, must be available to the Core logician. Nothing about an attempted solution to a logico-semantic paradox should imply that the trouble it reveals has its origin in operator-involving moves of classical (or even intuitionistic) reasoning that lie beyond the confines of Core Logic.

The *methodological* adequacy of Core Logic—its sufficiency for the hypothetico-deductive method of scientific theory-testing—was demonstrated in Tennant [1997], building on earlier results concerning Minimal Logic in Tennant [1985]. All these results now flow easily as corollaries of the central results in Tennant [2012a] and Tennant [2015b] for Core Logic and its classical extension, respectively. Core Logic matches Intuitionistic Logic on logical theorems, on inconsistencies, and on deducibility from consistent sets of premises. An analogous result holds in the classical case for the system of Core Logic extended by a classical rule of negation. There is really nothing further that we can demand from a system of logic.

2.3.7 Other important advantages of Core Logic

The advantages for *computational logic* (or automated deduction) of using systems such as Core Logic, in which all proofs are in a very exigent kind of normal form (without loss of completeness), were set out in Tennant

---

[9] This thesis was first advanced in Tennant [2015a], Tennant [under submission] and Tennant [forthcoming a].

[1992]. Indeed, Core Logic owes its genesis to accommodating the demands placed on proof theory by the need for efficient but completeness-conserving constraints on proof search in automated deduction. The rules of inference in proof theory, which afford an inductive, 'top down' *construction* of proofs, are reformulated in such a way that the proofs that result can also be thought of as the natural termini of efficient 'bottom up' *search*.[10]

Core Logic, as already intimated above, enjoys the *proof-theoretic* distinction that its proofs have exactly the same structure whether they are presented as natural deductions or as sequent proofs (Tennant [2012a]). More pithily: 'ND ≅ SC'. This is an immediate consequence of the fact that the ND- and SC-formulations meet the aforementioned intrinsic normality requirement.

The reader will recall that on the ND side, the requirement is that all elimination rules be stated in so-called parallelized form,[11] and the natural deductions of Core Logic be defined in such a way that major premises of eliminations (MPEs) always *stand proud* with no proof work above them.[12] This of course ensures that all proofs (as natural deductions) are in normal form. Isomorphism between natural deductions in Core Logic and the corresponding cut-free (and thinning-free) sequent proofs of Core Logic is then immediate.

This isomorphism property would be enjoyed by any logical system that resembled Core Logic in the key respect of satisfying the intrinsic normality requirement.[13] But to the best of my knowledge there is no extant rival to Core Logic that matches it in this respect *and* avoids the First Lewis Paradox in both its positive and its negative form.

---

[10] Here the terms of art 'top down' and 'bottom up' are best understood when one is working within a sequent calculus rather than within the corresponding system of natural deduction. The directions 'up' and 'down' are to be understood in relation to the finished proof as a tree-like structure, with its root node at the bottom and its leaf nodes at the top, at tips of branches.

[11] Thorough parallelization of elimination rules was achieved by Schroeder-Heister [1984].

[12] This double whammy of parallelized forms *plus* MPEs standing proud was introduced in Tennant [1992]. The aim was to make natural deductions in normal form isomorphic to cut-free, thinning-free sequent proofs. The proofs that resulted were called *hybrid* proofs in that book. This was because they combined the economy of macro-structure of proof trees to be found in the sequent calculus with the economy of sentence-labeling (rather than sequent-labeling) of nodes within the trees, which is the economical feature of natural deduction. Hybrid proofs we now call *core proofs*.

[13] The systems of Classical, Intuitionistic, and Minimal logic can all be formulated in the way just described, but never have been. See Chapter 6 for these systems' reformulation along these lines.

2.3.8   Core Logic's eschewal of thinning and cut as rules of the proof system

In Tennant [2016] we set out the reasons why the structural rules of Thinning and Cut should not be rules *of* the system. Their presence destroys relevance and forfeits the prospect of making important epistemic gains.

The two rules of Thinning and Cut in standard presentations of the sequent calculus are, we argue, an unwitting source of unwanted irrelevancies in deductive reasoning. One can do without them. Gentzen's famous *Hauptsatz* (the 'Cut-Elimination Theorem') tells one how to eliminate cuts within a proof. So why not by analogy (following up on the topic mentioned in §1.1 for an imaginary discussion with Gentzen) establish a *Thinning-Elimination Theorem*? This would tell one how to eliminate thinnings within any cut-free proof. Gentzen, with his formulation of sequent rules, could not do this. In the system of Core Logic, with its subtly different rules, however, one can; and the result of eliminating all cuts *and* thinnings is a core proof. The rub, of course, is that upon eliminating thinnings one would in general end up with a cut-free, thinning-free proof of *some subsequent of* the overall sequent 'for which' one seeks such a proof. But *that* is a welcome *epistemic gain*!—by eschewing the cuts and thinnings, one might end up with a proof of a *logically stronger* result than one took oneself originally to have proved (by means of those cuts and thinnings, now eliminated). Gentzen himself seems to have missed this crucial point, with its implications for epistemic gain, in his initial study (Gentzen [1932]) of the structural rules (see Tennant [2015c]).

In that study, Gentzen laid an important propaedeutic part of his foundations for modern proof theory. He examined *only the structural rules* of what was to become his sequent-calculus approach to the standard systems of Intuitionistic and Classical Logic. The rules in question were Reflexivity, Thinning and Cut. He proved the following normalization theorem: Any proof constructed by means of his structural rules can be transformed into a proof of the same result, in which the only application of Thinning occurs at the final step.

This result, on reflection, is really quite remarkable. For it basically assures the system's user that there really is no point at all in having a rule of Thinning in the first place. If one can make do with thinnings only as terminal steps of proofs, then why not avoid them altogether? After all, the premise sequent for any application of Thinning is logically *at least as strong* and potentially even *stronger than* the final sequent that results from the thinning in question.

It was only in his next publications that Gentzen furnished the so-called Right and Left sequent rules for the logical operators.[14] Here I ask the reader to consider how best Gentzen might have built on his earlier success. Should not the following have occurred to him?:

> With proofs constructed by means of only the structural rules, I was able to show that all thinnings could be made terminal. Now, here I am, devising logical rules for a more detailed sequent calculus. Wouldn't it be wonderful if I could generalize that result to the full calculus? How might I formulate the Right and Left rules for these logical operators so as still to be able to prove that every proof, if it has a thinning at all, can be rewritten so as to have it only at the very end?

Had this thought occurred to Gentzen, he would have discovered Core Logic.

2.3.9   Core Logic's shedding of other old saws

We have seen that the Core logician refuses to have either Thinning or Cut as a rule of the system. We have not yet, however, explicity pointed out a couple of other old saws that are pretty toothless. The first is the so-called Deduction Theorem. For the systems **M**, **I**, **C**, and **R** we have the following:[15]

$$\Delta, \varphi \vdash \psi \quad \Leftrightarrow \quad \Delta \vdash \varphi \to \psi.$$

The Core logician is of course happy with the implication

$$\Delta, \varphi \vdash_C \psi \quad \Rightarrow \quad \Delta \vdash_C \varphi \to \psi,$$

since that holds simply by applying the rule $\to$I. But the converse implication

$$\Delta \vdash_C \varphi \to \psi \quad \Rightarrow \quad \Delta, \varphi \vdash_C \psi$$

is a different matter altogether. It would certainly go through on the further condition that Δ, φ is consistent. But that further condition is not explicitly built in to the statement of the Deduction Theorem. So the problematic direction succumbs to the dispositive counterexample

$$\neg\varphi \vdash_C \varphi \to \psi \ ; \text{ but } \neg\varphi, \varphi \nvdash_C \psi.$$

---

[14] Gentzen [1934, 1935].

[15] The result is straightforward for **M**, **I**, and **C**. The non-trivial direction for **R**, which is from left to right, was established by Church [1951] and Shaw-Kwei [1950]. See also Kron [1973].

And this is how it should be. We have

$$\neg\varphi \vdash_C \varphi \to \psi$$

because it is part of the meaning of $\to$ that $\varphi \to \psi$ is rendered true by the falsity of its antecedent. That inferential transition is *relevant* as well, since φ is a subformula of both the premise and the conclusion.[16] *But*

$$\neg\varphi, \varphi \nvdash_C \psi.$$

This is because ψ could in general be completely irrelevant to φ. The two sentences could be talking about completely different subject matters, and have no extralogical vocabulary in common at all. So of course one cannot expect the inference to go through in general.

A second old saw that the Core logician is happy to give up is the Replacement Principle (see §12.2). This is the principle that interdeducible sentences should be replaceable *salva veritate* in any context of deducibility. In the standard trio of logical systems (**M**, **I**, and **C**), as well as in the system **R** of Anderson and Belnap, the interdeducibility of any two sentences φ and ψ guarantees their interreplaceability, *salva veritate*, in all statements of deducibility:

$$\frac{\varphi \dashv\vdash \psi \qquad \theta_1, \ldots, \theta_n, \vdash \chi}{\theta'_1, \ldots, \theta'_n, \vdash \chi'} \quad \text{where } (\theta_1, \ldots, \theta_n : \chi)^\varphi_\psi = (\theta'_1, \ldots, \theta'_n : \chi')^\varphi_\psi$$

But this is very much by courtesy of *unrestricted* Cut! The failure of the Replacement Principle for Core Logic is of course of a piece with the failure of *unrestricted* transitivity. Consider, for example, the two core-interdeducible sentences

$$\psi \vee (\varphi \wedge \neg\varphi) \dashv\vdash \psi.$$

In Core Logic we have

$$\varphi \wedge \neg\varphi \vdash \psi \vee (\varphi \wedge \neg\varphi);$$

but emphatically, once again,

$$\varphi \wedge \neg\varphi \nvdash \psi.$$

So Replacement fails for Core Logic.

The Core logician's response to the demand that Replacement should hold is simply to challenge interdeducibility as an adequate explication of the *synonomy* of two sentences. Let us make the following abbreviation.

---

[16] Indeed, it is a *negative* subformula of both of them—that is, it occurs in both of them 'with the same parity'. See Chapter 10 for further development of this and related ideas.

Definition 1. $\varphi \sim \psi \equiv_{df} \varphi$ *is synonymous with* ψ.

We stress that the notion of synonymy is pre-formal, or intuitive; and we are simply introducing the symbol '$\sim$' as shorthand for it. Intuitively, the following inference holds:

$$\frac{\varphi \sim \psi \quad \theta_1, \ldots, \theta_n, \vdash \chi}{\theta'_1, \ldots, \theta'_n, \vdash \chi'} \quad \text{where } (\theta_1, \ldots, \theta_n : \chi)_\psi^\varphi = (\theta'_1, \ldots, \theta'_n : \chi')_\psi^\varphi$$

That the displayed inference holds when the first premise is $\varphi \sim \psi$, but not (for Core Logic) when it is $\varphi \dashv\vdash \psi$, simply shows that, though $\varphi \sim \psi$ implies $\varphi \dashv\vdash \psi$, the converse implication fails to hold. That is, *interdeducibility is an inadequate explication of the notion of sentence synonymy*. This was the lesson of the classic study of Smiley [1962], which revealed that for many a logical system, interdeducibility is *not* sufficient for interreplaceability, *salva veritate*, in all statements of deducibility. For Smiley, as for us, synonymy is to be explicated as such interreplaceability. Systems like **M**, **I**, **C**, and **R** happen to serve up interdeducibility as a sufficient condition for synonymy. Core Logic, however, does not. But that does not affect its methodological adequacy one iota.

2.4 Why 'core'?

Core Logic has recently emerged in a new role, in a way that both explains and justifies my preference for its new label. An argument was presented in Tennant [2012b] for the following revision-theoretic thesis:

> Core Logic is the minimal inviolable core of logic without any part of which one would not be able to establish the rationality of belief-revision.

Revising one's belief system with respect to a proposition p involves *contracting* the system with respect to $\neg p$, and then *expanding* the resulting system with respect to p. We showed [*op. cit.*] that the reasoning involved (at the meta-level) in carrying out all the necessary adjustments requires no more than, but all of, Core Logic. Thus no principle of Core Logic could ever be given up in the course of a process of rational belief-revision, on pain of not being able to carry out that very process. This endows Core Logic with a certain *reflexive stability* enjoyed by no other logic.

This provides yet another 'endogenous' argument for Core Logic, showing that $\mathbb{C}$ is a body of principles exactly responding to 'pressures from within'. These pressures make the system-builder *commit to* certain logical principles, rather than eschew them; but they do no more.

The fact that such disparate perspectives on the problem of optimal choice of a logical system all converge on this single system of Core Logic is an indication that it captures a very stable, central, and robust notion of logical deducibility. Indeed, one can go so far as to claim: Core Logic is *the* formal system that correctly explicates the notion of *a priori because analytically valid* logical inferences. We need to reconsider, on the totality of the formal evidence, the existing 'asymmetry' involved in regarding Core Logic as some sort of hobbled residue after attempts to effect controversial reforms of Classical Logic.

Matters should, rather, be seen the other way round. The asymmetry consists in Classical Logic having been allowed to *accrete* problematic forms of inference in addition to the logically licit ones of Core Logic, and thereby to become *overblown* in relation to genuine methodological needs.

How has it come to pass that this beautiful logical kernel (Core Logic)—a fragment of which Aristotle took off the Platonic shelf (see Tennant [2014a])—acquired both the pith and the husk that turn it into Classical Logic? The explanation envisaged at present is two-pronged.

First, there is the slide to a conception of verification-transcendent truth, and accompanying toleration of non-constructive forms of classical reasoning. This is a matter of making one's logic conform to a (mistaken) *semantic realist* conception of reality as determinate in crucial regards (namely, the horns for any application of Classical Dilemma). Inferences that are *classically* valid without exploiting irrelevance, but are not part of Core Logic, are *synthetic a priori*, if they are valid at all. Along the lines of Tennant [1996b]: Any strictly classical logical moves governing negation should be understood not as justified by meanings—for Core Logic exhausts all logical connections forged by meanings—but rather as *expressing a 'semantic realist' attitude* as to the determinacy of truth values of the 'litmus sentences' to which those rules are applied.

An example of such a litmus sentence is φ below, which is assumed in the positive horn of a Classical Dilemma:

$$
\begin{array}{cc}
(i)\underline{} & \underline{}(i) \\
\varphi & \neg\varphi \\
\vdots & \vdots \\
\psi & \psi \\
\hline
& \quad(i) \\
\psi
\end{array}
$$

The semantic realist's attitude within the realm of mathematics has been vigorously disavowed by intuitionists ever since Brouwer. Anti-realists in general refuse to accept that every declarative sentence of one's language *has a determinate truth value even if* it lies beyond our powers, as rational thinkers, to discover what that truth value is. This Principle of Bivalence was identified by Dummett as *constitutive* of the view he called semantic realism. Where we differ from Dummett is that we do not share his view that subscribing to the Principle of Bivalence (and therefore adopting classical rules of inference, such as Dilemma) somehow imposes upon the logical operators 'inchoate classical meanings'. Rather, as argued in Tennant [1996b], the adoption of strictly classical rules manifests one's attitude of realism precisely because the logical operators involved in the litmus sentence in question are endowed *only with their core meanings*.

The second prong of explanation of how Core Logic might have become bloated into Classical Logic is as follows. There is the desire to enshrine a rule of Cut *in* the system, rather than having Cut merely be admissible *for* the system. This leads to overreach in our conception of deductive transitivity, and opens the door to the (uncomfortably tolerated) paradoxes of relevance. This is a matter of making one's logic conform to a (mistaken) *methodological convention* that one should round out and simplify, rather than be astute and discerning about what exactly follows from what. Even if strictly classical inferences should be permitted, the *relevantized* system $\mathbb{C}^+$ of Classical Core Logic provides all the patterns of deductive reasoning that provably suffice for science and mathematics, but which at present have to be surgically extracted from within the unnecessarily bloated system **C** of Classical Logic, leaving a great deal of fat behind.

Chapter 3

The Logic of Evaluation

Abstract

§3.1 introduces the standard logical operators of first-order logic.

§3.2 explains why T and F are the only truth values even for the intuitionist. It prepares the ground for the inferentialist's derivation of the truth tables. The notion of a verification of a sentence, and the complementary notion of a falsification, are introduced.

§3.3 explains how sentences of propositional logic may be verified or falsified with respect to an *atomic basis*—the inferentialist's analog of a truth-value assignment to atomic sentences. Basis-relative verifications and falsifications are co-inductively defined. It emerges that the rules by means of which one constructs such evaluations allow one to derive the truth tables row by row, reading them from left to right. These rules determine what we call the system $\mathbb{T}$ of Truth-Table Logic.

§3.4 extends the treatment to first order, using the notion of saturated terms and formulae. We introduce the important notion of *model-relative* rules that verify universals or falsify existentials. These can be responsible for 'infinite sideways branchings' within verifications and falsifications. We establish a metatheorem to the effect that if a sentence can be furnished with both a verification and a falsification relative to a set of literals, then one can track down a particular atom which occurs along with its negation in that set. This presages a stronger result later on about how any 'failure' of transitivity with any pair of core proofs is compensated for by their revelation of an inconsistency in their combined premises.

§3.5 considers more general atomic bases that allow for the expression of conceptual inclusions and/or contrarieties. The resulting choice of rules of verification and falsification determines what we call the system $\mathbb{E}$, the Logic of Evaluation. The previous co-inductive definition of verifications and falsifications of individual *sentences* now becomes the definition of a *determination* of a *sequent* with respect to a general atomic basis. The sequents can have *atomic hypotheses* on the left. The stage is finally set for the ushering in of the rules of inference that determine the deductive system $\mathbb{C}$ of Core Logic. To anticipate: this will be done by (i) allowing for *complex* hypotheses on the left,

(ii) allowing sentences to replace the absurdity symbol in conclusion-positions, and (iii) *voiding the basis*, so that deducibility becomes a model-invariant matter of *form*, not of *content*.

3.1 The logical operators

We shall be concerned here with the following logical expressions of English, and the corresponding logical symbols:

Logical symbol	Logical operation/relation	English correlate
¬	Negation	it is not the case that ...
∨	Disjunction	either ... or ... (or both)
∧	Conjunction	... and ...
→	Implication	... only if ... / if ... then ...
$\exists x$	Existential quantification	there exists x such that ... x ...
$\forall x$	Universal quantification	for every x ... x ...
=	Identity	... is identical to ...

A system of logic should emerge as a canon of inferences warranted by the meanings of these logical words, and nothing else. That is, logic should be a body of *analytically valid* inferences. But on the view being advanced here, those meanings are conferred or acquired by virtue of those words being subject to certain rules of inference. The crucial questions to confront, then, are: *What exactly are these rules? And what should we be taking into consideration when we try to frame them?*

3.2 Our inferentialism delivers the truth tables

Let us take the well-known *truth tables for the truth-functional connectives* as our expository starting point. And let us see exactly how to generate and justify them, from an *inferentialist*'s point of view. The results are instructive.

But before we embark on this investigation, let us make an important point clear in advance. There is an issue of conceptual or logical priority to clarify, lest the reader form a mistaken impression about our approach. We are not taking the truth tables as an exercise in meaning-specifying formal semantics. We are not laying down the truth tables as enjoying foundational priority, or as accomplishing some prior fixing of operator meanings to which rules of inference would subsequently have to be held responsible. Rather, we are taking the rules of inference that we lay down as in themselves meaning-specifying; and we shall be demonstrating how an extremely modest set of rules serves to generate the familiar truth tables row by row, reading them from left to right.

The truth tables should really be called *truth-value* tables, because they involve two truth values, T and F, with F figuring just as importantly within them as T. We need to bear that overriding consideration in mind, and not let it be eclipsed by our preference, when forming beliefs or making assertions, for true propositions over false ones.

The intuitionist, it should be noted, agrees with the classicist that T and F are the only *definite* values that any proposition can enjoy. For the classicist there can be no other values; every proposition is held to be either true or false. *For the intuitionist*, too, there can be no other values—but for a different reason. The intuitionist is concerned only to *prove* propositions, or to *refute* them. Those that are provable are true; those that are refutable, false. But the intuitionist refuses to contemplate assigning any *other* possible value (or values) 'in between' truth and falsity as 'the' value of a proposition for which we have, at present, neither proof nor refutation.

The reason for this refusal is simple. Let us introduce some notation. For any truth value X, and truth-value assignment τ that assigns X to P, we shall write $\tau(P)=X$. We can then abbreviate this even further by writing X(P), suppressing mention of the assignment τ. Now suppose, for *reductio ad absurdum*, that there *were* some other truth value besides T and F that, according to a (would-be) intuitionist, some proposition P enjoyed. Call such a value V. Note that we (and the would-be intuitionist) would have the following primitive (metalinguistic) inferences, registering the pairwise distinctness of the three truth values in play:

$$\frac{T = F}{\bot} \qquad \frac{T = V}{\bot} \qquad \frac{F = V}{\bot}$$

whence we would have also the inferences

$$\frac{\tau(P)=T \quad \tau(P)=F}{\bot} \qquad \frac{\tau(P)=T \quad \tau(P)=V}{\bot} \qquad \frac{\tau(P)=F \quad \tau(P)=V}{\bot}$$

The latter inferences can be abbreviated even further:

$$\frac{T(P) \quad F(P)}{\bot} \qquad \frac{T(P) \quad V(P)}{\bot} \qquad \frac{F(P) \quad V(P)}{\bot}$$

All intuitionists—genuine and would-be—subscribe to the inferential principle

$$\frac{}{\tau(\varphi)=\mathrm{T}}{\scriptstyle(i)} \qquad\qquad \frac{}{\mathrm{T}(\varphi)}{\scriptstyle(i)}$$

$$\vdots \qquad \text{, i.e.} \qquad \vdots$$

$$\frac{\bot}{\tau(\varphi)=\mathrm{F}}{\scriptstyle(i)} \qquad\qquad \frac{\bot}{\mathrm{F}(\varphi)}{\scriptstyle(i)}$$

This is because one shows a proposition to be *false* by *refuting* the assumption that it is *true*. That is to say: if its truth is impossible *modulo* other
assumptions, then it is false (given those assumptions). This is the essence
of intuitionistic negation too. Indeed the rule (¬I) of Negation Introduction,

$$\frac{}{\varphi}{\scriptstyle(i)}$$

$$(\neg\mathrm{I}) \qquad\qquad \vdots$$

$$\frac{\bot}{\neg\varphi}{\scriptstyle(i)}$$

which is framed wholly within the object language, mirrors at that level
what the metalinguistic principle diplayed above (whether in equational or
predicational form) conveys at the metalevel.

A basic Tarskian principle, acceptable to the intuitionist, is that if a
sentence is not true, then its negation is:[1]

$$\frac{}{\mathrm{T}(\varphi)}{\scriptstyle(i)}$$

$$(\mathrm{T}\neg) \qquad\qquad \vdots$$

$$\frac{\bot}{\mathrm{T}(\neg\varphi)}{\scriptstyle(i)}$$

It is easy to see that (T¬) can be derived using (¬I) in the metalanguage,

---

[1] This is one of the basic rules of the inferential theory of truth presented in Tennant
[1987b], at pp. 70–75.

followed by a step of T-Introduction:

$$
\begin{array}{c}
\dfrac{\rule{1.2cm}{0.4pt}}{\varphi}\!^{(i)} \\[4pt]
\overline{T(\varphi)} \\[4pt]
\vdots \\[4pt]
\dfrac{\perp}{\neg\varphi}\!^{(i)} \\[4pt]
\overline{T(\neg\varphi)}
\end{array}
$$

Conversely, $(\neg I)$ can be derived using $(T\neg)$, by employing a final step of T-Elimination:

$$
\begin{array}{c}
\dfrac{\rule{1.5cm}{0.4pt}}{T(\varphi)}\!^{(i)} \\[4pt]
\varphi \\[4pt]
\vdots \\[4pt]
\dfrac{\perp}{T(\neg\varphi)}\!^{(i)} \\[4pt]
\neg\varphi
\end{array}
$$

We can now show in short order that it is impossible for any proposition P to be assigned a truth value V distinct from both T and F:

$$
\cfrac{V(P) \quad \cfrac{\rule{1.6cm}{0.4pt}}{T(P)}\!^{(1)}}{\cfrac{V(P) \quad \cfrac{\perp}{F(P)}\!^{(1)}}{\perp}}
$$

There is also an interesting technical result vindicating the intuitionist's refusal to countenance any possible truth values distinct from T and F, which was established by Gödel [1932]: it is impossible to devise, using finitely many truth values, truth-value tables (so-called matrices) for the connectives that are characteristic for intuitionistic propositional logic. Gödel published his result before Gentzen furnished us with natural-deduction systems. The latter afford, as we have seen, a knock-down inferentialist refutation of the possibility of truth values *sustaining a sensible philosophical interpretation* other than T and F.

The stress on the need for the truth values to sustain a sensible philosophical interpretation is only underscored if it turns out that the 'many-valuedness' advocate has to retreat to an infinitude of 'truth values', in light of Gödel's result. There are in fact at least three *prima facie* different ways in which Intuitionistic Logic could *technically* be viewed as *uncountably infinitely many-valued*. But none of these ways would sustain a sensible philosophical interpretation for any of the truth values other than the ones among them that would count as T and F. The three treatments in question have been provided by McKinsey and Tarski [1948], Shoesmith and Smiley [1971], and McCarty [2005].[2] We leave to the reader the casual inspection that would suffice to convince one that the sheer technical artificiality of these 'truth values', ingenious though their furnishing has been, prevents them from sustaining any sensible philosophical interpretation. Even though there are continuum-many of them, they do not—unlike, say, probability values in the real interval [0,1]—afford any kind of operationalization in terms of dispositions to behavior on the part of rational agents; or explanation of their practices of assertion, denial, conjecture, and supposition; or their *valuing* propositions or statements enjoying such values, in any significantly differentiating way.

This brief survey of attempts to render Intuitionistic Logic as a many-valued system shows, in fact, that as far as Intuitionistic Logic is concerned the whole formal-semantical approach based on *assigning values* to proposi-

---

[2] *This long footnote may be skipped by the non-technical reader.*

See McKinsey and Tarski [1948], Theorem 4.3 at p. 12, for the result that a propositional formula is an intuitionistic theorem if and only if it vanishes identically in every Brouwerian algebra (upon any assignments of elements of the algebra to atoms, and with connectives interpreted by particular algebraic operations). Theorem 1.19 on p. 134 of McKinsey and Tarski [1946] states

> The algebra of closed sets of a topological space, and every subalgebra of this algebra, is a Brouwerian algebra. Conversely, every Brouwerian algebra is isomorphic to a subalgebra of the algebra of closed sets of a topological space.

In particular, the closed sets of any Euclidean space form a Brouwerian algebra with uncountably many elements. And these are to serve as 'truth values' in this particular 'formal semantics' for Intuitionistic Logic.

See Shoesmith and Smiley [1971], Theorem 1, especially the definition of the matrix for the extended calculus M (with the matrix *elements*—the 'truth values'—being the *wffs* of M), and what counts as an assignment of such values, at p. 611 *infra*. The extended calculus is obtained from the original unextended one by supplying new propositional variables takng the form of *ordered pairs* consisting of a propositional variable of the unextended calculus, along with a *set of wffs* of the same. This is what leads to there being an uncountable infinity of wffs in M, hence also of 'truth values'.

See McCarty [2005] at p. 361–2 for the suggestion that intuitionists take for their 'space' of truth values the enigmatic Prop, defined as the set of all subsets of the singleton of 0, and which McCarty proves to be uncountable and indecomposable (Corollary 7), and a Dedekind set, i.e. neither finite nor infinite (Corollary 8).

tions is ill-conceived. That is the deep reason why the first really illuminating formal semantics for intuitionism, namely Kripke's possible-worlds (or states-of-knowledge) semantics succeeded whereas 'assignment-of-values' semantics had failed. For the fundamental notion in Kripke's formal semantics is that of a *node* (or *index*, or *world*) *forcing* a sentence. One can think of that as entitling any agent apprised of what has been established 'within' that node or state-of-knowledge to *assert* the sentence in question—to *hold it as true*. And being entitled to *deny* a sentence is equivalent to being able to *assert its negation*—which of course one can do just in case one is able to *refute* the assumption that it might ever, in the future, prove to be true (which is to say: *be proved* to be true).

We have digressed, however instructively; let us now return to the main thread. The familiar truth values T and F are all that the constructivist *needs* in order to attain the eventual enlightenment afforded by Core Logic.

We shall use

as a complex symbol representing a *verification* of the sentence φ (of a formal language) by appeal to some collection Λ of 'basic facts' assumed fixed and available;[3] and we shall use

as a complex symbol representing a *falsification* of the sentence ψ by appeal to the collection Λ. The symbol $\bot$ represents *absurdity*, or *constant falsehood*. It is, if you like, the inferentialist's analog of the semantic value F, deployed within certain kinds of complex, proof-like constructions (including proofs themselves). We shall be explaining these presently.

It is important to appreciate that the complex entities represented by these complex symbols are not necessarily proofs. They need not be effectively surveyable; they need not be finite. We shall on occasion use the following synonyms for the respective notions:

verification; truthmaker; evaluation 'proof'

falsification; falsitymaker; evaluation 'disproof'

---

[3] We choose the Greek letter Λ because it suggests, alliteratively, a set of *l*iterals.

and will often drop the scare quotes even in the infinite, unsurveyable cases. In the finite, surveyable cases, however, the metaphysical and epistemic guises coincide: verifications are indeed proofs, and falsifications disproofs. It is only in the infinite case—when we move to first-order logic—that the two notions come apart. More on that anon. For the time being, while we are dealing with propositional logic and the connectives $\neg$, $\wedge$, $\vee$ and $\rightarrow$, the constructions that we call verifications and falsifications will be special kinds of proof and disproof respectively. And the rules by whose applications we can construct them will be like logical rules of inference.

As will emerge in due course, verifications and falsifications are inductively defined, hence complex, abstract objects. They involve sentences of the object language; but the inductive definition is of course given in the metalanguage. Like all inductive definitions, that of verifications and falsifications requires *basis clauses*—stipulations of what are to count as the *simplest* objects of the kinds being defined.

3.3   Atomic determinations

3.3.1   Using the simplest kind of atomic basis

We shall begin our investigation of verifications and falsifications (truthmakers and falsitymakers) within the simplified confines of propositional logic, which involves only the connectives.

At the propositional level, the simplest constructions of the kinds being defined are the following *atomic determinations* (for atomic sentences A):

$$\frac{}{A} \qquad \frac{A}{\bot}$$

These correspond, respectively, to saying 'A is true' or 'A is false'. We shall be dealing in general with finite sets Λ of such atomic determinations—containing, for each atomic sentence A, at most one of the atomic determinations just displayed. Such a set Λ is called an *atomic basis*. It is the rule equivalent of an *assignment of truth values* to the atomic sentences involved. The requirement that *at most one* of these two possibilities be included in Λ (for each atomic sentence A) is simply the requirement, standard in the semantical treatment of propositional languages, that such an assignment be single-valued where defined—hence, semantically *coherent*.

3.3.2 Co-inductive definition of Λ-relative verifications and falsifications

Equipped with this *very simple* conception of Λ,[4] it is now a straightforward matter to state the basis clauses in the co-inductive definition of Λ-*relative* verifications and falsifications. They are as follows.

$$\overline{A} \;\text{ is a verification of } A \text{ relative to } \left\{\overline{A}\right\}$$

$$\frac{A}{\perp} \;\text{ is a falsification of } A \text{ relative to } \left\{\frac{A}{\perp}\right\}$$

More formally:

$$\overline{\mathcal{V}\left(\overline{A}, A, \left\{\overline{A}\right\}\right)} \qquad \overline{\mathcal{F}\left(\frac{A}{\perp}, A, \left\{\frac{A}{\perp}\right\}\right)}$$

So much for the basis clauses in the inductive definition of Λ-relative verifications and falsifications. Let us now investigate exactly what the inductive-definitional clauses are.

Verifying and falsifying negations

The truth table for negation is as follows.

φ	$\neg\varphi$
T	F
F	T

One reads the truth table row by row, from left to right. The first row of the truth table for negation says, in effect,

Any verification of φ immediately yields a falsification of $\neg\varphi$.

---

[4] In the simplest case under consideration here, the atomic basis Λ has only members of the two forms

$$\overline{A} \qquad \frac{A}{\perp}$$

In due course we shall be considering (still at the propositional level) atomic bases of a more general form. One can allow, for example, rules with more than one premise, and whose conclusions are atoms (or $\perp$). One can also allow premises in the rules to be literals (atoms or negations of atoms). For the time being, however, we stay with the simplest case of an atomic basis.

So the first row of the truth table for negation tells us that

Any verification

$$\Lambda$$
$$\vdots$$
$$\varphi$$

of φ relative to Λ immediately yields a falsification

$$\overbrace{\neg\varphi\,,\,\Lambda}$$
$$\vdots$$
$$\bot$$

of $\neg\varphi$ relative to Λ.

Now the inferentialist can easily justify *this* claim, because he has a rule for the immediate construction of *falsifications from verifications* as follows:

If $\begin{matrix}\Lambda\\V\\\varphi\end{matrix}$ is a verification of φ relative to Λ, then $\dfrac{\begin{matrix}\Lambda\\V\\\neg\varphi\qquad\varphi\end{matrix}}{\bot}$ is a

falsification of $\neg\varphi$ relative to Λ.

So the inferentialist can justify the *first* row of the truth table for negation. The *second* row says

Any falsification of φ relative to Λ immediately yields a verification of $\neg\varphi$ relative to Λ.

Or, in slightly more expansive terms:

Any falsification $\quad\begin{matrix}\overbrace{\varphi\,,\,\Lambda}\\\vdots\\\bot\end{matrix}\quad$ of φ relative to Λ immediately yields

a verification $\quad\begin{matrix}\Lambda\\\vdots\\\neg\varphi\end{matrix}\quad$ of $\neg\varphi$ relative to Λ.

Once again the inferentialist can justify *this* claim because he has a rule for the immediate construction of *verifications from falsifications* as follows:

$$\text{If} \quad \overbrace{\varphi\,,\,\Lambda}^{} \atop \dfrac{F}{\bot} \quad \text{is a falsification of } \varphi \text{ relative to } \Lambda, \text{ then}$$

$$(i)\text{---} \atop \overbrace{\varphi\,,\,\Lambda}^{} \atop F \atop \dfrac{\bot}{\neg\varphi}\,(i)$$

is a verification of $\neg\varphi$ relative to Λ.

(Italic F stands for a falsification; roman F stands for the truth value False.) Note the 'discharge' notation. Within the falsification F, the conclusion $\bot$ is derived from the *assumption* φ, *modulo* the basic information Λ. The purpose of the discharge stroke over that assumption occurrence of φ is to emphasize that in the resulting verification of $\neg\varphi$, the newly verified conclusion $\neg\varphi$ does *not* depend on φ; it depends only on Λ.

The inferentialist's two rules of immediate construction can be expressed more compactly by the following *meta*linguistic conditionals:

$$\mathcal{V}(V,\varphi,\Lambda) \;\Rightarrow\; \mathcal{F}\left(\dfrac{\neg\varphi\;\;V}{\bot}\,,\neg\varphi,\Lambda\right)$$

$$\mathcal{F}(F,\varphi,\Lambda) \;\Rightarrow\; \mathcal{V}\left(\dfrac{F}{\neg\varphi}\,,\neg\varphi,\Lambda\right)$$

Alternatively, they can be expressed by the following (metalinguistic) *inductive-definitional rules*:

$$\dfrac{\mathcal{V}(V,\varphi,\Lambda)}{\mathcal{F}\left(\dfrac{\neg\varphi\;\;V}{\bot}\,,\neg\varphi,\Lambda\right)} \qquad \dfrac{\mathcal{F}(F,\varphi,\Lambda)}{\mathcal{V}\left(\dfrac{F}{\neg\varphi}\,,\neg\varphi,\Lambda\right)}$$

We see now that the inductive definition is in fact *co-inductive*. The two ternary notions $\mathcal{V}$ and $\mathcal{F}$ *intercalate* in the formulation of the co-inductive definitional rules for the unary connective $\neg$. This will happen also with the co-inductive definitional rules corresponding to the binary connective $\rightarrow$, as we shall presently see.

Verifying and falsifying conjunctions

The truth table for conjunction is

φ_1	φ_2	$\varphi_1 \wedge \varphi_2$
T	T	T
T	F	F
F	T	F
F	F	F

Once again one reads the truth table row by row, from left to right. The first row of the truth table for conjunction says, in effect,

> Any verification of φ_1 relative to Λ_1 together with any verification of φ_2 relative to Λ_2 immediately yield a verification of $\varphi_1 \wedge \varphi_2$ relative to $\Lambda_1 \cup \Lambda_2$.

The third and fourth rows together reveal that a falsification of the conjunct φ_1 immediately yields a falsification of the conjunction:

> Any falsification of φ_1 relative to Λ immediately yields a falsification of $\varphi_1 \wedge \varphi_2$ relative to Λ.

The second and fourth rows together reveal that a falsification of the conjunct φ_2 likewise immediately yields a falsification of the conjunction:

> Any falsification of φ_2 relative to Λ immediately yields a falsification of $\varphi_1 \wedge \varphi_2$ relative to Λ.

These observations are captured by the following co-inductive definitional rules.

$$\frac{\mathcal{V}(V_1, \varphi_1, \Lambda_1) \qquad \mathcal{V}(V_2, \varphi_2, \Lambda_2)}{\mathcal{V}\left(\begin{array}{cc} V_1 & V_2 \\ \hline \varphi_1 \wedge \varphi_2 \end{array} , \varphi_1 \wedge \varphi_2 , \Lambda_1 \cup \Lambda_2 \right)}$$

$$\frac{\mathcal{F}(F, \varphi_1, \Lambda)}{\mathcal{F}\left(\begin{array}{cc} \varphi_1 \wedge \varphi_2 & F \\ \hline \bot \end{array} , \varphi_1 \wedge \varphi_2 , \Lambda \right)} \qquad \frac{\mathcal{F}(F, \varphi_2, \Lambda)}{\mathcal{F}\left(\begin{array}{cc} \varphi_1 \wedge \varphi_2 & F \\ \hline \bot \end{array} , \varphi_1 \wedge \varphi_2 , \Lambda \right)}$$

It follows that the inferentialist can justify the left-to-right reading of each row of the truth table for $\wedge$.

Verifying and falsifying disjunctions

The truth table for disjunction is

φ_1	φ_2	$\varphi_1 \vee \varphi_2$
T	T	T
T	F	T
F	T	T
F	F	F

Once again one reads the truth table row by row, from left to right. The fourth row of the truth table for disjunction says, in effect,

Any falsification of φ_1 relative to Λ_1 together with any falsification of φ_2 relative to Λ_2 immediately yield a falsification of $\varphi_1 \vee \varphi_2$ relative to $\Lambda_1 \cup \Lambda_2$.

The first and second rows together reveal that a verification of the disjunct φ_1 immediately yields a verification of the disjunction:

Any verification of φ_1 relative to Λ immediately yields a verification of $\varphi_1 \vee \varphi_2$ relative to Λ.

The first and third rows together reveal that a verification of the disjunct φ_2 likewise immediately yields a verification of the disjunction:

Any verification of φ_2 relative to Λ immediately yields a verification of $\varphi_1 \vee \varphi_2$ relative to Λ.

These observations are captured by the following co-inductive definitional rules.

$$\frac{\mathcal{V}(V, \varphi_1, \Lambda)}{\mathcal{V}\left(\dfrac{V}{\varphi_1 \vee \varphi_2}, \varphi_1 \vee \varphi_2, \Lambda\right)} \qquad \frac{\mathcal{V}(V, \varphi_2, \Lambda)}{\mathcal{V}\left(\dfrac{V}{\varphi_1 \vee \varphi_2}, \varphi_1 \vee \varphi_2, \Lambda\right)}$$

$$\frac{\mathcal{F}(F_1, \varphi_1, \Lambda_1) \qquad \mathcal{F}(F_2, \varphi_2, \Lambda_2)}{\mathcal{F}\left(\dfrac{\varphi_1 \vee \varphi_2 \quad F_1 \quad F_2}{\bot}, \varphi_1 \vee \varphi_2, \Lambda_1 \cup \Lambda_2\right)}$$

It follows that the inferentialist can justify the left-to-right reading of each row of the truth table for $\vee$.

Verifying and falsifying conditionals

The object-language conditional, represented by $\rightarrow$, has the following truth table, widely agreed upon as capturing the truth-value combinatorics of the 'if ... then ...' locution in mathematical reasoning:

φ_1	φ_2	$\varphi_1 \rightarrow \varphi_2$
T	T	T
T	F	F
F	T	T
F	F	T

The first and third rows tell us that

> Any verification of φ_2 immediately yields a verification of $\varphi_1 \to \varphi_2$.

The third and fourth rows tell us that

> Any falsification of φ_1 immediately yields a verification of $\varphi_1 \to \varphi_2$.

The second row tells us that

> Any verification of φ_1 relative to Λ_1 and falsification of φ_2 relative to Λ_2 together immediately yield a falsification of $\varphi_1 \to \varphi_2$ relative to $\Lambda_1 \cup \Lambda_2$.

These observations are captured by the following co-inductive definitional rules, intercalating once more for $\mathcal{V}$ and $\mathcal{F}$.

$$\frac{\mathcal{V}(V, \varphi_2, \Lambda)}{\mathcal{V}\left(\dfrac{V}{\varphi_1 \to \varphi_2}, \varphi_1 \to \varphi_2, \Lambda\right)} \qquad \frac{\mathcal{F}(F, \varphi_1, \Lambda)}{\mathcal{V}\left(\dfrac{F}{\varphi_1 \to \varphi_2}, \varphi_1 \to \varphi_2, \Lambda\right)}$$

$$\frac{\mathcal{V}(V, \varphi_1, \Lambda_1) \qquad\qquad \mathcal{F}(F, \varphi_2, \Lambda_2)}{\mathcal{F}\left(\dfrac{\varphi_1 \to \varphi_2 \quad V \quad F}{\bot}, \varphi_1 \to \varphi_2, \Lambda_1 \cup \Lambda_2\right)}$$

It follows that the inferentialist can justify the left-to-right reading of each row of the truth table for $\to$.

3.3.3   Rules of verification and falsification, in graphic form

Here we gather up, provide convenient labels for, and restate in a more accessible *graphic* form the rules stated above for the construction of verifications and/or falsifications of propositional compounds relative to the simplest kind of atomic basis Λ. These graphic forms will look familiar to any logician well-versed in the natural-deduction systems of Gentzen and Prawitz—albeit with some unusual differences of detail, since we are dealing with evaluations, not with deductive proofs in general. The reader should, however, pay careful attention to the following important features of the rules when stated in the forms they enjoy below.

1. A **verification** of a sentence φ (relative to an interpretation M) establishes φ as its *conclusion*:

2. A **falsification** of a sentence φ (relative to an interpretation M) establishes absurdity ($\bot$) from φ as its *major premise*:

Inspection of the rules will reveal that the MPE-occurrence of φ is at the terminal step, and that it does *not* itself stand as the conclusion of a verification. In other words, the MPE-occurrence of φ has no 'evaluation work' above it. It *stands proud.*

3. Verifications and falsifications are 'relative to an interpretation M' in the sense that the premises of a verification of φ, and the premises (other than φ itself) of a falsification of φ, are *literals* (or *rules of atomic determination*, as described in §3.2) from the 'atomic diagram' of M. Hence the notation Λ_M.

Here now are the rules in graphic form. By means of the boxes subscripted to the discharge-strokes, we intend to remind the reader that the assumptions in question really have been used, and are available to be discharged, upon applying the rule in question. Also, the absence of any vertically descending dots above the major premises of the falsification rules indicates that those premises stand proud, with no 'evaluation work' above them.

$(\neg\mathcal{V})$

$$
\underbrace{\square\!\!-\!\!(i) \atop \varphi\,,\,\Lambda}
$$
$$
\vdots
$$
$$
\frac{\bot}{\neg\varphi}(i)
$$

$(\neg\mathcal{F})$

$$
\Lambda
$$
$$
\vdots
$$
$$
\frac{\neg\varphi \qquad \varphi}{\bot}
$$

$(\wedge\mathcal{V})$

$$
\begin{array}{cc} \Lambda_1 & \Lambda_2 \\ \vdots & \vdots \\ \varphi_1 & \varphi_2 \end{array}
$$
$$
\frac{}{\varphi_1 \wedge \varphi_2}
$$

$(\wedge\mathcal{F})$

$$
\frac{\varphi_1 \wedge \varphi_2 \qquad \underbrace{\square\!\!-\!\!(i) \atop \varphi_1\,,\,\Lambda} \atop \vdots \atop \bot}{\bot}(i)
\qquad\qquad
\frac{\varphi_1 \wedge \varphi_2 \qquad \underbrace{\square\!\!-\!\!(i) \atop \varphi_2\,,\,\Lambda} \atop \vdots \atop \bot}{\bot}(i)
$$

$(\vee\mathcal{V})$

$$
\begin{array}{cc} \Lambda & \Lambda \\ \vdots & \vdots \\ \varphi_1 & \varphi_2 \end{array}
$$
$$
\frac{\varphi_1}{\varphi_1 \vee \varphi_2} \qquad \frac{\varphi_2}{\varphi_1 \vee \varphi_2}
$$

$(\vee\mathcal{F})$

$$
\frac{\varphi_1 \vee \varphi_2 \qquad \underbrace{\square\!\!-\!\!(i) \atop \varphi_1\,,\,\Lambda_1} \atop \vdots \atop \bot \qquad \underbrace{\square\!\!-\!\!(i) \atop \varphi_2\,,\,\Lambda_2} \atop \vdots \atop \bot}{\bot}(i)
$$

$(\to \mathcal{V})$

$$
\begin{array}{cc}
\begin{array}{c}
\Box\!\!-\!\!(i) \\
\overbrace{\varphi_1 \, , \, \Lambda} \\
\vdots \\
\dfrac{\bot}{\varphi_1 \to \varphi_2}(i)
\end{array}
&
\begin{array}{c}
\Lambda \\
\vdots \\
\dfrac{\varphi_2}{\varphi_1 \to \varphi_2}
\end{array}
\end{array}
$$

$(\to \mathcal{F})$

$$
\dfrac{\quad\begin{array}{cc}
\begin{array}{c}
\Lambda_1 \\
\vdots \\
\varphi_1 \to \varphi_2
\end{array}
&
\begin{array}{c}
\quad\quad \Box\!\!-\!\!(i) \\
\varphi_1 \quad\quad \overbrace{\varphi_2 \, , \, \Lambda_2} \\
\vdots \\
\bot
\end{array}
\end{array}\quad (i)}{\bot}
$$

Note that the atomic bases Λ, Λ_1 and Λ_2 featuring in these rules are strictly *cumulative* as one 'goes down' a verification or falsification constructed in accordance with the rules. *No member of these sets ever gets discharged.* They simply 'parametrize' the various rule applications, so that one can tell, by the end of the verification or falsification in question, what atomic rules have been applied in its construction. Any 'assumptions' that come to be discharged within any verification or falsification are sentence occurrences standing as major premises for applications of $\mathcal{F}$-rules.

The foregoing rules comprise the propositional system of *Truth-Table Logic*, which we shall call $\mathbb{T}$.[5] It is generated by the simplest kinds of atomic determination—namely, those dealing with but a single atom at a time. If we make atomic determinations by means of atoms A or their negations $\neg A$ (instead of the corresponding rules of inference $\overline{A}$ or $\frac{A}{\bot}$), then the sets Λ will be (non-empty) sets of *literals*. These will be the *sentential* (as opposed to *inferential*) versions of the simplest kinds of atomic determination of which we have been speaking.[6]

By inspection of the rules one readily sees, by an easy mathematical induction, that every provable sequent in the logic $\mathbb{T}$ is of one of the forms

$$\Lambda : \varphi \quad\quad \text{or} \quad\quad \varphi, \Lambda : \bot$$

where φ is any sentence and Λ is a non-empty set of literals involving only such atoms as occur in φ. So the proving or refuting of sentences φ that

---

[5] See Tennant [1989].

[6] As indicated in footnote 4, there are more complex kinds of atomic determination, which we shall be describing in §3.5.

can be accomplished within $\mathbb{T}$ is, respectively, always from, or *modulo*, a non-empty set Λ of *literals*.

Note that we impose no requirement that Λ not contain any atom along with its negation. So, for example, the $\mathbb{T}$-proof

$$\frac{A \quad \neg A}{A \wedge \neg A}$$

establishes the sequent $A, \neg A : A \wedge \neg A$, which is of the form $\Lambda : \varphi$ with (obviously) inconsistent Λ. But it is clear that there is no $\mathbb{T}$-proof of the 'positive version' of the notorious First Lewis Paradox $A, \neg A : B$. Nor is there any $\mathbb{T}$-proof of the 'negative version' $A, \neg A : \neg B$. This is enough to qualify $\mathbb{T}$ as a relevance logic; and indeed even more can be said about variable-sharing in $\mathbb{T}$ that would clinch its status as a relevant logic. Chapter 10 will establish a 'strongest possible' variable-sharing result for Classical Core Logic. Since Classical Core Logic includes $\mathbb{T}$, the result in question holds for $\mathbb{T}$ as well.

The Double Negation inference $\neg\neg A : A$ is of neither of the two forms displayed above; so it is not a provable sequent in $\mathbb{T}$. For the same reason, neither is the Law of Excluded Middle, $\emptyset : A \vee \neg A$. So, in delivering the left–right action of the truth tables, the inferentialist using $\mathbb{T}$ makes no commitment at all to a realist outlook about the determinacy of truth.

We have shown, then, that *truth-value determination* for any complex sentence φ, relative to a truth-value assignment, can be accomplished by means of inferences that are both *constructive* and *relevant*. Such inferences suffice to mimic any calculation of the truth value of a compound sentence once given truth values for its atomic constituents. There is nothing strictly *classical* about any aspect of the way in which a Classical logician works out the truth value of a propositional compound when given an assignment of truth values to its atoms. Nor is there anything irrelevant.

3.4 Extension to first order: saturated terms and formulae

Thus far we have been treating only the connectives of propositional logic. We shall now develop some ideas that will be useful when treating the semantics of *first-order* languages, which allow one to make relational predications, identity statements, and universally or existentially quantified statements.

In general, terms and formulae of the object language may contain free variables. If they do, then they are called *open*. Those that are not open are called *closed*. A closed formula is called a *sentence*. A closed term may be called a (simple or complex) *name*. The semantic value of a name, when it has one, is an *individual*, which the name is then said to *denote*. The semantic value of a sentence, when it has one, is a *truth value*, and the sentence is said to be *true* or *false* according to whether that value is T or F.

Closing an open term or formula involves substituting closed terms (of the object language) for free occurrences of variables. Thus one could substitute the object-linguistic name j for the free occurrence of the variable x in the open term $f(x)$, to obtain the closed term $f(j)$ ('the father of John'). Or, to complicate the example slightly, one could substitute the closed object-linguistic term $m(j)$ for that free occurrence, to obtain the closed term $f(m(j))$ ('John's maternal grandfather'). Likewise, an open formula, say $L(x,y)$, may be closed by substituting closed terms for its free occurrences of variables. One such closing would result in the sentence $L(m(f(j)), f(j))$ ('John's paternal grandmother loves his father').

Here we shall introduce an operation on open terms and formulae analogous to the operation of closing, but importantly different from it. The new operation will be called *saturation*. Like the operation of closing, the operation of saturation gets rid of all free occurrences of variables within an object-linguistic term or formula. But the way it does so is importantly different. Instead of substituting closed object-linguistic terms for free occurrences of variables, saturation is effected by substituting *individuals from the domain* for those free occurrences. Thus if α and β are individuals from the domain, one saturation of the open formula $L(x,y)$ would be $L(\alpha,\beta)$. Another one would be $L(m(f(\alpha)),f(\alpha))$, where the saturation is effected by substituting the saturated terms $m(f(\alpha))$ and $f(\alpha)$ for the free occurrences of the variables x and y respectively.

When the domain D supplies all the individuals involved in a saturation operation, the resulting saturated terms are called *saturated D-terms*, and the resulting saturated formulae are called *saturated D-formulae*.

Saturated terms and formulae are object-linguistic and metalinguistic hybrids. But, as mathematical objects, they are well defined. When one treats, in standard Tarskian semantics, of assignments of individuals to variables, one is assuming such well-defined status for ordered pairs of the form $\langle x,\alpha \rangle$, where x is a free variable of the object language, and α is an individual from the domain of discourse.[7] Since standard semantics is already

---

[7] This is true not only of Tarski's original treatment Tarski [1956], which invoked *infinite sequences* of individuals correlated with object-linguistic variables, but also of the treatment (in Tennant [1978]) of Tarski's approach that appeals, more modestly, to *finitary* assignments of individuals to the free variables in a formula.

committed to the use of such hybrid entities, it may as well take advantage of similar hybrid entities such as saturated terms and formulae. Formal semantics seeks, after all, to describe how the sentences of a language succeed in saying things about the objects in its domain of discourse.

We shall be taking advantage of such hybrid entities by having them feature in atomic rules of inference that constitute the 'atomic diagram' of a model. Saturated formulae will also feature importantly in the rules of verification and falsification (relative to a model M) for *primitive* saturated formulae. These rules are stated in §3.4.1.

We shall now examine what an inferentialist might call the 'rules of evaluation' by means of which *quantified* sentences are determined as true or as false under a given interpretation. Interpretations, or models, are now construed as consisting of domains of individuals, with denotations supplied for names, and with extensions supplied for predicates.

3.4.1 Rules of verification and of falsification at first order

The simplest verifications (relative to an interpretation M of a first-order language) are of the form

$$\frac{}{P(\alpha_1,\ldots,\alpha_n)}M$$

where in M the predicate P holds of $\alpha_1,\ldots,\alpha_n$. This is a verification of $P(\alpha_1,\ldots,\alpha_n)$ relative to M.

The simplest falsifications (relative to an interpretation M) are of the form

$$\frac{P(\alpha_1,\ldots,\alpha_n)}{\bot}M$$

where in M the predicate P fails to hold of $\alpha_1,\ldots,\alpha_n$. This is a falsification of $P(\alpha_1,\ldots,\alpha_n)$ relative to M.

We have also the following rules enabling formation of verifications, and of falsifications, of primitive saturated formulae relative to a model M.

$$\frac{A(\alpha_1,\ldots,\alpha_n) \quad t_1 = \alpha_1 \quad \ldots \quad t_n = \alpha_n}{A(t_1,\ldots,t_n)}$$

where A is a primitive n-place predicate, $t_1,\ldots,t_n$ are saturated D-terms, and $\alpha_1,\ldots,\alpha_n$ are individuals in the domain.

$$\dfrac{\overline{A(\boldsymbol{\alpha_1}, \ldots, \boldsymbol{\alpha_n})}^{(i)}}{\bot}_M \qquad \dfrac{t_1 = \boldsymbol{\alpha_1} \quad \ldots \quad t_n = \boldsymbol{\alpha_n} \qquad A(t_1, \ldots, t_n)}{\bot}^{(i)}$$

where A is a primitive n-place predicate, $t_1, \ldots, t_n$ are saturated D-terms, and $\boldsymbol{\alpha_1}, \ldots, \boldsymbol{\alpha_n}$ are individuals in the domain.

$$\dfrac{f(\boldsymbol{\alpha_1}, \ldots, \boldsymbol{\alpha_n}) = \boldsymbol{\alpha} \qquad u_1 = \boldsymbol{\alpha_1} \qquad \ldots \qquad u_n = \boldsymbol{\alpha_n}}{f(u_1, \ldots u_n) = \boldsymbol{\alpha}}$$

where f is a primitive n-place function sign, $u_1, \ldots, u_n$ are saturated D-terms and $\boldsymbol{\alpha}, \boldsymbol{\alpha_1}, \ldots, \boldsymbol{\alpha_n}$ are individuals in the domain.

$$\dfrac{\overline{f(\boldsymbol{\alpha_1}, \ldots, \boldsymbol{\alpha_n}) = \boldsymbol{\alpha}}^{(i)}}{\bot}_M \qquad \dfrac{u_1 = \boldsymbol{\alpha_1} \quad \ldots \quad u_n = \boldsymbol{\alpha_n} \qquad f(u_1, \ldots, u_n) = \boldsymbol{\alpha}}{\bot}^{(i)}$$

where f is a primitive n-place function sign, $u_1, \ldots, u_n$ are saturated D-terms, and $\boldsymbol{\alpha}, \boldsymbol{\alpha_1}, \ldots, \boldsymbol{\alpha_n}$ are individuals in the domain.

$$\overline{\boldsymbol{\alpha} = \boldsymbol{\alpha}} \qquad \text{where } \boldsymbol{\alpha} \text{ is an individual in the domain.}$$

Evaluation rules are rules of verification or falsification. More complex verifications and falsifications (of complex saturated formulae, relative to M) are built up in the first-order case not only by means of our evaluation rules in §3.3.3 for the connectives, but also by means of evaluation rules for the quantifiers. Here we address the question of evaluation rules for the quantifiers, and present them similarly in graphic form. The reader should not lose sight of the fact, however, that these graphic forms are just handy and familiar-looking abbreviations for more fully detailed formal co-inductive-definitional clauses for the two notions

$$V \text{ is an } M\text{-relative verification of } \begin{bmatrix} \exists x \varphi \\ \forall x \varphi \end{bmatrix} \text{ relative to } \Lambda_M$$

and

$$F \text{ is an } M\text{-relative falsification of } \begin{bmatrix} \exists x \varphi \\ \forall x \varphi \end{bmatrix} \text{ relative to } \Lambda_M.$$

Here, then, are the graphic forms of the evaluation rules in question.

$(\exists\mathcal{V})$
$$\begin{array}{c} \Lambda \\ \vdots \\ \dfrac{\varphi^x_\alpha}{\exists x \varphi} \end{array}$$
where α is any individual in the domain

$(\exists\mathcal{F})$
$$\cfrac{\exists x \varphi \qquad \underbrace{\begin{array}{c} \square\!\!-\!\!\!-(i) \\ \varphi^x_{\alpha_1} \,, \, \Lambda_1 \end{array}}_{} \quad \begin{array}{c} \vdots \\ \perp \end{array} \quad \cdots \quad \underbrace{\begin{array}{c} \square\!\!-\!\!\!-(i) \\ \varphi^x_{\alpha_n} \,, \, \Lambda_n \end{array}}_{} \quad \begin{array}{c} \vdots \\ \perp \end{array} \quad \cdots}{\perp}(i)\, M$$
where $\alpha_1, \ldots, \alpha_n, \ldots$ are all the individuals in the domain

$(\forall\mathcal{V})$
$$\cfrac{\begin{array}{c} \Lambda_1 \\ \vdots \\ \psi(\alpha_1) \end{array} \quad \cdots \quad \begin{array}{c} \Lambda_n \\ \vdots \\ \psi(\alpha_n) \end{array} \quad \cdots}{\forall x \psi(x)}\, M$$
where $\alpha_1, \ldots, \alpha_n, \ldots$ are all the individuals in the domain

$(\forall\mathcal{F})$
$$\cfrac{\forall x \psi(x) \qquad \underbrace{\begin{array}{c} \square\!\!-\!\!\!-\!\!\!-(i) \\ \psi(\alpha) \,, \, \Lambda \end{array}}_{} \begin{array}{c} \vdots \\ \perp \end{array}}{\perp}(i)$$
where α is any individual in the domain

Note how the two rules $(\exists\mathcal{F})$ and $(\forall\mathcal{V})$ have their inference strokes labeled with M. This is in order to emphasize that applications of these rules are *M-relative*. For it is the domain of M that determines exactly which immediate sub-evaluations have to be constructed for the application of the rule in question. $(\exists\mathcal{F})$ calls for a falsification of each instance; whereas $(\forall\mathcal{V})$ calls for a verification of each instance.

3.4.2 Further features of first-order verifications and falsifications relative to the simplest kind of atomic basis

With the move to first order, it is worth repeating our past mention of notable features of verifications and falsifications, and expanding on them as appropriate for the first-order case.

1. Verifications and falsifications are *relative to an interpretation (that is, model)* M. They 'depend' *only on M-literals* as their 'undischarged assumptions'. But they 'depend' also on those applications of *M-relative* evaluation rules for the quantifiers, namely $(\forall \mathcal{V})$ and $(\exists \mathcal{F})$, that require domain-many sub-evaluations for all the individual instances from the domain.

2. Verifications and falsifications are, mathematically speaking, (labeled) trees all of whose branches are finite. But, if the domain of M is infinite, then these trees have infinite branchings at applications of the M-relative rules $(\forall \mathcal{V})$ and/or $(\exists \mathcal{F})$.

3. The 'premises' of a verification or falsification are 'atomic facts' of M. The set of such facts is consistent: for no atomic A does the set contain both

$$\overline{A} \quad \text{and} \quad \frac{A}{\bot} \ .$$

4. Any verification of a complex sentence φ has φ as the conclusion of its last step, which is an application of the $\mathcal{V}$-rule for the operator dominant in φ, that is the operator applied last in constructing φ.

5. Any falsification of a complex sentence φ has φ as the major premise of its last step, which is an application of the $\mathcal{F}$-rule for the operator dominant in φ.

6. Note that in any verification or falsification, every major premise for the application of an $\mathcal{F}$-rule *stands proud*, in the sense that it is not the conclusion of any verification. Moreover, any such major premise is *discharged* at the very next step. Hence all undischarged major premises φ are those of terminal applications of $\mathcal{F}$-rules in falsifications of φ.

7. Suppose the domain of M is finite. Then every M-relative verification or falsification is finite and effectively surveyable. Moreover, one can

in principle decide, in a finite time, whether a given construction is indeed an *M*-relative verification or falsification; and, if so, of which sentence φ.

It is a straightforward matter to prove (4) and (5) simultaneously by induction.

Finally, a historical note by way of further explanation. I am grateful to an anonymous referee who suggested that my truthmakers might be 'cousins', or 'twin brothers', of Brouwer's 'fully analyzed proofs'. (See p. 394, fn. 8 of Brouwer [1927].) But infortunately this is to misrepresent their structure. In truthmakers and falsitymakers alike, even in ones relative to models whose domains are infinite, there is a *uniform upper bound* on the lengths of all branches. This is *not* the case with Brouwer's fully analyzed proofs. Branches in *those* might well be finite, but in general they can be *arbitrarily* (albeit finitely) long. So the best analogy one is entitled to suggest is that my truthmakers are like 'cousins, *infinitely many times removed*' from Brouwer's fully analyzed proofs.

Nor, for similar reasons, are my truthmakers to be identified with proofs in Peano Arithmetic extended with Carnap's infinitary ω-rule.

Also, my truthmakers are *entirely general*: for any model *M*, and for any sentence φ that is true in *M*, there is at least one *M*-relative truthmaker for φ (and in general there are many).

3.4.3   On the soundness of the evaluation rules

Evaluations constructed in accordance with the foregoing rules have as their premises (i.e. undischarged assumptions) certain *literals* that are used to codify an assignment of truth values to saturated atomic formulae. We need assurance that it is impossible for any saturated formula to receive the value *T* according to one evaluation, but the value *F* according to some other. That assurance is provided by the following result. It says that evaluative contradictions can always be 'atomically localized'.

Metatheorem 1. *Suppose that V is a verification of θ modulo the set τ of literals, and that F is a falsification of θ modulo the set σ of literals. Suppose moreover that in both V and F all steps of $(\exists\text{-}\mathcal{F})$ and/or $(\forall\text{-}\mathcal{V})$ are with respect to the same individuals $\alpha_1, \ldots, \alpha_n, \ldots$.*

Then $\tau \cup \sigma$ is 'literally incoherent', i.e. $\tau \cup \sigma$ contains, for some atomic saturated literal A, both $\dfrac{}{A}$ and $\dfrac{A}{\bot}$.

Proof. By induction on θ.[8]

Basis. The co-inductive definition of verifications and of falsifications provides for verifications or falsifications of *atomic* saturated literals A only via the following basis clauses:

(i) $\overline{A}$ is a verification of A modulo $\left\{\overrightarrow{A}\right\}$.

(ii) $\frac{A}{\perp}$ is a falsification of A modulo $\left\{\frac{A}{\perp}\right\}$.

We note that $\left\{\overrightarrow{A}\right\} \cup \left\{\frac{A}{\perp}\right\}$ is literally incoherent.

Inductive Hypothesis (IH). Suppose the result holds for the immediate saturated subformulae of θ.

Inductive Step. We consider θ by cases, according to whether θ has the form $\neg\varphi$, $\varphi \wedge \psi$, $\varphi \vee \psi$, $\varphi \rightarrow \psi$, $\exists x\varphi$ or $\forall x\varphi$; and where, by IH, the result holds for φ and for ψ in the connected cases, and for each individual instance φ_α^x in the quantified cases.

Case 1: $\theta = \neg\varphi$. The supposed verification V and falsification F have the respective forms

$$
\begin{array}{c}
\square\!\!-\!\!(i) \\
\underbrace{\varphi, \tau} \\
F' \\
\dfrac{\perp}{\neg\varphi}(i)
\end{array}
\qquad \text{and} \qquad
\begin{array}{c}
\sigma \\
V' \\
\dfrac{\neg\varphi \quad \varphi}{\perp}
\end{array} .
$$

By IH applied to φ, the verification V', and the falsification F', the result holds.

---

[8] This induction proceeds on the obvious transitive well-founded relation 'θ_1 is simpler than θ_2' among saturated formulae, according to which any saturated instance θ_1 of a quantified formula θ_2 is simpler than θ_2, and any constituent of a saturated formula is simpler than it.

Case 2: $\theta = \varphi \wedge \psi$. The supposed verification V and falsification F have the respective forms

$$
\begin{array}{cc}
\tau_1 & \tau_2 \\
V_1 & V_2 \\
\varphi & \psi \\
\hline
\multicolumn{2}{c}{\varphi \wedge \psi}
\end{array}
\qquad \text{and one of}
$$

$$
\begin{array}{c}
\square\!\!-\!\!(i) \\
\overbrace{\varphi \,,\, \sigma} \\
F_1 \\
\dfrac{\varphi \wedge \psi \qquad \bot}{\bot}(i)
\end{array}
\qquad \text{or} \qquad
\begin{array}{c}
\square\!\!-\!\!(i) \\
\overbrace{\psi \,,\, \sigma} \\
F_2 \\
\dfrac{\varphi \wedge \psi \qquad \bot}{\bot}(i)
\end{array} .
$$

By IH applied to φ, the verification V_1, and the falsification F_1, the set $\tau_1 \cup \sigma$ is literally incoherent. Likewise, by IH applied to ψ, the verification V_2, and the falsification F_2, the set $\tau_2 \cup \sigma$ is literally incoherent. Either way, the set $(\tau_1 \cup \tau_2) \cup \sigma$ is literally incoherent.

Case 3: $\theta = \varphi \vee \psi$. The supposed verification V and falsification F have the respective forms

$$
\begin{array}{c}
\tau \\
V_1 \\
\varphi \\
\hline
\varphi \vee \psi
\end{array}
\qquad \text{or} \qquad
\begin{array}{c}
\tau \\
V_2 \\
\psi \\
\hline
\varphi \vee \psi
\end{array} ,
\qquad \text{and}
$$

$$
\dfrac{\varphi \vee \psi \qquad
\begin{array}{c}\square\!\!-\!\!(i)\\ \overbrace{\varphi \,,\, \sigma_1}\\ F_1 \\ \bot\end{array}
\qquad
\begin{array}{c}\square\!\!-\!\!(i)\\ \overbrace{\psi \,,\, \sigma_2}\\ F_2 \\ \bot\end{array}}{\bot}(i) .
$$

By IH applied to φ, the verification V_1, and the falsification F_1, the set $\tau \cup \sigma_1$ is literally incoherent. Likewise, by IH applied to ψ, the verification V_2, and the falsification F_2, the set $\tau \cup \sigma_2$ is literally incoherent. Either way, the set $\tau \cup (\sigma_1 \cup \sigma_2)$ is literally incoherent.

Case 4: $\theta = \varphi \rightarrow \psi$. The supposed verification V and falsification F have the respective forms

$$
\begin{array}{c}
\square\!\!-\!\!(i)\\
\overbrace{\varphi \,,\, \tau}\\
F_1 \\
\dfrac{\bot}{\varphi \rightarrow \psi}(i)
\end{array}
\qquad \text{or} \qquad
\begin{array}{c}
\tau \\
V_1 \\
\psi \\
\hline
\varphi \rightarrow \psi
\end{array} ,
\qquad \text{and}
$$

$$
\dfrac{\varphi \rightarrow \psi \qquad
\begin{array}{c}\sigma_1\\ V_2 \\ \varphi\end{array}
\qquad
\begin{array}{c}\square\!\!-\!\!(i)\\ \overbrace{\psi \,,\, \sigma_2}\\ F_2 \\ \bot\end{array}}{\bot}(i) .
$$

By IH applied to φ, the verification V_2, and the falsification F_1, the set $\tau \cup \sigma_1$ is literally incoherent. Likewise, by IH applied to ψ, the verification V_1 and the falsification F_2, the set $\tau \cup \sigma_2$ is literally incoherent. Either way, the set $\tau \cup (\sigma_1 \cup \sigma_2)$ is literally incoherent.

Case 5: $\theta = \exists x \varphi$. The supposed verification V and falsification F have the respective forms

$$
\begin{array}{c}
\tau \\
V' \\
\varphi^x_{\alpha_k} \\
\hline
\exists x \varphi
\end{array}
\quad \text{and} \quad
\begin{array}{c}
\overset{\Box \text{---}(i)}{\underbrace{\varphi^x_{\alpha_1} , \sigma_1}} \\
F_1 \\
\bot
\end{array}
\;\cdots\;
\begin{array}{c}
\overset{\Box \text{---}(i)}{\underbrace{\varphi^x_{\alpha_n} , \sigma_n}} \\
F_n \\
\bot
\end{array}
\;\cdots\;
\begin{array}{c}
\\
\\
\hline
\bot
\end{array}
\begin{array}{c}
\exists x \varphi \\
\\
\hline
(i)
\end{array}
.
$$

By IH applied to $\varphi^x_{\alpha_k}$, the verification V', and the falsification F_k, the set $\tau \cup \sigma_k$ is literally incoherent. Hence the set $\tau \cup (\bigcup_n \sigma_n)$ is literally incoherent.

Case 6: $\theta = \forall x \varphi$. The supposed verification V and falsification F have the respective forms

$$
\begin{array}{ccc}
\tau_1 & & \tau_n \\
V_1 & \cdots & V_n \\
\psi(\alpha_1) & & \psi(\alpha_n) \\
\hline
\multicolumn{3}{c}{\forall x \psi(x)}
\end{array}
\quad \text{and} \quad
\begin{array}{c}
\overset{\Box \text{---}(i)}{\underbrace{\psi(\alpha_k) , \sigma}} \\
F' \\
\hline
\bot
\end{array}
\begin{array}{c}
\forall x \psi(x) \\
\bot \\
\hline
(i)
\end{array}
.
$$

By IH applied to $\psi(\alpha_k)$, the verification V_k and the falsification F', the set $\tau_n \cup \sigma$ is literally incoherent. Hence the set $(\bigcup_n \tau_n) \cup \sigma$ is literally incoherent. $\qquad\square$

We indicate the end of any proof of a metatheorem (as we have just done here) by employing the familiar end-of-proof marker '$\square$' according to the usual conventions of rigorous but informal mathematical proof.

Let us write
$$ \tau \Vdash \varphi $$
to state that there exists a verification

$$
\begin{array}{c}
\tau \\
V \;; \\
\varphi
\end{array}
$$

and write

$$\varphi, \tau \Vdash \bot$$

to state that there exists a falsification

$$\underbrace{\varphi, \tau}_{F}\ .$$
$$\bot$$

Metatheorem 1 can be construed as saying that

if $\tau \Vdash \varphi$ and $\varphi, \sigma \Vdash \bot$, then $\tau, \sigma \Vdash \bot$

—where the consequent, moreover, is true by virtue of inferential literals of the form $\dfrac{}{A}$ and $\dfrac{A}{\bot}$ in $\tau \cup \sigma$. This is a prototypical statement of *Cut* for the 'deducibility' relation $\Vdash$ —the kind of deducibility that is involved in verifying and/or falsifying saturated formulae. This version of Cut says, roughly, that if one uses a verification and a falsification of any sentence to cut through to absurdity, then the overall set of premise-literals reduces, for some atom A, to $\{A, \neg A\}$. Note that there is epistemic gain involved— the Cut-elimination process codified by the recursive function $[V, F]$ to be defined below ends up identifying the offending pair $\{A, \neg A\} \subseteq (\tau \cup \sigma)$, where τ is the set of literals used as premises in the verification (or 'eval-uation proof') V, and σ is the set of premises used in the falsification (or evaluation 'disproof') F. Note that $\{A, \neg A\} : \bot$ is a *subsequent* of $\tau \cup \sigma : \bot$.

Cut in this primitive form, concerning *evaluations*, is therefore a precur-sor for the more general version that can be established for *proofs* in Core Logic, to wit:

There is a computable function $[\Pi, \Sigma]$ such that

for any core proofs $\begin{array}{c}\Delta\\ \Pi\\ \varphi\end{array}$ and $\begin{array}{c}\overbrace{\varphi, \Gamma}\\ \Sigma\\ \psi\end{array}$, $[\Pi, \Sigma]$ is a core proof of some

subsequent of $\Delta \cup \Gamma : \psi$.

The proof of Metatheorem 1 can be reworked in an illuminating way, as furnishing an inductive definition of a certain two-place 'normalization' or 'Cut-elimination' function[9] $[V, F]$ on any verification V and falsification F

---

[9] The function can be called either a normalization function or a Cut-elimination func-tion because in Core Logic natural deductions are structurally isomorphic to the corre-sponding sequent proofs.

of some same saturated formula. The reader will recognize how in the following definition of $[V, F]$ we appear, in the inductive clauses, to be applying *reduction procedures* in order to eliminate what look like *maximal sentence occurrences*. The complex sentence in question stands as the conclusion of the terminal 'introduction step' in the verification V, and as the major premise of the terminal 'elimination step' in the falsification F. It is only when we reach the more general setting of Cut for Core Logic that the need arises for operations other than such reductions. See Chapter 6 for details.

Definition 2. *Let the binary operation* $[V, F]$ *on verifications V and falsifications F be defined by the following basis clause and inductive clauses. The usual closure clause applies.*

Basis clause:

$$\left[\frac{}{A}\,, \frac{A}{\bot} \right] = \frac{\overline{A}}{\bot}$$

Inductive clauses:

$$\left[\begin{array}{cc} \begin{array}{c} \square\!\!-\!\!(i) \\ \underbrace{\varphi,\ \tau} \\ F' \\ \dfrac{\bot}{\neg\varphi}(i) \end{array} & , & \dfrac{\begin{array}{cc} & \sigma \\ \neg\varphi & \varphi \end{array}}{\bot}\ V' \end{array} \right] = \left[\begin{array}{cc} \dfrac{\sigma}{\varphi}\ V' & , & \begin{array}{c} \square\!\!-\!\!(i) \\ \underbrace{\varphi,\ \tau} \\ F' \\ \bot \end{array} \end{array} \right]$$

$$\left[\begin{array}{cc} \begin{array}{cc} \tau_1 & \tau_2 \\ V_1 & V_2 \\ \varphi & \psi \\ \hline \multicolumn{2}{c}{\varphi \wedge \psi} \end{array} & , & \dfrac{\varphi \wedge \psi \quad \begin{array}{c} \square\!\!-\!\!(i) \\ \underbrace{\varphi,\ \sigma} \\ F' \\ \bot \end{array}}{\bot}(i) \end{array} \right] = \left[\begin{array}{c} \tau_1 \\ V_1 \\ \varphi \end{array}\ , \begin{array}{c} \square\!\!-\!\!(i) \\ \underbrace{\varphi,\ \sigma} \\ F' \\ \bot \end{array} \right]$$

$$\left[\begin{array}{cc} \begin{array}{cc} \tau_1 & \tau_2 \\ V_1 & V_2 \\ \varphi & \psi \\ \hline \multicolumn{2}{c}{\varphi \wedge \psi} \end{array} & , & \dfrac{\varphi \wedge \psi \quad \begin{array}{c} \square\!\!-\!\!(i) \\ \underbrace{\psi,\ \sigma} \\ F' \\ \bot \end{array}}{\bot}(i) \end{array} \right] = \left[\begin{array}{c} \tau_2 \\ V_2 \\ \psi \end{array}\ , \begin{array}{c} \square\!\!-\!\!(i) \\ \underbrace{\psi,\ \sigma} \\ F' \\ \bot \end{array} \right]$$

$$
\left[\;
\begin{array}{c}
\tau \\ V' \\ \dfrac{\varphi}{\varphi \vee \psi}
\end{array}
\;,\;
\varphi \vee \psi
\quad
\dfrac{
\overset{\square\!-\!\!(i)}{\overbrace{\varphi,\sigma_1}}\\ F_1 \\ \bot
\qquad
\overset{\square\!-\!\!(i)}{\overbrace{\psi,\sigma_2}}\\ F_2 \\ \bot
}{\bot}{\scriptstyle(i)}
\;\right]
=
\left[\;
\begin{array}{cc}
\tau & \overset{}{\overbrace{\varphi,\sigma_1}} \\
V' \;,\; & F_1 \\
\varphi & \bot
\end{array}
\;\right]
$$

$$
\left[\;
\begin{array}{c}
\tau \\ V' \\ \dfrac{\psi}{\varphi \vee \psi}
\end{array}
\;,\;
\varphi \vee \psi
\quad
\dfrac{
\overset{\square\!-\!\!(i)}{\overbrace{\varphi,\sigma_1}}\\ F_1 \\ \bot
\qquad
\overset{\square\!-\!\!(i)}{\overbrace{\psi,\sigma_2}}\\ F_2 \\ \bot
}{\bot}{\scriptstyle(i)}
\;\right]
=
\left[\;
\begin{array}{cc}
\tau & \overset{}{\overbrace{\psi,\sigma_2}} \\
V' \;,\; & F_2 \\
\psi & \bot
\end{array}
\;\right]
$$

$$
\left[\;
\begin{array}{c}
\overset{\square\!-\!\!(i)}{\overbrace{\varphi,\tau}}\\ F_1 \\ \dfrac{\bot}{\varphi \to \psi}{\scriptstyle(i)}
\end{array}
\;,\;
\varphi \to \psi
\quad
\dfrac{
\begin{array}{c}\sigma_1\\ V_2\\ \varphi\end{array}
\qquad
\overset{\square\!-\!\!(i)}{\overbrace{\psi,\sigma_2}}\\ F_2 \\ \bot
}{\bot}{\scriptstyle(i)}
\;\right]
=
\left[\;
\begin{array}{cc}
\sigma_1 & \overset{}{\overbrace{\varphi,\tau}} \\
V_2 \;,\; & F_1 \\
\varphi & \bot
\end{array}
\;\right]
$$

$$
\left[\;
\begin{array}{c}
\tau \\ V_1 \\ \dfrac{\psi}{\varphi \to \psi}
\end{array}
\;,\;
\varphi \to \psi
\quad
\dfrac{
\begin{array}{c}\sigma_1\\ V_2\\ \varphi\end{array}
\qquad
\overset{\square\!-\!\!(i)}{\overbrace{\psi,\sigma_2}}\\ F_2 \\ \bot
}{\bot}{\scriptstyle(i)}
\;\right]
=
\left[\;
\begin{array}{cc}
\tau & \overset{}{\overbrace{\psi,\sigma_2}} \\
V_1 \;,\; & F_2 \\
\psi & \bot
\end{array}
\;\right]
$$

$$
\left[\;
\begin{array}{c}
\tau \\ V' \\ \dfrac{\varphi^{x}_{\alpha_k}}{\exists x \varphi}
\end{array}
\;,\;
\exists x \varphi
\quad
\dfrac{
\overset{\square\!-\!\!(i)}{\overbrace{\varphi^{x}_{\alpha_1},\sigma_1}}\\ F_1 \\ \bot
\;\cdots\;
\overset{\square\!-\!\!(i)}{\overbrace{\varphi^{x}_{\alpha_n},\sigma_n}}\\ F_n \\ \bot
\;\cdots
}{\bot}{\scriptstyle(i)}
\;\right]
=
\left[\;
\begin{array}{cc}
\tau & \overset{\square\!-\!\!(i)}{\overbrace{\varphi^{x}_{\alpha_k},\sigma_k}} \\
V' \;,\; & F_k \\
\varphi^{x}_{\alpha_k} & \bot
\end{array}
\;\right]
$$

$$
\left[
\begin{array}{ccc}
\tau_1 & \tau_n & \\
V_1 & \cdots & V_n & \cdots \\
\psi(\alpha_1) & \psi(\alpha_n) \\
\hline
\forall x\psi(x)
\end{array}
\right. ,
\quad
\begin{array}{c}
\Box\!\!-\!\!\!-\!\!\!-(i) \\
\overbrace{\psi(\alpha_k)\,,\ \sigma} \\
\hline
F' \\
\forall x\psi(x) \qquad \bot \\
\hline
\bot
\end{array}\!\!-(i)
\left.\vphantom{\begin{array}{c}a\\a\\a\\a\end{array}}\right]
=
\left[
\begin{array}{cc}
\tau_k & \overbrace{\psi(\alpha_k)\,,\ \sigma} \\
V_k & \\
\psi(\alpha_k) & F' \\
 & \bot
\end{array}
\right]
$$

Corollary 1. *Let a model M be defined on domain D, and let θ be a saturated formula all of whose extra-logical expressions receive interpretation in M. Then it is not the case both that θ has an M-relative verification and that θ has an M-relative falsification.*

Proof. For no atomic saturated A does M contain both $\dfrac{\ \ }{A}$ and $\dfrac{A}{\bot}$. $\square$

3.4.4 A note on nomenclature

We now echo what we did earlier in the propositional case, adapting it straightforwardly here to the first-order case. For models M, saturated formulae φ, and constructions V, we stipulate that the following three forms of expression will be equivalent:

> V is a verification of φ relative to M;
> V is an evaluation proof of φ relative to M;
> V is an M-relative truthmaker for φ.

Likewise, for models M and saturated formulae φ again, and constructions F, the following three will be equivalent:

> F is a falsification of φ relative to M;
> F is an evaluation disproof of φ relative to M;
> F is an M-relative falsitymaker for φ.

Note that we are taking the liberty of dropping the square quotes around 'proof' and 'disproof'. They need to be borne tacitly in mind, though, in the case where the constructions in question happen to be infinitary as a result of the domain of M being infinite. The choice of form of expression will be determined by context. For example, in discussions of truthmaking, it will be natural to speak of truthmakers and falsitymakers.[10] In a treatment of

---

[10] We prefer the term 'falsitymaker' to the term 'false-maker' of Armstrong [2004], §1.4; in contrast with 'truth' we need the noun 'falsity' rather than the adjective 'false'.

the problem of how one might state formal criteria of cognitive significance (as in Tennant [1997], Chapter 11), the use of 'verification' and 'falsification' is more natural, tying, as it does, into the terminological worlds of the likes of Ayer, Popper, and Carnap, who wrestled with that particular problem of demarcation in the philosophy of science. Finally, in more proof-theoretically flavored discussions, which treat of the normalizability of proofs and similar constructions, it might be more appropriate to use the terms 'evaluation proof' and 'evaluation disproof'.

Whichever terminology one chooses to adopt, this much is clear: one is speaking of the same kind of formally (inductively) definable abstract object. We are *reifying* 'ways of being true' (or false) under a particular interpretation or model M. Truth-in-M consists in the *existence* of such an M-relative truthmaker. It is worth remarking also that in game-theoretic semantics for first-order logic—see Tennant [1978]—an M-relative truthmaker for the sentence φ codifies a winning strategy for the player who starts as Player T in the game on φ, against the background of the model M. See §8.3 for further details.

3.5 More general atomic bases

Thus far our discussion of verifications and falsifications relative to a model, even at first order, has presupposed that the atomic determinations are of the simplest possible forms, (coherently) determining atoms (or saturated atomic formulae) as true or as false. We indicated in §3.3.1 the possibility of a more general kind of atomic determination, which it is our concern now to elaborate. Our atomic bases Λ are about to be 'beefed up'—but in ways that do justice to the idea that it is only the extralogical primitives that are being interpreted.

Note that we have allowed all along for any given atom A to be *undetermined*, or *not interpreted*, by a particular (simple) atomic basis Λ. This would simply involve the *absence* from Λ of both

$$\overline{A} \qquad \text{and} \qquad \frac{A}{\bot} \ .$$

Even in the absence of a determination of an atom A occurring in a sentence φ, it might be possible nevertheless to evaluate φ by attending to available determinations of its other atoms. But we are allowing, in principle, for this *not* to be the case; relative to certain bases Λ, the truth value of certain sentences φ might be undefined, or not determined. Nothing in the 'logic of truth-tabular computation' demands or implies that every sentence φ must receive a truth value under every interpretation Λ.

We preserve this guiding constructivist thought as we consider now the ways in which atomic bases might be enriched while still deserving to be called 'atomic'.

3.5.1 Inclusions and contrarieties

Two very simple ways in which atomic bases can more accurately reflect the 'atomic knowledge' possessed by speakers of a language is by registering *inclusions* and/or *contrarieties* among atomic sentences. Take, for example, the following two patterns of inference:

$$\frac{t \text{ is red}}{t \text{ is colored}} \qquad \frac{t \text{ is red} \quad t \text{ is green}}{\bot}$$

Anyone who knows English is prepared to grant the validity of these inferences without knowing the color of the object t in question or even whether it has a color at all. The validity of these particular inferences involving color predicates is *analytic*. They are fully justified by *conceptual* connections understood by anyone who has mastered those predicates.

The suggestion now is that we would do well to consider atomic bases that contain not only atomic determinations of the forms

$$\frac{}{A} \quad \text{and} \quad \frac{A}{\bot}$$

(for particular sentences A), but also atomic rules of inference of the forms

$$\frac{A}{B} \quad \text{and} \quad \frac{A \quad B}{\bot}$$

(for particular sentences A and B, not placeholders or propositional variables). *Analytic* inclusions and contrarieties would of course be eligible for inclusion in these expanded atomic bases. But we could also include particular inferences of the foregoing general forms that are *synthetic*, and indeed even *contingent*. (We have of course already allowed for contingent postive or negative atomic determinations in our atomic bases.)

For our purposes in attaining to Core Logic, *this will be generalization enough* of our atomic bases Λ consisting only of literals. Yet it is only a first tentative step whose further generalizations should be rather obvious. One could consider, for example, finitary atomic rules of the forms

$$\frac{A_1 \ \ldots \ A_n}{B} \quad \text{and} \quad \frac{A_1 \ \ldots \ A_n}{\bot} \ .$$

Then one could countenance the possibility of involving as premises not just positive literals (i.e. atoms), but also negative ones (i.e. negations of atoms). This, however, can be effected without introducing negation explicitly. The general forms allowing for both positive and negative premises are

$$
\begin{array}{ccc}
 & \overline{}^{(i)} & \overline{}^{(i)} \\
 & B_1 & B_m \\
 & \vdots & \cdots \quad \vdots \qquad \text{where } n + m \geq 0 \\
A_1 \ \ldots \ A_n & \bot & \bot \\
\cline{1-3}
 & \multicolumn{2}{c}{C}^{(i)}
\end{array}
$$

and

$$
\begin{array}{ccc}
 & \overline{}^{(i)} & \overline{}^{(i)} \\
 & B_1 & B_m \\
 & \vdots & \cdots \quad \vdots \qquad \text{where } n + m > 0 \\
A_1 \ \ldots \ A_n & \bot & \bot \\
\cline{1-3}
 & \multicolumn{2}{c}{\bot}^{(i)}
\end{array}
$$

As an instance of the first form, we have, when $n + m = 0$, the atomic determination

$$\overline{C}$$

As an instance of the second form, we have, when $n = 1$ and $m = 0$, the atomic determination

$$\frac{A}{\bot}$$

As another instance of the first form, we have, when $n = 1$ and $m = 0$, the atomic inclusion rule

$$\frac{A}{C}$$

And as another instance of the second form, we have, when $n = 2$ and $m = 0$, the atomic rule of contrariety

$$\frac{A_1 \qquad A_2}{\bot}$$

Thus both inclusions and contrarieties are covered by the general form—which is all that concerns us here.

Inclusions and contrarieties, it turns out, represent enough of a departure to occasion an important extension to our inductive definition of the sorts of constructions that count as verifications and/or falsifications relative to an atomic basis Λ. Note that when Λ was of the simplest kind, involving only atomic determinations, we were able to define verifications and falsifications of a complex sentence φ relative to Λ *without any need to keep track of hypothetical assumptions other than φ itself.*

This feature now changes when we countenance inclusions and contrarieties. First, suppose that we have no atomic determination for either the atom A or the atom B. If, however, we have the atomic inclusion rule

$$\frac{A}{B}$$

then any instance of that very rule counts as a verification of B relative to the atomic basis

$$\left\{ \frac{A}{B} \right\}$$

and the set

$$\{A\}$$

of *as-yet-undischarged* atomic hypotheses. (Note: there is no line over A here; it is being treated as a hypothesis, not as a model-relative atomic fact.)

Second (still supposing that we have no atomic determination for either the atom A or the atom B), any instance of the atomic rule of contrariety

$$\frac{A \qquad B}{\perp}$$

counts as a falsification of B relative to the atomic basis

$$\left\{ \frac{A \qquad B}{\perp} \right\}$$

and the set $\{A\}$ of as-yet-undischarged atomic hypotheses. It *also* counts as a falsification of A relative to the same atomic basis, but with $\{B\}$ as its set of as-yet-undischarged atomic hypotheses.

Suppose an atomic basis $\mathcal{B}$ contains just the rules

$$\frac{A \qquad B}{C} \qquad \text{and} \qquad \frac{C \qquad D}{E}$$

It is clear that, even though it is not *in* the basis $\mathcal{B}$, the rule

$$\frac{A \qquad B \qquad D}{E}$$

is justified relative to $\mathcal{B}$. The justification is

$$\frac{\dfrac{A \qquad B}{C} \qquad D}{E}$$

This is a determination, relative to the basis given, of the *atomic sequent* $A, B, D : E$.

3.5.2 The notion of a determination relative to an atomic basis

Before, when dealing with bases that contained only atomic determinations

$$\frac{}{A} \qquad \text{and} \qquad \frac{A}{\bot} \; ,$$

we could construct determinations of sequents involving only atomic sentences only if those sequents were of the forms

$$\emptyset : A \qquad A : \emptyset$$

The 'base construction' A was a verification of A relative to $\{\,\overline{A}\,\}$. We wrote this first basis clause of the co-inductive definition of verifications and falsifications as

$$\mathcal{V}(A, A, \{\overline{A}\})$$

Likewise the 'base construction' $\dfrac{A}{\bot}$ was a falsification of A relative to $\left\{\dfrac{A}{\bot}\right\}$. We wrote this second basis clause of the co-inductive definition of verifications and falsifications as

$$\mathcal{F}\left(\dfrac{A}{\bot}, A, \left\{\dfrac{A}{\bot}\right\}\right)$$

In speaking (even in the base cases) of verifications and falsifications of *sentences* we lose sight of the fact that such verifications and falsifications are *determinations of sequents*. A base-case *verification* 'of A' determines the sequent $\emptyset : A$; while a base-case *falsification* 'of A' determines the sequent $A : \emptyset$. Thus instead of writing

$$\mathcal{V}(A, A, \{\overline{A}\})$$

we could have written

$$\mathcal{D}(A; \, \emptyset {:} A \, ; \{\overline{A}\});$$

and instead of writing

$$\mathcal{F}\left(\frac{A}{\perp} \, , A, \left\{\frac{A}{\perp}\right\}\right)$$

we could have written

$$\mathcal{D}\left(\frac{A}{\perp} \, ; \, A {:} \emptyset \, ; \left\{\frac{A}{\perp}\right\}\right).$$

We use semicolons to separate the three arguments of the ternary predicate $\mathcal{D}$ in order to avoid confusion with commas that may appear within complex specifications of the sequent (among its premises) occupying the second argument place.

It is worth rewriting our earlier $\mathbb{T}$-rules of verification and falsification in this new notation, so that it will be easier to appreciate exactly how the rules will be generalized once we broaden our conception of the atomic bases relative to which sentences may be evaluated. The curious-seeming occurrences of $\emptyset$ in the following restatement of the $\mathbb{T}$-rules are inserted at this stage with an eye to how they will be replaced, eventually, by possibly non-empty sets $\mathcal{A}$ of undischarged but dischargeable atomic assumptions. This is called for only when the very simple bases Λ are themselves replaced by the more general atomic bases $\mathcal{B}$. The simple bases Λ (as special cases of $\mathcal{B}$) only ever give rise to $\emptyset$ as a possibility for $\mathcal{A}$. In normal circumstances, mention of the empty set would be suppressed in graphic statements of rules of inference. But it appears here for good expository reasons, as the path to the Logic of Evaluation $\mathbb{E}$ unfolds.

Graphic form of the rules of $\mathbb{T}$

$(\neg \mathcal{V})$

$$\underbrace{\begin{array}{c}\square\!\!-\!\!(i)\\ \varphi\,,\,\emptyset\,;\,\Lambda\end{array}}$$

$$\vdots$$

$$\cfrac{\bot}{\neg\varphi}(i)$$

$(\neg \mathcal{F})$

$$\underbrace{\emptyset\,;\,\Lambda}$$

$$\vdots$$

$$\cfrac{\neg\varphi \qquad \varphi}{\bot}$$

$(\wedge \mathcal{V})$

$$\underbrace{\emptyset\,;\,\Lambda_1}\qquad\underbrace{\emptyset\,;\,\Lambda_2}$$

$$\quad\vdots\qquad\qquad\vdots$$

$$\cfrac{\varphi_1 \qquad\qquad \varphi_2}{\varphi_1 \wedge \varphi_2}$$

$(\wedge \mathcal{F})$

$$\underbrace{\begin{array}{c}\square\!\!-\!\!(i)\\ \varphi_1\,,\,\emptyset\,;\,\Lambda\end{array}}$$

$$\vdots$$

$$\cfrac{\varphi_1 \wedge \varphi_2 \qquad \bot}{\bot}(i)$$

$$\underbrace{\begin{array}{c}\square\!\!-\!\!(i)\\ \varphi_2\,,\,\emptyset\,;\,\Lambda\end{array}}$$

$$\vdots$$

$$\cfrac{\varphi_1 \wedge \varphi_2 \qquad \bot}{\bot}(i)$$

$(\vee \mathcal{V})$

$$\underbrace{\emptyset\,;\,\Lambda}\qquad\underbrace{\emptyset\,;\,\Lambda}$$

$$\quad\vdots\qquad\qquad\vdots$$

$$\cfrac{\varphi_1}{\varphi_1 \vee \varphi_2}\qquad\cfrac{\varphi_2}{\varphi_1 \vee \varphi_2}$$

$(\vee\mathcal{F})$

$$\cfrac{\cfrac{\overbrace{\substack{\Box\!\!-\!\!(i)\\ \varphi_1\,,\,\emptyset\,;\,\Lambda_1}}\\ \vdots\\ \varphi_1 \vee \varphi_2 \qquad \bot}{} \qquad \cfrac{\overbrace{\substack{\Box\!\!-\!\!(i)\\ \varphi_2\,,\,\emptyset\,;\,\Lambda_2}}\\ \vdots\\ \bot}{}}{\bot}(i)$$

$(\to\mathcal{V})(a)$

$$\cfrac{\overbrace{\substack{\Box\!\!-\!\!(i)\\ \varphi_1\,,\,\emptyset\,;\,\Lambda}}\\ \vdots\\ \cfrac{\bot}{\varphi_1 \to \varphi_2}(i)}{}$$

$(\to\mathcal{V})(b)$

$$\cfrac{\overbrace{\substack{\emptyset\,;\,\Lambda}}\\ \vdots\\ \cfrac{\varphi_2}{\varphi_1 \to \varphi_2}}{}$$

$(\to\mathcal{F})$

$$\cfrac{\varphi_1 \to \varphi_2 \qquad \cfrac{\overbrace{\substack{\emptyset\,;\,\Lambda_1}}\\ \vdots\\ \varphi_1}{} \qquad \cfrac{\overbrace{\substack{\Box\!\!-\!\!(i)\\ \varphi_2\,,\,\emptyset\,;\,\Lambda_2}}\\ \vdots\\ \bot}{}}{\bot}(i)$$

Atomic determinations upon generalizing Λs to $\mathcal{B}$s

We are now considering adding more complex 'within-basis constructions' as determinations of sequents of either one of the two atomic forms

$$A_1, \ldots, A_n : A$$

and

$$A_1, \ldots, A_n, A : \emptyset$$

(Remember that in this latter context $\emptyset$ plays the role of $\bot$.)

A determination of

$$A_1, \ldots, A_n : A$$

will be called a verification of A relative to *both* the set $\{A_1, \ldots, A_n\}$ $(=\mathcal{A})$ of undischarged atomic assumptions *and* the basis $\mathcal{B}$ of atomic rules used in its construction. Thus, our earlier example

$$\cfrac{\cfrac{A \qquad B}{C} \qquad D}{E}$$

is a verification of E relative to the set $\{A, B, D\}$ of undischarged atomic assumptions *and* the basis

$$\left\{ \frac{A \quad B}{C}, \frac{C \quad D}{E} \right\}$$

consisting of the two atomic rules that were used in its construction.

A determination of

$$A_1, \ldots, A_n, A : \emptyset$$

will be called a falsification of A relative to *both* the set $\{A_1, \ldots, A_n\}(= \mathcal{A})$ of undischarged atomic assumptions *and* the basis $\mathcal{B}$ of atomic rules used in its construction.

Atomic determinations within the basis $\mathcal{B}$ of course help to determine the membership of the set $\mathcal{A}$ of undischarged atomic assumptions. Thus if the basis $\mathcal{B}$ consists of the rules

$$\frac{}{A} \quad \frac{}{B} \quad \frac{}{D} \quad \frac{A \quad B}{C} \quad \frac{C \quad D}{E}$$

then the determination

$$\Pi \ = \ \frac{\dfrac{\overline{A} \quad \overline{B}}{C} \quad \overline{D}}{E}$$

is a verification of E relative to both $\emptyset$ (the empty set of undischarged assumptions) and $\mathcal{B}$. Or, more succinctly:

$$\mathcal{D}(\Pi; \emptyset : E ; \mathcal{B})$$

And if the basis $\mathcal{B}$ consists of the rules

$$\frac{}{B} \quad \frac{A \quad B}{C} \quad \frac{C \quad D}{\perp}$$

then the determination

$$\Xi \ = \ \frac{\dfrac{A \quad \overline{B}}{C} \quad D}{\perp}$$

is a falsification of A relative to both the set $\{D\}$ of undischarged assumptions and $\mathcal{B}$. Or, more succinctly:

$$\mathcal{D}(\Xi;\, D, A:\emptyset;\mathcal{B})$$

Note that this succinct form covers also the case where we wish to say that Ξ is a falsification of D relative to the set $\{A\}$ of undischarged assumptions and $\mathcal{B}$. So it is more economical to deal with the ternary form

$$\mathcal{D}(\Pi;\, \tau:\emptyset;\, \mathcal{B})$$

than a form of expression that requires one to focus arbitrarily on a single member A of τ as 'the one' that is being falsified *modulo* the set $\tau \setminus \{A\}$ of the other undischarged assumptions. (The Quine–Duhem problem makes certain choices of formal relations not worth defining here.)

What emerges from our discussion is the important lesson that atomic inferential bases that provide for inclusions and contrarieties motivate the choice of this new ternary relation $\mathcal{D}$ in our theory of inferential semantics. We shall henceforward use the inductively definable relation $\mathcal{D}$ holding among constructions Π, certain sequents S, and atomic inferential bases $\mathcal{B}$:

$$\mathcal{D}(\Pi; S; \mathcal{B})$$

The sequents S will be allowed to take either one of two forms. First, we shall allow (possibly complex) sentences φ to be verified *modulo* a set of undischarged *atomic* assumptions *and* an atomic basis $\mathcal{B}$. In such cases the sequent S will be of the form

$$A_1, \ldots A_n : \varphi.$$

Second, we shall allow (possibly complex) sentences φ to be falsified *modulo* a set of undischarged *atomic* assumptions *and* an atomic basis $\mathcal{B}$. In such cases the sequent S will be of the form

$$\varphi, A_1, \ldots A_n : \emptyset.$$

Thus we shall have instances

$$\mathcal{D}(\Pi;\, A_1, \ldots, A_n:\varphi;\, \mathcal{B})$$

of our new relation, as well as instances

$$\mathcal{D}(\Pi;\, \varphi, A_1, \ldots, A_n:\emptyset;\, \mathcal{B})$$

Note how this generalizes smoothly from the earlier case, where the atomic basis Λ (a special case of $\mathcal{B}$) consisted of atomic determinations only, and we had no need to keep track of undischarged atomic assumptions (whence n was 0—or, equivalently, $\mathcal{A}$ was empty). In the earlier case we were dealing, in effect, with the simpler instances

$$\mathcal{D}(\Pi;\, \emptyset : \varphi;\, \Lambda)$$

(expressed in terms of our new relation), as well as instances

$$\mathcal{D}(\Pi;\, \varphi : \emptyset;\, \Lambda).$$

For

$$\mathcal{D}(\Pi;\, \emptyset : \varphi;\, \Lambda)$$

we wrote, even more simply,

$$\mathcal{V}(\Pi, \varphi, \Lambda);$$

and for

$$\mathcal{D}(\Pi;\, \varphi : \emptyset;\, \Lambda)$$

we likewise wrote

$$\mathcal{F}(\Pi, \varphi, \Lambda).$$

Such was life without undischarged atomic assumptions to track. But, with such assumptions in the picture, we attain a new (part of the) rule for verifying conditionals.

3.5.3 Broadening the atomic basis affords us a new way to verify conditionals

For the inferentialist, the 'if …then …' construction affords a convenient way to encapsulate, within a *single* sentence, a transition in thought that would otherwise be expressed as an inference between *two* sentences. So, if one's atomic basis contains the inference rule $\dfrac{A}{B}$ governing the particular sentences A and B, then one should be justified in asserting $A \to B$ outright. Put in terms of our new formal notation:

$$\mathcal{D}\left(\frac{A}{B}\,;\, A{:}B;\, \left\{\frac{A}{B}\right\}\right) \;\Longrightarrow\; \mathcal{D}\left(\frac{\dfrac{\overline{A}^{(1)}}{B}}{A \to B}{}^{(1)}\,;\, \emptyset{:}A \to B;\, \left\{\frac{A}{B}\right\}\right)$$

Note that the double arrow is *metalinguistic* here. We could, if we wished, be 'thoroughly inferentialist' by rendering this conditional as the metalinguistic *inference*

$$\dfrac{\mathcal{D}\left(\dfrac{A}{B} \; ; \; A\!:\!B \; ; \; \left\{\dfrac{A}{B}\right\}\right)}{\mathcal{D}\left(\dfrac{\overline{}^{(1)}\!\!\!\!\dfrac{A}{B}}{A \to B}{}_{(1)} \; ; \; \emptyset\!:\!A \to B \; ; \; \left\{\dfrac{A}{B}\right\}\right)}$$

More generally, where $\mathcal{A}$ is a set of atoms other than A:

$$\mathcal{D}(\Pi; \, \mathcal{A}, A\!:\!B; \, \mathcal{B}) \implies \mathcal{D}\left(\dfrac{\Pi}{A \to B} \; ; \; \mathcal{A}\!:\!A \to B; \, \mathcal{B}\right) ;$$

or again as a metalinguistic rule of inference:

$$\dfrac{\mathcal{D}(\Pi; \, \mathcal{A}, A\!:\!B; \, \mathcal{B})}{\mathcal{D}\left(\dfrac{\Pi}{A \to B} \; ; \; \mathcal{A}\!:\!A \to B; \, \mathcal{B}\right)}$$

This is one of the inductive clauses in the definition of the ternary relation $\mathcal{D}(\Pi, S, \mathcal{B})$. It can be expressed in graphic form as follows:

$$\dfrac{\underbrace{\mathcal{B} \; , \; \mathcal{A} \; , \; \overset{\overline{}^{(i)}}{A}}_{\Pi}}{\dfrac{\overline{}^{(i)}\,B}{A \to B}}$$

where it is to be understood that A is not a member of the set $\mathcal{A}$ of undischarged atomic assumptions.

All our earlier definitional rules for the formation of verifications and falsifications can now be re-expressed (see §3.5.4) as definitional rules for the formation of $\mathcal{B}$-relative *determinations of sequents* of either one of the forms

$$\mathcal{A} : \varphi$$

or

$$\varphi, \mathcal{A} : \emptyset$$

where $\mathcal{B}$ is an atomic inferential basis, $\mathcal{A}$ is a set of atoms, and the sentence φ, if atomic, is not in $\mathcal{A}$. The assumptions in $\mathcal{A}$ are potentially dischargeable upon further applications of rules. The rules in $\mathcal{B}$, which are not discharge-able, serve to interpret the atoms of the language. Determinations of φ as, respectively, true or false, are once again relative to that interpretation. The broader atomic basis $\mathcal{B}$ here takes over from our former Λ; and $\mathcal{A}$ is about to be put in place of $\emptyset$ in the graphic rules for $\mathbb{T}$ stated in §3.5.2. What we have here is an ever-so-slight inferentialist extension of the standard, but rather limited, idea of an interpretation consisting only of an assignment of truth values to atoms. We are now allowing for an interpretation to involve also a specification of 'analytic' *inclusions* and *contrarieties* holding among the particular atomic sentences that are being interpreted; and accordingly allowing for the possibility of accumulating some undischarged atomic as-sumptions as evaluations proceed.[11]

3.5.4   Rules of determination relative to an atomic basis

Definition 3. *Let Π_1 be a finite tree of sentences with an occurrence of A as its root. Let Π_2 be a finite tree of sentences with at least one leaf occurrence of A. Then $\Pi_2[A/\Pi_1]$ is the tree that results by grafting, at each leaf occurrence of A in Π_2, a copy of Π_1 so that the root occurrence of A in that copy of Π_1 takes over from that leaf occurrence of A within Π_2.*

Example. Let Π_1 and Π_2 be

$$\frac{\dfrac{B \quad C}{G \quad\quad H}}{A} \qquad \text{and} \qquad \frac{\dfrac{A \quad D}{A \quad\quad E}}{F}$$

respectively. Then $\Pi_2[A/\Pi_1]$ is

$$\frac{\dfrac{\dfrac{B \quad C}{G \quad H}}{A} \quad\quad\quad \dfrac{\dfrac{B \quad C}{G \quad H}}{A} \quad\quad D}{F}$$

Here now are the basis and inductive clauses in the definition of our new ternary relation $\mathcal{D}(\Pi; S; \mathcal{B})$. They are the rules of $\mathbb{E}$ in *metalinguistic inferential form*, expressing clauses in the metalinguistic inductive definition of the

---

[11] The notion of an atomic inferential basis can be generalized even further, so as to include atomic rules whose applications discharge (atomic) assumptions. See, for example, Sandqvist [2015]. For our expository purposes, however, we have incorporated just enough complexity into atomic bases, and we do not need to consider generalizing them any further.

ternary relation $\mathcal{D}$. Note how once again the bases $\mathcal{B}$ *merely accumulate.* Their members are never discharged. Remember that we write $\bot$ interchangeably with $\emptyset$ as a succedent of a sequent.

$$\mathcal{D}\left(\frac{A_1,\ldots,A_n}{A} \; ; \; \{A_1,\ldots,A_n\}:A \; ; \; \left\{ \frac{A_1,\ldots,A_n}{A} \right\} \right)$$

$$\mathcal{D}\left(\frac{A_1,\ldots,A_n}{\bot} \; ; \; \{A_1,\ldots,A_n\}:\bot \; ; \; \left\{ \frac{A_1,\ldots,A_n}{\bot} \right\} \right)$$

$$\frac{\mathcal{D}(\Pi_1 \; ; \; \mathcal{A}_1:A \; ; \; \mathcal{B}_1) \qquad \mathcal{D}(\Pi_2 \; ; \; A,\mathcal{A}_2:C \; ; \; \mathcal{B}_2)}{\mathcal{D}(\Pi_2[A/\Pi_1] \; ; \; \mathcal{A}_1,\mathcal{A}_2:C \; ; \; \mathcal{B}_1,\mathcal{B}_2)}$$

$$\frac{\mathcal{D}(\Pi; \; \varphi, \mathcal{A}:\bot \; ; \; \mathcal{B})}{\mathcal{D}\left(\frac{\Pi}{\neg\varphi} \; ; \; \mathcal{A}:\neg\varphi; \; \mathcal{B} \right)} \qquad \frac{\mathcal{D}(\Pi; \; \mathcal{A}:\varphi \; ; \; \mathcal{B})}{\mathcal{D}\left(\frac{\neg\varphi \quad \Pi}{\bot} \; ; \; \neg\varphi, \mathcal{A}:\bot \; ; \; \mathcal{B} \right)}$$

$$\frac{\mathcal{D}(\Pi_1; \; \mathcal{A}_1:\varphi_1; \; \mathcal{B}_1) \qquad \mathcal{D}(\Pi_2; \; \mathcal{A}_2:\varphi_2; \; \mathcal{B}_2)}{\mathcal{D}\left(\frac{\Pi_1 \quad \Pi_2}{\varphi_1 \wedge \varphi_2} \; ; \; \mathcal{A}_1 \cup \mathcal{A}_2:\varphi_1 \wedge \varphi_2; \; \mathcal{B}_1 \cup \mathcal{B}_2 \right)}$$

$$\frac{\mathcal{D}(\Pi; \; \mathcal{A}, \varphi_i:\bot; \; \mathcal{B})}{\mathcal{D}\left(\frac{\varphi_1 \wedge \varphi_2 \quad \Pi}{\bot} \; ; \; \mathcal{A}, \varphi_1 \wedge \varphi_2:\bot; \; \mathcal{B} \right)} \qquad i = 1,2$$

$$\frac{\mathcal{D}(\Pi; \; \mathcal{A}:\varphi_i \; ; \; \mathcal{B})}{\mathcal{D}\left(\frac{\Pi}{\varphi_1 \vee \varphi_2} \; ; \; \mathcal{A}:\varphi_1 \vee \varphi_2; \; \mathcal{B} \right)} \qquad i = 1,2$$

$$\frac{\mathcal{D}(\Pi_1; \; \mathcal{A}_1, \varphi_1:\bot; \; \mathcal{B}_1) \qquad \mathcal{D}(\Pi_2; \; \mathcal{A}_2, \varphi_2:\bot; \; \mathcal{B}_2)}{\mathcal{D}\left(\frac{\varphi_1 \vee \varphi_2 \quad \Pi_1 \quad \Pi_2}{\bot} \; ; \; \mathcal{A}_1 \cup \mathcal{A}_2, \varphi_1 \vee \varphi_2:\bot; \; \mathcal{B}_1 \cup \mathcal{B}_2 \right)}$$

$$\frac{\mathcal{D}(\Pi;\ \mathcal{A}:\varphi_2;\ \mathcal{B})}{\mathcal{D}\left(\dfrac{\Pi}{\varphi_1\to\varphi_2}\ ;\ \mathcal{A}:\varphi_1\to\varphi_2;\ \mathcal{B}\right)}\qquad \frac{\mathcal{D}(\Pi;\ \mathcal{A},\varphi_1:\varphi_2;\ \mathcal{B})}{\mathcal{D}\left(\dfrac{\Pi}{\varphi_1\to\varphi_2}\ ;\ \mathcal{A}:\varphi_1\to\varphi_2;\ \mathcal{B}\right)}\qquad \frac{\mathcal{D}(\Pi;\ \mathcal{A},\varphi_1:\bot;\ \mathcal{B})}{\mathcal{D}\left(\dfrac{\Pi}{\varphi_1\to\varphi_2}\ ;\ \mathcal{A}:\varphi_1\to\varphi_2;\ \mathcal{B}\right)}$$

$$\frac{\mathcal{D}(\Pi;\ \mathcal{A}_1:\varphi_1;\ \mathcal{B}_1)\qquad\qquad \mathcal{D}(\Pi;\ \mathcal{A}_2,\varphi_2:\bot;\ \mathcal{B}_2)}{\mathcal{D}\left(\dfrac{\varphi_1\to\varphi_2\ \ \Pi\ \ \Pi}{\bot}\ ;\ \mathcal{A}_1\cup\mathcal{A}_2,\varphi_1\to\varphi_2:\bot;\ \mathcal{B}_1\cup\mathcal{B}_2\right)}$$

The foregoing rules comprise the propositional *Logic of Evaluation*, which we shall call $\mathbb{E}$. It is generated by considering what is minimally involved in determining the truth value of a sentence relative to an interpretation $\mathcal{B}$ that might not only assign truth values directly to atoms, but also constrain those assignments by means of atomic rules of inclusion and contrariety.

By inspection of the rules one readily sees, via an easy mathematical induction, that every conclusion that they permit one to derive is of one of the forms

$$\mathcal{D}(\Pi; \mathcal{A}:\varphi;\ \mathcal{B})\qquad\text{or}\qquad \mathcal{D}(\Pi; \varphi, \mathcal{A}:\bot;\ \mathcal{B}),$$

so that every sequent provable in the logic $\mathbb{E}$ *modulo* atomic rules $\mathcal{B}$ is of one of the forms

$$\mathcal{A}:\varphi\qquad\text{or}\qquad \varphi,\mathcal{A}:\bot,$$

where φ is any sentence and $\mathcal{A}$ is a non-empty set of atoms occurring in φ. So the proving or refuting of sentences φ that can be accomplished within $\mathbb{E}$ (relative to an atomic basis $\mathcal{B}$) is, respectively, always from, or *modulo*, a (possibly empty) set $\mathcal{A}$ of *atoms*.

The primitive rules of inference of $\mathbb{E}$ concerning the logical operators have just been stated as clauses in the definition of the ternary relation $\mathcal{D}(\Pi; S; \mathcal{B})$. We now restate them in the more familiar graphic form, and invite the reader to compare them with the graphic rules for $\mathbb{T}$ stated in §3.5.2. We retain the former labeling of these rules as rules of verification or falsification ($\mathcal{V}$ or $\mathcal{F}$), because that is still apposite relative to our broadened conception of the atomic bases $\mathcal{B}$ that can serve to interpret the atoms.

Remember: the $\mathcal{A}$s are finite sets of atomic hypotheses; while the $\mathcal{B}$s are finite atomic bases, i.e. sets of atomic rules of inference constituting an interpretation.

Graphic form of the rules of $\mathbb{E}$

$(\neg\mathcal{V})$

$$
\underbrace{\overset{\square\!-\!(i)}{\varphi\,,\;\mathcal{A}\,;\,\mathcal{B}}}
$$
$$
\vdots
$$
$$
\frac{\bot}{\neg\varphi}\,\text{---}(i)
$$

$(\neg\mathcal{F})$

$$
\underbrace{\mathcal{A}\,;\,\mathcal{B}}
$$
$$
\vdots
$$
$$
\frac{\neg\varphi\qquad\varphi}{\bot}
$$

$(\wedge\mathcal{V})$

$$
\underbrace{\mathcal{A}_1\,;\,\mathcal{B}_1}\qquad\underbrace{\mathcal{A}_2\,;\,\mathcal{B}_2}
$$
$$
\vdots\qquad\qquad\vdots
$$
$$
\frac{\varphi_1\qquad\qquad\varphi_2}{\varphi_1\wedge\varphi_2}
$$

$(\wedge\mathcal{F})$

$$
\underbrace{\overset{\square\!-\!(i)}{\varphi_1\,,\;\mathcal{A}\,;\,\mathcal{B}}}\qquad\qquad\underbrace{\overset{\square\!-\!(i)}{\varphi_2\,,\;\mathcal{A}\,;\,\mathcal{B}}}
$$
$$
\vdots\qquad\qquad\qquad\qquad\vdots
$$
$$
\frac{\varphi_1\wedge\varphi_2\qquad\bot}{\bot}(i)\qquad\qquad\frac{\varphi_1\wedge\varphi_2\qquad\bot}{\bot}(i)
$$

$(\vee\mathcal{V})$

$$
\underbrace{\mathcal{A}\,;\,\mathcal{B}}\qquad\underbrace{\mathcal{A}\,;\,\mathcal{B}}
$$
$$
\vdots\qquad\qquad\vdots
$$
$$
\frac{\varphi_1}{\varphi_1\vee\varphi_2}\quad\frac{\varphi_2}{\varphi_1\vee\varphi_2}
$$

$(\lor \mathcal{F})$

$$
\begin{array}{cc}
\overset{\square\text{—}(i)}{\underbrace{\varphi_1 \ , \ \mathcal{A}_1 \ ; \ \mathcal{B}_1}} & \overset{\square\text{—}(i)}{\underbrace{\varphi_2 \ , \ \mathcal{A}_2 \ ; \ \mathcal{B}_2}} \\
\vdots & \vdots \\
\end{array}
$$

$$
\varphi_1 \lor \varphi_2 \qquad \bot \qquad\qquad \bot
$$
$$
\frac{\hspace{6cm}}{\bot}{\scriptstyle(i)}
$$

$(\rightarrow \mathcal{V})$

$$
\begin{array}{ccc}
\overset{\square\text{—}(i)}{\underbrace{\varphi_1 \ , \ \mathcal{A} \ ; \ \mathcal{B}}} & \underbrace{\mathcal{A} \ ; \ \mathcal{B}} & \overset{\square\text{—}(i)}{\underbrace{\varphi_1 \ , \ \mathcal{A} \ ; \ \mathcal{B}}} \\
\vdots & \vdots & \vdots \\
\dfrac{\bot}{\varphi_1 \rightarrow \varphi_2}{\scriptstyle(i)} & \dfrac{\varphi_2}{\varphi_1 \rightarrow \varphi_2} & \dfrac{\varphi_2}{\varphi_1 \rightarrow \varphi_2}{\scriptstyle(i)}
\end{array}
$$

$(\rightarrow \mathcal{F})$

$$
\begin{array}{cc}
\underbrace{\mathcal{A}_1 \ ; \ \mathcal{B}_1} & \overset{\square\text{—}(i)}{\underbrace{\varphi_2 \ , \ \mathcal{A}_2 \ ; \ \mathcal{B}_2}} \\
\vdots & \vdots \\
\end{array}
$$

$$
\varphi_1 \rightarrow \varphi_2 \qquad \varphi_1 \qquad\qquad \bot
$$
$$
\frac{\hspace{6cm}}{\bot}{\scriptstyle(i)}
$$

Note that the second two parts of $(\rightarrow \mathcal{V})$ could be combined as follows:

$$
\overset{\lozenge\text{—}(i)}{\underbrace{\varphi_1 \ , \ \mathcal{A} \ ; \ \mathcal{B}}}
$$
$$
\vdots
$$
$$
\dfrac{\varphi_2}{\varphi_1 \rightarrow \varphi_2}{\scriptstyle(i)}
$$

Here the diamond affixed to the discharge stroke indicates that the subordinate verification of φ_2 need not have φ_1 among its undischarged assumptions; but, if it does, then all assumption occurrences of φ_1 are discharged upon applying the rule.

Finally, the $\mathcal{V}$-rules and $\mathcal{F}$-rules for sentences with quantifiers dominant can now be written in the way that is set out below. (Note that we are now also allowing for the model M to have an infinite domain.) Bear in mind that the various atomic bases $\mathcal{B}$, $\mathcal{B}_i$ that are mentioned in the rules (and which are included within the broader 'atomic diagrams' now being countenanced for models M) are strictly cumulative 'down through' the construction of any

M-relative evaluation; none of their members is ever discharged. Matters are different, however, with the sets $\mathcal{A}$, $\mathcal{A}_i$ of undischarged, but potentially dischargeable, assumptions.

$(\forall\mathcal{V})$

$$\frac{\overbrace{\mathcal{A}_1 \; ; \; \mathcal{B}_1}\qquad\qquad\overbrace{\mathcal{A}_n \; ; \; \mathcal{B}_n}}{\begin{array}{cc} \vdots & \cdots \quad \vdots \quad \cdots \\ \psi(\alpha_1) & \psi(\alpha_n) \end{array}}$$
$$\frac{}{\forall x \psi(x)} \; M$$

where $\alpha_1, \ldots, \alpha_n \ldots$ are all the individuals in the domain

$(\forall\mathcal{F})$

$$\frac{\forall x\psi(x) \qquad \underbrace{\dfrac{\overline{\qquad}^{(i)}}{\psi(\alpha)} \; , \; \mathcal{A} \; ; \; \mathcal{B}}}{}$$
$$\frac{\forall x\psi(x) \qquad \begin{array}{c} \vdots \\ \bot \end{array}{}^{(i)}}{\bot}$$

where α is an individual in the domain

$(\exists\mathcal{V})$

$$\frac{\overbrace{\mathcal{A} \; ; \; \mathcal{B}}}{\begin{array}{c}\vdots\\ \psi(\alpha)\end{array}}$$
$$\frac{\psi(\alpha)}{\exists x\psi(x)}$$

where α is an individual in the domain

$(\exists\mathcal{F})$

$$\frac{\exists x\psi(x) \qquad \underbrace{\dfrac{\overline{\qquad}^{(i)}}{\psi(\alpha)} \; , \; \mathcal{A}_1 \; ; \; \mathcal{B}_1}\quad \bot \qquad \cdots \qquad \underbrace{\dfrac{\overline{\qquad}^{(i)}}{\psi(\alpha)} \; , \; \mathcal{A}_n \; ; \; \mathcal{B}_n}\quad \bot}{\bot} {}^{(i)} \, M$$

where $\alpha_1, \ldots, \alpha_n \ldots$ are all the individuals in the domain

Chapter 4

From the Logic of Evaluation to the Logic of Deduction

Abstract

§4.1 describes the process whereby one morphs smoothly from the verification and falsification rules of the *model-relative* Logic of Evaluation $\mathbb{E}$ to the *model-invariant*, deductive system $\mathbb{C}$ of Core Logic. This is done by (i) allowing *complex* sentences in the antecedents of sequents, (ii) allowing *complex sentences* to replace the absurdity symbol as consequents of sequents, and (iii) *voiding the basis of atomic rules*, so that deducibility becomes a model-invariant matter of *form*, not of *content*.

§4.2 sets out in graphic form the rules of Core Logic as they result from the morphing process that begins with the $\mathcal{V}$- and $\mathcal{F}$-rules of the Logic of Evaluation.

§4.3 explains reasons for preferring the parallelized forms of certain elimination rules in natural deduction (the ones for $\wedge$, $\rightarrow$, and $\forall$) to their more conventional serial forms.

§4.4 explains how $\bot$ (absurdity) can make its way into proofs as a conclusion, as required for applications of ($\neg$I). We proceed to introduce the notion of harmony between introduction and elimination rules, focusing on $\neg$ for the purposes of illustration. This heralds the full treatment of reduction procedures for the logical operators that will be provided in Chapter 6.

§4.5 examines Disjunctive Syllogism, revisiting Core Logic's liberalized rule of proof by cases in order to stress its virtues.

§4.6 explains in detail how the rules of natural deduction are to be construed as inductive clauses within the inductive definition of the notion of proof.

4.1 From evaluation to deduction

We saw in Chapter 3 how to furnish interpretations (in the form of atomic bases $\mathcal{B}$) that can not only make outright atomic determinations, but also impose constraints of inclusion and contrariety among extralogical primitives. Relative to such an interpretation $\mathcal{B}$, one seeks to determine the truth value of any sentence involving those primitives. In the first-order case there is the need, of course, to specify the domain of individuals, since we need verifications and falsifications of *instances* of quantifications in order to evaluate them.

There are two rules for each logical operator—a $\mathcal{V}$-rule and an $\mathcal{F}$-rule—enabling one to evaluate as *True* or *False* any sentence φ containing occurrences of those operators, in any model interpreting the primitive, non-logical expressions occurring in φ. These $\mathcal{V}$-and $\mathcal{F}$-rules comprise the Logic of Evaluation $\mathbb{E}$.

Verifications have possibly complex conclusions φ; while falsifications have conclusions $\bot$. Moreover, both kinds of construction have as undischarged but dischargeable assumptions only *atoms*. These form the set $\mathcal{A}$ of assumptions for a 'conditional' evaluation (verification or falsification) of φ. If a verification or falsification Π of φ has $\mathcal{A}$ empty, then it verifies or falsifies φ *outright*, relative to the atomic basis $\mathcal{B}$ whose rules have found application within Π.

But what happens when one does not have full enough access to the basic (atomic) facts about a situation? One cannot always be in a position to work out for oneself the truth value of a given sentence in any situation in which one finds oneself. Often one has the benefit of some sincere assertions from fellow speakers, in a situation of which one is not fully apprised. One needs, under such circumstances, to be able to work out at least some of the major *consequences* of what one has been told. That is, one needs to be able to *deduce* conclusions from *complex* premises in general, and not only from *atomic* ones. (The conclusions may be complex, or atomic, or $\bot$.) How might that be done?

We offer for consideration the following methodological suggestion. Let the rules governing the connectives in the Logic of Evaluation $\mathbb{E}$ (see §3.5.4) morph smoothly into rules of general deduction among complex propositional premises and conclusions. This will happen by systematically tweaking the $\mathbb{E}$-rules in order to attain the degree of generality that a deductive system must enjoy. We shall find that the $\mathcal{V}$-rules of $\mathbb{E}$ for the connectives will become Introduction rules; and the $\mathcal{F}$-rules of $\mathbb{E}$ for the connectives will become Elimination rules. The systematic tweaks are as follows:

1. Let the sets $\mathcal{A}$ of *atoms*, in both the $\mathcal{V}$-rules and the $\mathcal{F}$-rules, be replaced by sets Δ of *possibly complex premises*. (Sets of atoms will of course still be allowed as special cases of such Δ.)

2. Let the atomic bases $\mathcal{B}$ be empty, so that all mention of them can be suppressed.

Nothing more remains to be done in order to generalize the $\mathcal{V}$-rules for the connectives into Introduction rules. In order to finalize the generalization of the $\mathcal{F}$-rules for the connectives into Elimination rules,

3. Let θ appear in place of $\bot$ as a placeholder for possibly complex conclusions. (Conclusion occurrences of $\bot$ will of course still be allowed as special cases of such θ.)

Under these suggested modifications, *the $\mathcal{V}$-rules of* $\mathbb{E}$ *for the connectives morph into their Introduction rules in* $\mathbb{C}$, *and their $\mathcal{F}$-rules morph into their Elimination rules in* $\mathbb{C}$:

$$\left.\begin{array}{l}(@\text{-}\mathcal{V})\\(@\text{-}\mathcal{F})\end{array}\right\} \rightsquigarrow \left\{\begin{array}{l}(@\text{I})\\(@\text{E})\end{array}\right.$$
$$\mathcal{A}/\Delta$$
$$\mathcal{B}/\emptyset$$
$$\bot/\theta$$

An Introduction rule tells one under what circumstances one can *infer to* a conclusion of a particular form. The corresponding Elimination rule tells one what can be *inferred from* such a sentence, when it is used as the so-called *major premise*. The major premise sits *immediately above* the inference stroke. And, even though the new context is that of deductive transition in general, *major premises still stand proud*, with no proof work above them. They are *not* required or allowed to function in any other way; in particular, they are *not* going to give rise to any proofs not in normal form. All evaluations (truthmakers and falsitymakers) were in normal form; and so too will be deductive proofs.

The feature of targets for falsification standing proud is essential to the way we evaluate compounds. In passing to a more general deductive logic, in which we prove complex conclusions in general rather than just the absurdity sign, we make *minimal* changes to the rules of evaluation. The resulting rules of eliminative form then have their premises standing proud too. That is the essential feature of Left rules in the sequent calculus. With their MPEs standing proud, the natural-deduction elimination rules are finally in a form that allows for utterly natural isomorphism between natural deductions and their corresponding sequent proofs. This is an essential advance on Gentzenian proof theory's vexingly complicated treatment of translations between ND-proofs and SC-proofs in the decades immediately following Gentzen's pioneering contributions.

We are retaining the overall *form* of our $\mathcal{V}$- and $\mathcal{F}$-rules, but letting particular symbols therein admit of a more general interpretation. In particular—and this is crucial—we are retaining the *naturally forced relevance* of premises to conclusions that is readily manifest in the process of evaluating sentences as true or as false on the basis of similar evaluations of their constituents. And we are *not adding any other rules*. In particular, we are *not adding any structural rules*. We are merely modestly generalizing only the rules of $\mathbb{E}$, the logic of evaluation, so that they turn naturally into the rules of Core Logic $\mathbb{C}$. In this way, the meanings of the logical operators that arise from their role in *making sentences true or false under interpretation* are revealed as being able to sustain *all logically licit deductive transitions from complex premises to complex conclusions*.

4.2 Graphic rules of Core Logic $\mathbb{C}$

Here, then, are the Introduction and Elimination rules for the connectives that respectively result from making the changes just suggested to the foregoing $\mathcal{V}$- and $\mathcal{F}$-rules of the Logic of Evaluation $\mathbb{E}$. These are the rules for Core Logic $\mathbb{C}$. Both verifications and falsifications now become *proofs* or *deductions*, for which we shall use the symbol Π (with or without subscripts). Each Introduction–Elimination pair of rules will be accompanied by some brief remarks clarifying our implementation of the methodological suggestion above.

$$
(\neg\text{I}) \qquad
\begin{array}{c}
\square\!\!-\!\!(i) \\
\overbrace{\varphi \;,\; \Delta} \\
\Pi \\
\hline
\dfrac{\bot}{\neg\varphi}\!\!-\!\!(i)
\end{array}
\qquad
(\neg\text{E}) \qquad
\begin{array}{cc}
& \Delta \\
& \Pi \\
\neg\varphi & \varphi \\
\hline
\multicolumn{2}{c}{\bot}
\end{array}
$$

In the rules for negation ($\neg$), there is no reason to substitute θ for $\bot$.

$$(\wedge\text{I}) \quad \frac{\begin{matrix} \Delta_1 & \Delta_2 \\ \Pi_1 & \Pi_2 \\ \varphi_1 & \varphi_2 \end{matrix}}{\varphi_1 \wedge \varphi_2} \qquad (\wedge\text{E}) \quad \frac{\varphi_1 \wedge \varphi_2 \qquad \begin{matrix} \overset{\square \underline{\quad}(i)}{\overbrace{\varphi_1 \ , \ \Delta}} \\ \Pi \\ \theta \end{matrix}}{\theta}(i) \qquad \frac{\varphi_1 \wedge \varphi_2 \qquad \begin{matrix} \overset{\square \underline{\quad}(i)}{\overbrace{\varphi_2 \ , \ \Delta}} \\ \Pi \\ \theta \end{matrix}}{\theta}(i)$$

The Introduction rule for $\wedge$ is straightforward. The Elimination rule, in its present form, bears its 'imprint of origin' in the separate halves of the falsification rule $(\wedge\mathcal{F})$, according to which one could falsify a conjunction by falsifying either one of its conjuncts. The two halves of the Elimination rule just stated can now be combined into a *single* graphic rule that says that whatever conclusion θ can be inferred by using *either one or both* of the two assumptions φ and ψ can be inferred by using, instead, their conjunction $\varphi_1 \wedge \varphi_2$:

$$(\wedge\text{E}) \quad \frac{\varphi_1 \wedge \varphi_2 \qquad \begin{matrix} \overset{(i) \underline{\quad}\square\underline{\quad}(i)}{\overbrace{\varphi_1 \ , \ \varphi_2 \ , \ \Delta}} \\ \vdots \\ \theta \end{matrix}}{\theta}(i)$$

Here we use the box again to annotate the discharge strokes. This time it is to indicate that we must have used, in the subordinate proof of θ indicated by the vertical dots, at least one, possibly both, of the conjuncts φ_1, φ_2 as undischarged assumptions. For, if neither of them stood thus as undischarged assumptions, then there would be no point at all in 'eliminating' the conjunction $\varphi_1 \wedge \varphi_2$—we would already have a proof of θ from the available undischarged assumptions. The addition to the latter of the conjunction $\varphi_1 \wedge \varphi_2$ (as a result of the contemplated application of the rule) would unnecessarily bloat the overall set of undischarged assumptions, thereby weakening the logical result. Not only that, the conjunction in question could be *completely irrelevant* to the subject matter about which one is reasoning when carrying out the subproof indicated by the vertical dots. This is why we impose the very reasonable relevance requirement indicated by the box.

In the case of disjunction $(\vee)$ there are three distinct patterns in which it would be permissible to replace $\bot$ by the more general placeholder θ when formulating the Elimination rule from the falsification rule. We retain all three of these patterns. There is also a fourth pattern, where $\bot$ is not replaced as the conclusion of either of the case proofs. We state it explicitly and separately here as $(\vee E_{\perp\perp})$, even though it could be taken as a special instance of $(\vee E_{\theta\theta})$.

$$(\vee I) \qquad \dfrac{\begin{array}{c} \Delta \\ \Pi \\ \varphi_1 \end{array}}{\varphi_1 \vee \varphi_2} \qquad \dfrac{\begin{array}{c} \Delta \\ \Pi \\ \varphi_2 \end{array}}{\varphi_1 \vee \varphi_2}$$

$$(\vee E_{\theta\perp}) \qquad \dfrac{\varphi_1 \vee \varphi_2 \qquad \begin{array}{c} \square\!\!-\!\!(i) \\ \overbrace{\varphi_1 \;,\; \Delta_1} \\ \Pi_1 \\ \theta \end{array} \qquad \begin{array}{c} \square\!\!-\!\!(i) \\ \overbrace{\varphi_2 \;,\; \Delta_2} \\ \Pi_2 \\ \perp \end{array}}{\theta}\,(i)$$

$$(\vee E_{\perp\theta}) \qquad \dfrac{\varphi_1 \vee \varphi_2 \qquad \begin{array}{c} \square\!\!-\!\!(i) \\ \overbrace{\varphi_1 \;,\; \Delta_1} \\ \Pi_1 \\ \perp \end{array} \qquad \begin{array}{c} \square\!\!-\!\!(i) \\ \overbrace{\varphi_2 \;,\; \Delta_2} \\ \Pi_2 \\ \theta \end{array}}{\theta}\,(i)$$

$$(\vee E_{\theta\theta}) \qquad \dfrac{\varphi_1 \vee \varphi_2 \qquad \begin{array}{c} \square\!\!-\!\!(i) \\ \overbrace{\varphi_1 \;,\; \Delta_1} \\ \Pi_1 \\ \theta \end{array} \qquad \begin{array}{c} \square\!\!-\!\!(i) \\ \overbrace{\varphi_2 \;,\; \Delta_2} \\ \Pi_2 \\ \theta \end{array}}{\theta}\,(i)$$

$$(\vee E_{\perp\perp}) \qquad \dfrac{\varphi_1 \vee \varphi_2 \qquad \begin{array}{c} \square\!\!-\!\!(i) \\ \overbrace{\varphi_1 \;,\; \Delta_1} \\ \Pi_1 \\ \perp \end{array} \qquad \begin{array}{c} \square\!\!-\!\!(i) \\ \overbrace{\varphi_2 \;,\; \Delta_2} \\ \Pi_2 \\ \perp \end{array}}{\perp}\,(i)$$

The first part of $(\rightarrow\mathcal{V})$ keeps its conclusion $\perp$ in its subordinate proof when it morphs into the first part of $(\rightarrow I)$. For it is only in erstwhile $\mathcal{F}$-rules that we are substituting the more general placeholder θ for $\perp$.

$$(\rightarrow\text{I})(\text{a})\qquad \begin{array}{c} \square\!\!-\!\!(i) \\ \overbrace{\varphi_1\ ,\ \Delta} \\ \Pi \\ \underline{\qquad\bot\qquad}\!\!-\!\!(i) \\ \varphi_1 \rightarrow \varphi_2 \end{array}\qquad (\rightarrow\text{I})(\text{b})\qquad \begin{array}{c} \lozenge\!\!-\!\!(i) \\ \overbrace{\varphi_1\ ,\ \Delta} \\ \Pi \\ \underline{\qquad\varphi_2\qquad}\!\!-\!\!(i) \\ \varphi_1 \rightarrow \varphi_2 \end{array}$$

$$(\rightarrow\text{E})\qquad \begin{array}{ccc} & \Delta_1 & \begin{array}{c}\square\!\!-\!\!(i)\\ \overbrace{\varphi_2\ ,\ \Delta_2}\end{array} \\ & \Pi_1 & \Pi_2 \\ \varphi_1 \rightarrow \varphi_2 & \varphi_1 & \theta \\ \hline \multicolumn{3}{c}{\qquad\qquad\theta\qquad\qquad}\!\!-\!\!(i) \end{array}$$

The two inherited features of $\mathbb{E}$ that were stressed above—non-vacuous discharge of certain assumptions, and MPEs standing proud—are crucial in ensuring for the resulting deductive logic $\mathbb{C}$ that in any of its proofs, the premises (i.e. undischarged assumptions) are *relevant* to the conclusion. See Chapter 10 for a dispositive result that reveals the precise nature of such relevance.

The foregoing rules constitute (propositional) *Core Logic*. *The same method succeeds in generalizing the quantifier rules* from the evaluation setting to the deductive setting. There is, however, a further principled consideration in the case of the quantifiers, which we did not have to deal with in the case of the connectives.

Two of the quantifier rules make the transition smoothly from the case of evaluation to the more general case of deduction. The rule $(\exists\mathcal{V})$:

$$(\exists\mathcal{V})\qquad \begin{array}{c} \overbrace{\mathcal{A}\ ;\ \mathcal{B}} \\ \vdots \\ \underline{\psi(\alpha)} \\ \exists x\psi(x) \end{array}$$

becomes the rule of $\exists$-Introduction:

$$(\exists\text{I})\qquad \begin{array}{c} \Delta \\ \Pi \\ \underline{\psi^x_t} \\ \exists x\psi \end{array}$$

And the rule $(\forall\mathcal{F})$:

$(\forall\mathcal{F})$

$$\underbrace{\overline{\psi(\alpha)}^{(i)} , \mathcal{A} ; \mathcal{B}}$$
$$\vdots$$
$$\frac{\forall x\psi(x) \qquad \perp}{\perp}{}_{(i)}$$

becomes the rule of $\forall$-Elimination:

$(\forall\text{E})$

$$\underbrace{\overline{\psi^x_t}^{\square\_\_\_(i)} , \Delta}$$
$$\Pi$$
$$\frac{\forall x\psi \qquad \theta}{\theta}{}_{(i)}$$

By analogy with conjunction, this Elimination rule for $\forall$, which deals with a *single* counterinstance t, can be written equivalently and more economically (for the purposes of constructing proofs) as

$(\forall\text{E})$

$$\underbrace{\overset{(i)\_\_\_ \;\cdots\; \square \;\cdots\; \_\_\_(i)}{\psi^x_{t_1} , \quad \cdots \quad , \psi^x_{t_n}} , \Delta}$$
$$\Pi$$
$$\frac{\forall x\psi \qquad \theta}{\theta}{}_{(i)}$$

Here the box means that *at least one* instantiating assumption of the form $\psi^x_{t_i}$ must have been used, and be available for discharge.

In the transition from evaluation rules to rules of deduction in general, however, further changes have to be made in the case of our two 'M-relative' quantifier rules. When evaluating sentences in a model M, we required instance-by-instance verifications, or falsifications, depending on whether the rule being applied was $(\forall\mathcal{V})$ or $(\exists\mathcal{F})$ respectively. But *deductive* (as opposed to *evaluative*) reasoning does not concern any situation M in particular; rather, it is concerned with preserving truth-in-M, *whatever the model M may be*. We therefore replace all those 'instance' proofs (and disproofs) for individuals $\alpha_1, \ldots, \alpha_n \ldots$ with a single *proof template*, involving a *parameter* a in place of the various individuals α_i. Thus the M-relative rule $(\forall\mathcal{V})$—

$$(\forall\mathcal{V}) \quad \cfrac{\overbrace{\begin{array}{c}\mathcal{A}_1 \; ; \; \mathcal{B}_1\end{array}}\\ V_1 \\ \psi(\alpha_1)} \quad \cdots \quad \cfrac{\overbrace{\begin{array}{c}\mathcal{A}_n \; ; \; \mathcal{B}_n\end{array}}\\ V_n \\ \psi(\alpha_n)} \quad \cdots}{\forall x\psi(x)} \; M$$

where $\alpha_1, \ldots, \alpha_n \ldots$ are all the individuals in the domain

—becomes the rule of $\forall$-Introduction:

$$(\forall\text{I}) \quad \cfrac{\begin{array}{c}\Delta \\ \Pi \\ \psi(a)\end{array}}{\forall x\psi(x)}$$

where a does not occur in any sentence in Δ or in $\forall x\psi(x)$.

The subproof Π is the proof template just mentioned; it takes over from all the various V_i of the verification rule. The deductive reasoning has been *finitized*; we are on the way to having proofs be *effectively checkable*.

Likewise, the rule $(\exists\mathcal{F})$—

$$(\exists\mathcal{F}) \quad \cfrac{\exists x\psi(x) \qquad \cfrac{\overbrace{\overline{\psi(\alpha)}^{(i)} \;, \; \mathcal{A}_1 \; ; \; \mathcal{B}_1}}{\begin{array}{c}F_1 \\ \bot\end{array}} \quad \cdots \quad \cfrac{\overbrace{\overline{\psi(\alpha)}^{(i)} \;, \; \mathcal{A}_n \; ; \; \mathcal{B}_n}}{\begin{array}{c}F_n \\ \bot\end{array}} \quad \cdots}{\bot} \; {(i)\, M}$$

where $\alpha_1, \ldots, \alpha_n \ldots$ are all the individuals in the domain

—becomes the rule of $\exists$-Elimination:

$$(\exists\text{E}) \quad \cfrac{\exists x\psi(x) \qquad \cfrac{\overbrace{\Box\overline{}^{(i)}\\ \psi(a) \;, \; \Delta}}{\begin{array}{c}\Pi \\ \theta\end{array}}}{\theta} \; {(i)}$$

where a does not occur in any sentence in Δ or in $\exists x\psi(x)$ or in θ.

Once again, the proof template Π, with its parameter a, takes over from the various falsifications F_i of the falsification rule.

We now have in place all the rules of *first-order* Core Logic $\mathbb{C}$. They have, to be sure, been stated graphically. We set as a helpful exercise for the reader at this stage these rules' reformulation as clauses in the inductive definition of the notion 'Π is a core proof of conclusion φ from the set Δ of undischarged assumptions', which we abbreviate as $\mathcal{P}(\Pi, \varphi, \Delta)$. We need only point out that the sole basis clause will be $\mathcal{P}(\varphi, \varphi, \{\varphi\})$. (The reader who does not wish to undertake this exercise will find the answer in §4.6.)

4.3 Reasons for preferring parallelized to serial forms of Elimination rules

We have stated both $(\to E)$ and $(\wedge E)$ in so-called *parallelized form*, as opposed to the more usual *serial form* that is traditional with systems of natural deduction. The serial forms of these two rules are *Modus Ponens*:

$$\frac{\vdots \qquad \vdots}{\quad \dfrac{\varphi \to \psi \quad \varphi}{\psi}}$$

and

$$\frac{\vdots \qquad\qquad \vdots}{\dfrac{\varphi \wedge \psi}{\varphi} \qquad \dfrac{\varphi \wedge \psi}{\psi}} \cdot$$

These forms were given originally in Gentzen [1934, 1935] and were reprised in Prawitz [1965]. Both rules allow non-trivial proof work above their major premises.

The parallelized forms for these Elimination rules are greatly to be preferred. There are four reasons why this is so.

First, we achieve a pleasing uniformity of presentation, since the Elimination rule for $\vee$ (known as *Proof by Cases*) is always in parallelized form, and indeed cannot even be framed in serial form.[1]

Second, framing Elimination rules in parallelized form is the natural and obvious thing to do when one is concerned with a framework for efficient

---

[1] At first order, Gentzen and Prawitz have $(\exists E)$ likewise in parallelized form; and we shall be urging that their serial form of $(\forall E)$ be replaced by the parallelized form also.

proof search. When one is given a sequent to be proved (call this the 'original problem'), one has to break it down into sub-problems. One can do this by attending to the conclusion of the sequent, and framing the sub-problem as that of finding appropriate subordinate proofs that would enable one to apply the *Introduction* rule for the dominant operator in the conclusion. Alternatively, one can attend instead to one of the premises, and frame the sub-problem as that of finding appropriate subordinate proofs that would enable one to apply the *Elimination* rule for the dominant operator in the premise in question, which will then be the major premise for that elimination. The parallelized forms of the elimination rules serve this purpose perfectly.

Third, using Elimination rules in parallelized form makes for *shorter* formal proofs. This is because one can avoid the need to repeat whole fragments of proof every time one wishes to apply an Elimination rule. This makes automated proof search much more efficient.

These first three motivations for using parallelized forms of Elimination rules were treated in depth in Tennant [1992].

Fourth, and finally, using Elimination rules in parallelized form affords a solution to certain problems that would otherwise arise for the proof-theoretic criterion of paradoxicality to be discussed in Chapter 11.

The use of the *prima-facie* simpler and more appealing serial forms may well have impeded logicians from attaining the more uniform presentational point of view that is called for in order to appreciate the 'core nature' of Core Logic. In order to appreciate how Core Logic is the essential encapsulation of how the logical operators behave—independently of any considerations of dilution and/or transitivity—one needs to work with the *parallelized* forms of the Elimination rules. (The topics of dilution and transitivity will be discussed in due course.) Indeed, one needs to work with the parallelized forms subject to an important constraint: that major premises for elimination stand proud, with no proof work above them. The original source for the parallelized forms of Elimination rules (but without imposition of the aforementioned constraint) is, as pointed out in §2.3.7, footnote 11, Schroeder-Heister [1984]. The standing-proud feature was added by Tennant [1992].

4.4   Absurdity and contrarieties

We have seen how the absurdity symbol $\perp$ may be used as the conclusion of a falsification of a sentence φ relative to both a collection $\mathcal{A}$ of atomic assumptions and an atomic basis $\mathcal{B}$ of 'interpreting' rules of inference. That was within the context of the Logic of Evaluation $\mathbb{E}$. We then took an economically generalizing segue that turned the $\mathcal{V}$- and $\mathcal{F}$-rules of $\mathbb{E}$ into the Introduction and Elimination rules, respectively, of Core Logic $\mathbb{C}$.

4.4.1 Conclusion occurrences of absurdity ($\bot$)

The reader will be wondering how the absurdity symbol $\bot$ can ever make its way into a proof as a conclusion. For this is certainly needed in order for the rule ($\neg$I) to find application.

One simple way to produce $\bot$ as a conclusion is (as we have already seen) to have *rules of contrariety* within one's language or theory. Any pair of antonyms will suffice for this purpose. In Chapter 3 we advanced color-exclusion principles as good examples. If we understand 't is red (resp., green)' to mean 't is *monochromatically* red (resp., green) *all over, now*', (where t is any singular term denoting an object that can be colored) then we would have the following rule of contrariety in English:

$$\frac{t \text{ is red} \quad t \text{ is green}}{\bot}$$

Similar examples proliferate, by using any pair of English antonyms in place of 'red' and 'green'—pairs such as *friend–foe, rich–poor, hard–soft, nutritious–poisonous, tall–short, big–small,* etc.

Other ways to produce $\bot$ as a conclusion arise with the primitives of mathematical theories, such as geometry and arithmetic. We have the following rules expressing category-disjointness of geometrical primitives:

$$\frac{t \text{ is a point} \quad t \text{ is a line}}{\bot} \quad \frac{t \text{ is a point} \quad t \text{ is a plane}}{\bot} \quad \frac{t \text{ is a line} \quad t \text{ is a plane}}{\bot}$$

and we have the important axiom in arithmetic, that 0 is distinct from 1 (or: that 0 is distinct from *the successor of* 0), which can be expressed as the following rule of inference:

$$\frac{0 = s0}{\bot}$$

These examples should suffice to persuade the reader that there are ample ways of producing $\bot$ as a conclusion, so as to be able to apply ($\neg$I) subsequently. Here, for example, is a mini-proof exploiting the rules stated so far:

$$\frac{\overline{\rule{2cm}{0.4pt}}^{(1)}}{\begin{array}{cc} t \text{ is red} & t \text{ is green} \end{array}}$$

$$\frac{\bot}{\neg(t \text{ is red})}^{(1)}$$

This little proof shows that from the premise 't is green' the conclusion 'it is not the case that t is red' follows—at least, within a language in which we have the contrariety rule that has been applied at the first step. The second and final step of the proof is an application of the rule ($\neg$I) of Negation Introduction. Note how it discharges the assumption 't is red'. Actually, the construction just given qualifies also as a *verification* of $\neg(t$ is red), relative to both the set $\{t$ is green$\}$ of undischarged atomic assumptions, and the atomic basis $\left\{ \dfrac{t \text{ is red} \quad t \text{ is green}}{\bot} \right\}$. So it is a proof in the logic of evaluation $\mathbb{E}$.

Another way to produce $\bot$ as the conclusion of a proof is by applying the rule ($\neg$E) of Negation *Elimination*:

$$(\neg\text{E}) \qquad \frac{\neg\varphi \quad \overset{\vdots}{\varphi}}{\bot}$$

Recall that the major premise $\neg\varphi$ 'stands proud', with no proof work above it. The *minor* premise φ, however, *is* allowed to have proof work above it, as indicated by the vertical dots. In all our rule presentations of this graphic form, we have followed this convention. We remind the reader: the *absence* of vertical dots above a sentence occurrence means that it stands proud, with no proof work above it. By requiring major premises of eliminations to stand proud, one ensures that all proofs are in normal form.

4.4.2   Harmony between introduction and elimination rules

The Elimination rule for negation exactly balances the corresponding Introduction rule: The Introduction rule specifies that in order to infer $\neg\varphi$, one needs to reduce φ to absurdity; and the Elimination rule says that given $\neg\varphi$, one may infer from φ to absurdity. So the two rules are in harmony.

There is a formal way to depict the harmony between the Introduction and Elimination rules for any logical operator. It will be provided in due course in Chapter 6. Here we forge ahead, if somewhat prematurely, with an illustration of harmony using the negation rules.

Suppose one is given two core proofs

$$
\begin{array}{cc}
\Xi & \varphi, \Delta \\
\Pi & \Sigma \\
\varphi & \theta
\end{array}
\qquad \text{(where } \varphi \notin \Delta \text{ and } \Delta \text{ may be empty)}
$$

Here the sentence φ is called an *interpolant* between the combined set $\Xi \cup \Delta$ of premises, and the conclusion θ of Σ.

One can define an effective operation $[\Pi, \Sigma]$ that turns the two proofs Π and Σ into a proof of θ (or of the stronger conclusion $\bot$) from (some subset of) the combined set $\Xi \cup \Delta$ of premises. An important clause in the definition of this operation is the following *reduction procedure for negation*. It deals with the case where the interpolant φ is a negation, the proof Π ends with an application of Negation Introduction, and φ stands in Σ as the major premise of a terminal step of Negation Elimination (so that θ is $\bot$). In these circumstances there will be immediate subproofs Π' of Π, and Σ' of Σ, involved as shown:

$$
\left[
\begin{array}{c}
\overset{(i)}{\underline{}} \\
\varphi, \Xi \\
\Pi' \\
\dfrac{\bot}{\neg\varphi}(i)
\end{array}
\quad,\quad
\begin{array}{c}
\Delta \\
\Sigma' \\
\dfrac{\neg\varphi \quad \varphi}{\bot}(1)
\end{array}
\right]
\;=\;
\left[
\begin{array}{c}
\Delta \\
\Sigma' \\
\varphi
\end{array}
\quad,\quad
\begin{array}{c}
\varphi, \Xi \\
\Pi' \\
\bot
\end{array}
\right]
$$

The reduct on the right no longer contains the indicated occurrences of $\neg\varphi$, and has as its two arguments the simpler proofs Σ' and Π'. Because of the harmony between the Introduction and Elimination rules for negation, the displayed occurrences of negation on the left-hand side of the foregoing identity have been spirited away.

Our illustration of harmony in connection with negation is in one respect a special case, because the overall conclusion of the reduct is $\bot$ (and this in turn is because $\bot$ is the conclusion of the Negation Elimination rule). But with other Elimination rules (for other logical operators), the overall conclusion might be some sentence φ, rather than $\bot$. What would the reduct accomplish in such a case? It turns out that the test for harmony, which all Introduction–Elimination rule pairs will pass, is this: the reduct should establish some *subsequent* of the overall 'target' sequent that one would expect 'by transitivity'. And 'subsequent' here is to be understood generously, so as to allow (when Δ' is a subset of Δ) that the sequent $\Delta' : \bot$

counts as a subsequent of $\Delta : \varphi$.

The two rules of Negation Introduction and Elimination can also work together to produce logically complex relationships. A simple illustration is the proof that any sentence φ logically implies its own double negation $\neg\neg\varphi$:

$$
\begin{array}{c}
(1)\underline{\quad} \\
\neg\varphi \qquad \varphi \\
\hline
\underline{\quad\bot\quad}(1) \\
\neg\neg\varphi
\end{array}
$$

This can be extended to a proof that shows that any triple negation $\neg\neg\neg\varphi$ logically implies the single negation $\neg\varphi$:

$$
\begin{array}{c}
(1)\underline{\quad} \qquad \underline{\quad}(2) \\
\neg\varphi \qquad \varphi \\
\hline
\underline{\quad\bot\quad}(1) \\
\neg\neg\neg\varphi \qquad \neg\neg\varphi \\
\hline
\underline{\quad\bot\quad}(2) \\
\neg\varphi
\end{array}
$$

Let us now see how the rules for negation and conjunction can interact. First, we can prove the *Law of Non-Contradiction* (LNC), namely $\neg(\varphi \wedge \neg\varphi)$:

$$
\begin{array}{c}
(1)\underline{\quad} \qquad \underline{\quad}(1) \\
(2)\underline{\qquad\qquad} \qquad \neg\varphi \qquad \varphi \\
\varphi \wedge \neg\varphi \qquad\qquad \underline{\quad\bot\quad}(1) \\
\hline
\underline{\qquad\quad\bot\qquad\quad}(2) \\
\neg(\varphi \wedge \neg\varphi)
\end{array}
$$

Note that by the end of this proof, all its assumptions are discharged. The conclusion $\neg(\varphi \wedge \neg\varphi)$ has been proved *outright*. It rests on no assumptions whatsoever. For this reason it is called a logical *theorem*.

Second, here is a proof that the premise $\neg\psi$ logically implies the conclusion $\neg(\varphi \wedge \psi)$:

$$
\begin{array}{c}
\underline{\quad}(1) \\
(2)\underline{\qquad\qquad} \qquad \neg\psi \qquad \psi \\
\varphi \wedge \psi \qquad\qquad \underline{\quad\bot\quad}(1) \\
\hline
\underline{\qquad\quad\bot\qquad\quad}(2) \\
\neg(\varphi \wedge \psi)
\end{array}
$$

4.5 Disjunctive syllogism

Consider the following sequent:

$$\neg\varphi, \varphi \vee \psi : \psi.$$

It is called *Disjunctive Syllogism*. It is valid. Indeed, it is *perfectly* valid. For, by dropping any sentence on the left or on the right:

$$\varphi \vee \psi : \psi$$
$$\neg\varphi : \psi$$
$$\neg\varphi, \varphi \vee \psi : \bot$$

one obtains only *invalid* sequents.[2] Similar remarks hold for the other obvious form of Disjunctive Syllogism:

$$\neg\psi, \varphi \vee \psi : \varphi.$$

Disjunctive Syllogism is often encountered in mathematical reasoning. An example of a mathematical inference of this form is

$$x \not\leq y \,,\; x \leq y \vee x > y : x > y.$$

It is clear that we must be able to prove both forms of Disjunctive Syllogism, if our logic is to be adequate for mathematics. Since it has a premise of disjunctive form, we must ensure that the rule of Disjunction Elimination is formulated in a way that affords the sought proof.

4.5.1 Disjunction revisited

Recall this part of the rule of Disjunction Elimination that we called $(\vee E_{\bot\bot})$:

$$
(\vee E_{\bot\bot}) \quad
\begin{array}{ccc}
 & \Box\!\!-\!\!(i) & \Box\!\!-\!\!(i) \\
 & \varphi & \psi \\
 & \vdots & \vdots \\
\varphi \vee \psi & \bot & \bot \\
\hline
 & \bot &
\end{array}
\;(i)
$$

This finds expression as the fourth line of the truth table for $\vee$. This rule $(\vee E_{\bot\bot})$ limits one to overall conclusions of the form $\bot$. As has already been

---

[2] Note that the considerations entered here are entirely constructive, since the sets $\{\varphi \vee \psi\}$ and $\{\psi\}$ are finite and decidable.

indicated in §4.2, there is no compelling reason to be so restrictive when reasoning away from a major premise of disjunctive form. One way to achieve
greater generality, as indicated above, is to seek a sentential conclusion θ
in place of $\bot$ when applying elimination to a disjunction. Clearly the most
cautious and conservative generalization involves replacing *just one* of the
subordinate conclusions $\bot$ with θ. This gave rise to the two Elimination
rules $(\lor E_{\bot\theta})$ and $(\lor E_{\theta\bot})$ in §4.2. Together, these two rules say 'If you have
shown that one of the cases leads to absurdity, then the truth must lie with
the other case; so, you may bring down as the main conclusion the conclusion of that other case'. This thought respects relevance; it has no truck at
all with EFQ.

In the absence of a *reductio* of either disjunct, one wishes also to infer
from a disjunction any conclusion that one can infer from each of its disjuncts. This completes the process of generalizing $(\lor E_{\bot\bot})$:

$$
(\lor E_{\theta\theta}) \qquad
\begin{array}{cc}
\overset{\Box\text{———}(i)}{\varphi} & \overset{\Box\text{———}(i)}{\psi} \\[4pt]
\vdots & \vdots \\[2pt]
\end{array}
$$

$$
\cfrac{\varphi \lor \psi \qquad \theta \qquad \theta}{\theta}\,(i)
$$

The rule of Disjunction Elimination consists of the three parts just displayed:
$(\lor E_{\bot\theta})$, $(\lor E_{\theta\bot})$, and $(\lor E_{\theta\theta})$. The part we called $(\lor E_{\bot\bot})$ is now just a
special case of $(\lor E_{\theta\theta})$. If one wants a single graphic form of the rule of
Disjunction Elimination that covers exactly the official distinct parts of the
rule, one can write

$$
(\lor E) \qquad
\begin{array}{cc}
\overset{\Box\text{———}(i)}{\varphi} & \overset{\Box\text{———}(i)}{\psi} \\[4pt]
\vdots & \vdots \\[2pt]
\end{array}
$$

$$
\cfrac{\varphi \lor \psi \qquad \bot/\theta \qquad \bot/\theta}{\bot/\theta}\,(i)
$$

Disjunctive Syllogism tells one that if a disjunction is true, but one of
its disjuncts is not, then the truth must lie with the other disjunct. We are
now in a position to prove Disjunctive Syllogism, using only Introduction

and Elimination rules:

$$\cfrac{\varphi \vee \psi \qquad \cfrac{\neg\varphi \qquad \cfrac{}{\varphi}{\scriptstyle(1)}}{\bot} \qquad \cfrac{}{\psi}{\scriptstyle(1)}}{\psi}{\scriptstyle(1)}$$

The final step is an application of $(\vee E_{\bot\theta})$ (in which ψ is substituted for θ).
The other version of Disjunctive Syllogism has a similar proof:

$$\cfrac{\varphi \vee \psi \qquad \cfrac{}{\varphi}{\scriptstyle(1)} \qquad \cfrac{\neg\psi \qquad \cfrac{}{\psi}{\scriptstyle(1)}}{\bot}}{\varphi}{\scriptstyle(1)}$$

Here, the final step is an application of $(\vee E_{\theta\bot})$ (in which φ is substituted for θ).

4.6 The rules of natural deduction as clauses in the inductive definition of proof

We now define the notion $\mathcal{P}(\Pi, \varphi, \Delta)$: '$\Pi$ is a proof of the conclusion φ from the set Δ of undischarged assumptions'. This provides the solution to the exercise provided for the reader in §4.2.

We use a linear notation to denote the two-dimensional trees of sentence occurrences that we are calling proofs. We use square brackets to indicate the formation of a tree, with the convention that the first thing mentioned within the square brackets is the root node, and the remaining things within the square brackets are the immediate sub-trees, arranged from left to right. Thus the linear term

$$[\varphi, \Pi, \Sigma]$$

and the two-dimensional term

$$\cfrac{\Pi \quad \Sigma}{\varphi}$$

would denote the same abstract object, namely the tree with root node φ whose left-immediate subtree is Π and whose right-immediate subtree is Σ.

Definition 4 (Proof in natural deduction for Core Logic).
Basis clause: $\mathcal{P}(\varphi, \varphi, \{\varphi\})$

Inductive clauses follow, one for each rule of inference. Elimination rules are in parallelized form, with major premises standing proud (which is why they occur in the second argument place of the proof terms in the consequents of the inductive clauses in question: they are 'one-line' proofs). Note the membership requirements in the antecedents of certain inductive clauses; these ensure non-vacuous discharge of the assumptions in question.

$$(\neg I) \qquad \left. \begin{array}{c} \mathcal{P}(\Pi, \bot, \Gamma) \\ \varphi \in \Gamma \end{array} \right\} \quad \Rightarrow \quad \mathcal{P}([\neg\varphi, \Pi], \neg\varphi, \Gamma \setminus \{\varphi\})$$

In words: if Π proves $\bot$ from assumptions including φ, then $\dfrac{\Pi}{\neg\varphi}$ proves $\neg\varphi$ from those other assumptions.

$$(\neg E) \qquad \mathcal{P}(\Pi, \varphi, \Delta) \quad \Rightarrow \quad \mathcal{P}([\bot, \neg\varphi, \Pi], \bot, \{\neg\varphi\} \cup \Delta)$$

In words: If Π proves φ from Δ, then $\dfrac{\neg\varphi \quad \Pi}{\bot}$ proves $\bot$ from $\Delta, \neg\varphi$.

$$(\wedge I) \qquad \left. \begin{array}{c} \mathcal{P}(\Pi_1, \varphi_1, \Delta_1) \\ \mathcal{P}(\Pi_2, \varphi_2, \Delta_2) \end{array} \right\} \Rightarrow \mathcal{P}([\varphi_2 \wedge \varphi_2, \Pi_1, \Pi_2], \varphi_2 \wedge \varphi_2, \Delta_1 \cup \Delta_2)$$

In words: If Π_1 proves φ_1 from Δ_1, and Π_2 proves φ_2 from Δ_2, then $\dfrac{\Pi_1 \quad \Pi_2}{\varphi_1 \wedge \varphi_2}$ proves $\varphi_1 \wedge \varphi_2$ from $\Delta_1 \cup \Delta_2$.

$$(\wedge E) \qquad \left. \begin{array}{c} \mathcal{P}(\Pi, \theta, \Delta) \\ \varphi_1 \in \Delta \text{ or } \varphi_2 \in \Delta \end{array} \right\} \Rightarrow \mathcal{P}([\theta, \varphi_1 \wedge \varphi_2, \Pi], \theta, \{\varphi_1 \wedge \varphi_2\} \cup (\Delta \setminus \{\varphi_1, \varphi_2\}))$$

In words: If Π proves θ from assumptions including at least one of φ_1, φ_2, then $\dfrac{\varphi_1 \wedge \varphi_2 \quad \Pi}{\theta}$ proves θ from the assumptions of Π other than φ_1, φ_2, plus $\varphi_1 \wedge \varphi_2$.

$$(\vee I) \qquad \begin{array}{l} \mathcal{P}(\Pi, \varphi_1, \Delta) \Rightarrow \mathcal{P}([\varphi_1 \vee \varphi_2, \Pi], \varphi_1 \vee \varphi_2, \Delta); \\ \mathcal{P}(\Pi, \varphi_2, \Delta) \Rightarrow \mathcal{P}([\varphi_1 \vee \varphi_2, \Pi], \varphi_1 \vee \varphi_2, \Delta) \end{array}$$

In words: If Π proves φ_i from Δ ($i = 1, 2$), then $\dfrac{\Pi}{\varphi_1 \vee \varphi_2}$ proves $\varphi_1 \vee \varphi_2$ from Δ.

($\vee$E)

$$\left.\begin{array}{l} \mathcal{P}(\Pi_1, \theta, \Delta_1) \\ \mathcal{P}(\Pi_2, \theta, \Delta_2) \\ \quad \varphi_1 \in \Delta_1 \\ \quad \varphi_2 \in \Delta_2 \end{array}\right\} \Rightarrow \mathcal{P}([\theta, \varphi_1 \vee \varphi_2, \Pi_1, \Pi_2], \theta, \{\varphi_1 \vee \varphi_2\} \cup (\Delta_1 \backslash \{\varphi_1\}) \cup (\Delta_2 \backslash \{\varphi_2\}));$$

(Note that this covers the instance where θ is taken to be $\perp$.)

In words: If Π_i proves θ from certain assumptions including φ_i ($i = 1, 2$) then $\dfrac{\varphi_1 \vee \varphi_2 \quad \Pi_1 \quad \Pi_2}{\theta}$ proves θ from the remaining combined assumptions, plus $\varphi_1 \vee \varphi_2$.

$$\left.\begin{array}{l} \mathcal{P}(\Pi_1, \perp, \Delta_1) \\ \mathcal{P}(\Pi_2, \theta, \Delta_2) \\ \quad \varphi_1 \in \Delta_1 \\ \quad \varphi_2 \in \Delta_2 \end{array}\right\} \Rightarrow \mathcal{P}([\theta, \varphi_1 \vee \varphi_2, \Pi_1, \Pi_2], \theta, \{\varphi_1 \vee \varphi_2\} \cup (\Delta_1 \backslash \{\varphi_1\}) \cup (\Delta_2 \backslash \{\varphi_2\}));$$

$$\left.\begin{array}{l} \mathcal{P}(\Pi_1, \theta, \Delta_1) \\ \mathcal{P}(\Pi_2, \perp, \Delta_2) \\ \quad \varphi_1 \in \Delta_1 \\ \quad \varphi_2 \in \Delta_2 \end{array}\right\} \Rightarrow \mathcal{P}([\theta, \varphi_1 \vee \varphi_2, \Pi_1, \Pi_2], \theta, \{\varphi_1 \vee \varphi_2\} \cup (\Delta_1 \backslash \{\varphi_1\}) \cup (\Delta_2 \backslash \{\varphi_2\}))$$

In words, even more informally: if one has two case proofs Π_1, Π_2 respectively using φ_1, φ_2 as assumptions, and one of them has $\perp$ as its conclusion, then one may bring down the conclusion of the other one as the main conclusion of a proof by cases; that is, $\dfrac{\varphi_1 \vee \varphi_2 \quad \Pi_1 \quad \Pi_2}{\theta}$ proves θ from the remaining combined assumptions, plus $\varphi_1 \vee \varphi_2$.

($\rightarrow$I)(a) $\mathcal{P}(\Pi, \varphi_2, \Delta) \Rightarrow \mathcal{P}([\varphi_1 \rightarrow \varphi_2, \Pi], \varphi_1 \rightarrow \varphi_2, \Delta \backslash \{\varphi_1\})$;

In words, yet more informally:

If one has proved φ_2 without using φ_1 as an assumption, then one has thereby proved $\varphi_1 \rightarrow \varphi_2$ (from the same assumptions); *and* ...

If one has proved φ_2 from assumptions that include φ_1, then one has thereby proved $\varphi_1 \rightarrow \varphi_2$ (from the other assumptions).

(Vacuous discharge is *permitted.*)

$(\rightarrow I)(b)$ $\qquad \left. \begin{array}{c} \mathcal{P}(\Pi, \bot, \Delta) \\ \varphi_1 \in \Delta \end{array} \right\} \Rightarrow \mathcal{P}([\varphi_1 \rightarrow \varphi_2, \Pi], \varphi_1 \rightarrow \varphi_2, \Delta \setminus \{\varphi_1\})$

In words: If one has refuted the assumption φ_1, then one has thereby proved $\varphi_1 \rightarrow \varphi_2$ (from the remaining assumptions).

$(\rightarrow E)$

$\left. \begin{array}{c} \mathcal{P}(\Pi_1, \varphi_1, \Delta_1) \\ \mathcal{P}(\Pi_2, \theta, \Delta_2) \\ \varphi_2 \in \Delta_2 \end{array} \right\} \Rightarrow \mathcal{P}([\theta, \varphi_1 \rightarrow \varphi_2, \Pi_1, \Pi_2], \theta, \{\varphi_1 \rightarrow \varphi_2\} \cup \Delta_1 \cup (\Delta_2 \setminus \{\varphi_2\}))$

In words: If one has proved φ_1 from certain assumptions Δ_1, and has proved θ from assumptions that include φ_2, then one has proved θ from the latter assumptions other than φ_2, plus Δ_1 plus $\varphi_1 \rightarrow \varphi_2$.

$(\forall I)$ $\qquad \left. \begin{array}{c} \mathcal{P}(\Pi, \varphi_a^x, \Delta) \\ a \text{ does not occur in any member of } \Delta \end{array} \right\} \Rightarrow \mathcal{P}([\forall x \varphi, \Pi], \forall x \varphi, \Delta)$

In words: If one has proved an arbitrary instance φ_a^x (i.e. by making no assumptions about a), then one has proved $\forall x \varphi$ from the same assumptions.

$(\forall E)$ $\qquad \left. \begin{array}{c} \mathcal{P}(\Pi, \psi, \Delta) \\ \varphi_{t_1}^x, \ldots, \varphi_{t_n}^x \in \Delta \end{array} \right\} \Rightarrow \mathcal{P}([\psi, \forall x \varphi, \Pi], \psi, \{\forall x \varphi\} \cup (\Delta \setminus \{\varphi_{t_1}^x, \ldots, \varphi_{t_n}^x\}))$

In words: If one has proved ψ from certain assumptions among which are $\varphi_{t_1}^x, \ldots, \varphi_{t_n}^x$ $(n \geq 1)$, then one may conclude ψ from the remaining assumptions plus $\forall x \varphi$.

$(\exists I)$ $\qquad \mathcal{P}(\Pi, \varphi_t^x, \Delta) \Rightarrow \mathcal{P}([\exists x \varphi, \Pi], \exists x \varphi, \Delta)$

In words: If one has proved an instance φ_t^x from certain assumptions, then one has proved $\exists x \varphi$ from those assumptions.

$(\exists E)$ $\qquad \left. \begin{array}{c} \mathcal{P}(\Pi, \theta, \Delta \cup \{\varphi_a^x\}) \\ a \text{ does not occur in } \varphi \\ a \text{ does not occur in } \theta \\ a \text{ does not occur in any member of } \Delta \end{array} \right\} \Rightarrow \mathcal{P}([\theta, \exists x \varphi, \Pi], \theta, \{\exists x \varphi\} \cup \Delta)$

In words: If one has proved a conclusion θ not involving a, from assumptions among which φ_a^x is the only one involving a, and φ does not involve a, then one has proved θ from the remaining assumptions plus $\exists x \varphi$.

(=I) $\mathcal{P}(t = t, t = t, \emptyset)$

In words: Any sentence of the form $t = t$ is a proof of itself from no assumptions.

(=E) $\left. \begin{array}{c} \mathcal{P}(\Pi, \varphi, \Delta) \\ \varphi_u^t = \psi_u^t \end{array} \right\} \;\; \Rightarrow \;\; \mathcal{P}([\psi, t = u, \Pi], \psi, \{t = u\} \cup \Delta)$

In words: If one has proved φ from certain assumptions, and ψ results from φ by replacing at least one occurrence of t by one of u and/or vice versa, then one has proved ψ from those assumptions plus $t = u$.

Closure clause for system of core proof: if $\mathcal{P}(\Pi, \varphi, \Delta)$ in Core Logic, then this can be shown by appeal to the foregoing basis clause and inductive clauses.

Note: in ($\to$I)(a) there is no requirement that the antecedent φ_1 of the conditional be in Δ. This makes ($\to$I)(a) exactly like the usual rule of Conditional Proof. It is the added extra of ($\to$I)(b) as a new primitive rule (in the absence of EFQ) that makes Core Logic special, as far as the conditional is concerned. The combined effect of ($\to$I)(a) and ($\to$I)(b) is that premises remain relevant to conclusions, while the inferences in question yield the left-to-right readings of the three lines of the truth table for $\to$ that have T as their final entry. This ensures that Core Logic relevantizes *only at the level of the turnstile*. Relevance is a relation that holds between the *premises* of a proof and its *conclusion*.

Chapter 5

Motivating the Rules of Sequent Calculus

Abstract

§5.1 explains the natural correspondence between parallelized elimination rules in natural deduction, and Left rules in the sequent calculus. It presents an illustrative collection of natural deductions with their matching sequent proofs.

§5.2 states the rules of the sequent calculus for Core Logic.

§5.3 explains how these rules may be construed as inductive clauses in the inductive definition of the notion of sequent proof.

§5.4 describes the natural isomorphism between natural deductions in Core Logic, and the sequent proofs that correspond to them.

§5.5 examines the relations, between sequents, of concentration and dilution; and describes what it is for one sequent to strengthen another. We examine some possible global restrictions on proof formation, aimed at preventing proofs from proving dilutions of sequents already proved by a subproof. We establish the important result that the sequent rules of Core Logic maintain concentration, and we explain its importance for automated proof search.

5.1 Basic steps of inference

As noted in Chapter 4, the 'absolutely basic' steps of deduction are called *rules of inference*. We proceeded to motivate the statement of introduction and elimination rules for the usual connectives ($\neg$, $\wedge$, $\vee$, and $\rightarrow$), and the usual quantifiers ($\exists$ and $\forall$). These are the rules of so-called *natural deduction*. They allow one to construct a proof of an argument (or sequent)

$$\varphi_1, \ldots, \varphi_n : \psi$$

that takes the form of a tree, whose leaf nodes are labeled with the premises (or undischarged assumptions) $\varphi_1, \ldots, \varphi_n$ and whose root node is labeled with the conclusion ψ. Schematically, one can depict such a natural deduction Π as follows:

$$\underbrace{\varphi_1, \ldots, \varphi_n}$$
$$\Pi$$
$$\psi$$

The logical movement from premises to conclusion is represented as going *down* the page. Nodes of the proof tree are labeled by *sentences* (or by the absurdity symbol $\bot$). Each discrete step of inference is in accordance with one of the stated introduction or elimination rules.

The aim of this chapter is to present that alternative kind of logical system called the *sequent calculus*. It is an alternative only in the following respects:

1. As is the case with natural deduction, sequent proofs are in tree form; but each node in such a tree is labeled by the *sequent* established by that stage.

2. The simplest kind of natural deduction consists of a single occurrence of a sentence φ, and is a proof of φ from φ. The simplest kind of sequent proof, by contrast, consists of an 'initial sequent' of the form $\varphi : \varphi$.

3. The basic rules of inference in the sequent calculus allow one to introduce, on the right or on the left of the colon in a sequent, a new sentence with a particular logical operator @ dominant. Such rules are called the Right rule and the Left rule for @, and are suggestively labeled ':@' and '@:' respectively. The rule ':@' is the sequent form of the rule of @-Introduction in the system of natural deduction; and the rule '@:' is the sequent form of the rule of @-Elimination. The possible choices for @ (at the level of propositional logic) are the same as before: $\neg$, $\wedge$, $\vee$, and $\rightarrow$.

4. In natural deduction, assumptions that have been made for the sake of argument may come to be discharged by steps of inference taken 'lower down' in the proof. Such assumptions no longer count towards the set of undischarged assumptions of the proof. In sequent calculus, such 'discharge' of assumptions also takes place, but by the stratagem of removing a sentence from the left of a sequent when one applies a particular rule.

It is important to appreciate that the sequent calculus merely provides an *alternative form of presentation* of the proofs available in the system of natural deduction. It does not result in different sequents being provable; nor does it provide sequents with any new 'lines of proof'. Indeed, any 'line of proof' for a given sequent can be represented in either one of two essentially isomorphic ways: as a natural deduction; or as the corresponding sequent proof. (As we shall presently see, this is not true of Gentzen's sequent systems; but it is true of ours.)

Given a natural deduction in our sense, one can straightforwardly translate it into a sequent proof of the same result; and vice versa. The underlying 'trees' have their nodes labeled by *sentences* in the case of natural deductions, but by *sequents* in the case of sequent proofs. If one prunes away just those leaf nodes in a natural deduction that are labeled with major premises for eliminations, then the resulting tree (ignoring the labels) is isomorphic to the tree underlying the corresponding sequent proof. It involves the 'same primitive steps', arranged 'in the same constellation'.

This one–one correspondence between natural deductions and sequent proofs holds only by dint of the elimination rules having been *parallelized*, with major premises standing proud. For 'ordinary' natural deductions *à la* Gentzen and Prawitz, the correspondence is *not* one–one, as the following example shows. Corresponding to the single Gentzen–Prawitz natural deduction on the left (using *serial* $\wedge$-Elimination) are the *two* distinct sequent proofs on the right:

$$\frac{\dfrac{A \wedge B}{A}}{A \vee C} \qquad \frac{\dfrac{A : A}{A : A \vee C}}{A \wedge B : A \vee C} \qquad \frac{\dfrac{A : A}{A \wedge B : A}}{A \wedge B : A \vee C}$$

In the system of natural deduction with parallelized elimination rules, however, there is for each of the latter two sequent proofs a distinct natural deduction corresponding to it. These are, respectively,

$$\frac{A \wedge B \qquad \dfrac{\overline{A}^{(1)}}{A \vee C}}{A \vee C}{}^{(1)} \qquad \text{and} \qquad \frac{\dfrac{A \wedge B \qquad \overline{A}^{(1)}}{A}}{A \vee C}{}^{(1)}$$

Perhaps the best way to motivate the rules of the sequent calculus, and to underscore the points made above about the close correspondence between natural deductions and sequent proofs, is to render each of the natural deductions constructed in Chapter 4 as a sequent proof. After the reader has had the opportunity to examine how to translate each kind of proof into the other, we shall state the Right and Left rules in sequent calculus that correspond to the Introduction and Elimination rules, respectively, in natural deduction.

On the left are the simple natural deductions that we saw in Chapter 4, and to the right of each is its translation into sequent calculus.

$$\dfrac{\varphi \quad \psi}{\varphi \wedge \psi} \qquad\qquad \dfrac{\varphi : \varphi \quad \psi : \psi}{\varphi, \psi : \varphi \wedge \psi}$$

$$\dfrac{t \text{ is red} \quad t \text{ is red}}{\bot} \qquad\qquad \dfrac{}{t \text{ is red}, \ t \text{ is green} : \bot}$$

$$\dfrac{0 = s0}{\bot} \qquad\qquad \dfrac{}{0 = s0 : \bot}$$

$$\dfrac{\dfrac{}{t \text{ is red}}{}^{(1)} \quad t \text{ is green}}{\dfrac{\bot}{\neg(t \text{ is red})}{}^{(1)}} \qquad\qquad \dfrac{\dfrac{}{t \text{ is green}, \ t \text{ is red} : \bot}}{t \text{ is green} : \neg(t \text{ is red})}$$

$$\dfrac{\dfrac{}{\neg\varphi}{}^{(1)} \quad \varphi}{\dfrac{\bot}{\neg\neg\varphi}{}^{(1)}} \qquad\qquad \dfrac{\dfrac{\varphi : \varphi}{\varphi, \neg\varphi : \bot}}{\varphi : \neg\neg\varphi}$$

$$\dfrac{\dfrac{\dfrac{}{\neg\varphi}{}^{(1)} \quad \dfrac{}{\varphi}{}^{(2)}}{\dfrac{\bot}{\neg\neg\varphi}{}^{(1)}} \quad \neg\neg\neg\varphi}{\dfrac{\bot}{\neg\varphi}{}^{(2)}} \qquad\qquad \dfrac{\dfrac{\dfrac{\varphi : \varphi}{\varphi, \neg\varphi : \bot}}{\varphi : \neg\neg\varphi} \quad \varphi, \neg\neg\neg\varphi : \bot}{\neg\neg\varphi : \neg\varphi}$$

$$\dfrac{\dfrac{}{\varphi \wedge \neg\varphi}{}^{(2)} \quad \dfrac{\dfrac{}{\neg\varphi}{}^{(1)} \quad \dfrac{}{\varphi}{}^{(1)}}{\dfrac{\bot}{}{}^{(1)}}}{\dfrac{\bot}{\neg(\varphi \wedge \neg\varphi)}{}^{(2)}} \qquad\qquad \dfrac{\dfrac{\dfrac{\varphi : \varphi}{\varphi, \neg\varphi : \bot}}{\varphi \wedge \neg\varphi : \bot}}{\emptyset : \neg(\varphi \wedge \neg\varphi)}$$

$$
\begin{array}{c}
\dfrac{\quad}{\varphi \wedge \psi}(2) \qquad
\dfrac{\neg\psi \quad \overset{\displaystyle ——(1)}{\psi}}{\bot}(1)
\\[2ex]
\dfrac{\qquad\qquad\qquad\qquad}{\bot}(2)
\\[1ex]
\neg(\varphi \wedge \psi)
\end{array}
\qquad\qquad
\begin{array}{c}
\dfrac{\psi : \psi}{\dfrac{\neg\psi, \psi : \bot}{\dfrac{\neg\psi, \varphi \wedge \psi : \bot}{\neg\psi : \neg(\varphi \wedge \psi)}}}
\end{array}
$$

$$
\dfrac{\psi}{\varphi \to \psi}
\qquad\qquad
\dfrac{\psi : \psi}{\psi : \varphi \to \psi}
$$

$$
\dfrac{\neg\varphi \quad \overset{——(1)}{\varphi}}{\dfrac{\bot}{\varphi \to \psi}(1)}
\qquad\qquad
\dfrac{\varphi : \varphi}{\dfrac{\neg\varphi, \varphi : \bot}{\neg\varphi : \varphi \to \psi}}
$$

$$
\begin{array}{c}
\dfrac{\quad}{\varphi \to \psi}(2) \qquad \varphi \qquad
\dfrac{\neg\psi \quad \overset{——(1)}{\psi}}{\bot}(1)
\\[2ex]
\dfrac{\qquad\qquad\qquad\qquad\qquad\qquad}{\bot}(2)
\\[1ex]
\neg(\varphi \to \psi)
\end{array}
\qquad
\begin{array}{c}
\dfrac{\varphi : \varphi \quad \dfrac{\psi : \psi}{\neg\psi, \psi : \bot}}{\dfrac{\varphi, \neg\psi, \varphi \to \psi : \bot}{\varphi, \neg\psi : \neg(\varphi \to \psi)}}
\end{array}
$$

$$
\varphi \vee \psi \qquad
\dfrac{\neg\varphi \quad \overset{——(1)}{\varphi}}{\dfrac{\bot}{\psi}(1)} \qquad \overset{——(1)}{\psi}
\\[1ex]
\dfrac{\qquad\qquad\qquad\qquad\qquad\qquad}{\psi}(1)
\qquad\qquad
\dfrac{\dfrac{\varphi : \varphi}{\neg\varphi, \varphi : \bot} \quad \psi : \psi}{\varphi \vee \psi, \neg\varphi : \psi}
$$

5.2   Sequent calculus for Core Logic

Recall our conventions. We use Δ, Γ for (possibly empty) sets of sentences. *Sequents* are of the form $\Delta : \Gamma$, where Γ has at most one member. Instead of $\{\varphi\}$ we shall write φ when there can be no confusion. Instead of $\Delta \cup \Gamma$ we shall write Δ, Γ. Whenever we write $\Delta, \varphi_1, \dots \varphi_n : \Gamma$ $(n \geq 1)$ in a premise sequent, it is to be understood that $\varphi_i \notin \Delta$ $(1 \leq i \leq n)$. Instead of writing $\emptyset$ on the right of a colon, we simply leave a blank space.

The sequent rules for any logical operator @ are of two kinds: those for introducing a dominant occurrence of @ on the *right* of the colon of the sequent, and those for introducing it on the *left*. Right rules, denoted

$(:@)$, correspond to Introduction rules, denoted $(@\text{-I})$, in natural deduction. Left rules, denoted $(@:)$, correspond to Elimination rules, denoted $(@\text{-E})$, in natural deduction.

The reader should at this stage revisit §3.5.4, to check that the sequent rules below in effect simply extricate the middle arguments (the sequents) of the ternary relation $\mathcal{D}$ in the statement of verification and falsification rules for the logical operators in the Logic of Evaluation $\mathbb{E}$—but with the important changes that

(i) sets $\mathcal{A}$ (of atoms) in antecedents of sequents become sets Δ (of sentences in general), and

(ii) $\bot$ (i.e. $\emptyset$) as succedent becomes a singleton Γ—that is, either $\emptyset$ or $\{\theta\}$, for some possibly complex sentence θ.

$(:\neg)$ $\quad \dfrac{\Delta, \varphi:}{\Delta:\neg\varphi}$

$(\neg:)$ $\quad \dfrac{\Delta:\varphi}{\Delta,\neg\varphi:}$

$(:\wedge)$ $\quad \dfrac{\Delta_1:\varphi_1 \qquad \Delta_2:\varphi_2}{\Delta_1,\Delta_2:\varphi_1\wedge\varphi_2}$

$(\wedge:)$ $\quad \dfrac{\Delta,\varphi_i:\Gamma}{\Delta,\varphi_1\wedge\varphi_2:\Gamma} \quad$ for $i=1,2;$

$\left(\text{which affords also the more economical} \quad \dfrac{\Delta,\varphi,\psi:\Gamma}{\Delta,\varphi\wedge\psi:\Gamma}\right)$

$(:\vee)$ $\quad \dfrac{\Delta:\varphi_i}{\Delta:\varphi_1\vee\varphi_2} \quad$ for $i=1,2$

$(\vee:)$ $\quad \dfrac{\Delta_1,\varphi_1:\Gamma \quad \Delta_2,\varphi_2:\Gamma}{\Delta_1,\Delta_2,\varphi_1\vee\varphi_2:\Gamma} \quad \dfrac{\Delta_1,\varphi_1:\theta \quad \Delta_2,\varphi_2:}{\Delta_1,\Delta_2,\varphi_1\vee\varphi_2:\theta} \quad \dfrac{\Delta_1,\varphi_1: \quad \Delta_2,\varphi_2:\theta}{\Delta_1,\Delta_2,\varphi_1\vee\varphi_2:\theta}$

$(:\to)$ $\quad \dfrac{\Delta,\varphi_1:}{\Delta:\varphi_1\to\varphi_2} \quad \dfrac{\Delta:\varphi_2}{\Delta\setminus\{\varphi_1\}:\varphi_1\to\varphi_2}$

$(\to:)$ $\quad \dfrac{\Delta_1:\varphi_1 \qquad \Delta_2,\varphi_2:\Gamma}{\Delta_1,\Delta_2,\varphi_1\to\varphi_2:\Gamma}$

$(:\exists)\qquad \dfrac{\Delta:\psi^x_t}{\Delta:\exists x\psi}$

$(\exists:)\qquad \dfrac{\Delta,\psi^x_a:\Gamma}{\Delta,\exists x\psi:\Gamma}$ where the conclusion sequent has no occurrences of a

$(:\forall)\qquad \dfrac{\Delta:\psi}{\Delta:\forall x\psi^a_x}$ where a occurs in ψ but in no member of Δ

$(\forall:)\qquad \dfrac{\Delta,\psi^x_{t_1},\ldots,\psi^x_{t_n}:\Gamma}{\Delta,\forall x\psi:\Gamma}$

$(:=)\qquad \dfrac{}{\emptyset:t=t}$

$(=:)\qquad \dfrac{\Delta:\varphi}{\Delta,t=u:\psi}$ where $\varphi^t_u=\psi^t_u$

5.3 The sequent rules as clauses in the inductive definition of proof

We now define the notion $\mathcal{P}(\Pi,\Delta:\varphi)$—'$\Pi$ is a proof of the sequent $\Delta:\varphi$'. Note that this is a *binary* relation, not a ternary one as was the case with natural deduction. The second term of the relation is a *sequent*.

We use a linear notation again, this time to denote the two-dimensional trees of *sequent* occurrences that we are calling proofs. We use square brackets to indicate the formation of a tree, with the convention that the first thing mentioned within the square brackets is the root node (which now is always a sequent), and the remaining things within the square brackets are the immediate sub-trees, arranged from left to right. Thus the linear term

$$[\Delta:\varphi,\Pi,\Sigma]$$

and the two-dimensional term

$$\dfrac{\Pi\quad\Sigma}{\Delta:\varphi}$$

would denote the same abstract object, namely the tree with root node $\Delta : \varphi$ whose left-immediate subtree is Π and whose right-immediate subtree is Σ.

One drawback of this linear notation, in the context of the use we wish to make of it, is a certain prolixity. For we shall frequently wish to write statements of the form

$$\mathcal{P}([\Delta : \varphi, \Pi, \Sigma], \Delta : \varphi),$$

in which mention of the sequent $\Delta : \varphi$ is perforce repeated. A more compact notation for this would be

$$\mathcal{P}(\lfloor \Pi, \Sigma \rfloor; \Delta : \varphi).$$

The inward-pointing feet of the brackets are to remind the reader that the complex proof to be formed has Π as its left-immediate subtree, Σ as its right-immediate subtree, and the sequent $\Delta : \varphi$ as its conclusion. When there is only one immediate subtree Π, say, then of course we shall write

$$\mathcal{P}(\lfloor \Pi \rfloor; \Delta : \varphi).$$

Definition 5 (Proof in sequent calculus for Core Logic).

Basis clause: $\mathcal{P}(\varphi : \varphi ; \varphi : \varphi)$

Inductive clauses follow, one for each rule of inference. Note the membership requirements in the antecedents of certain inductive clauses; these ensure non-vacuous discharge of the assumptions in question.

$$(: \neg) \qquad \left. \begin{array}{c} \mathcal{P}(\Pi ; \Gamma : \bot) \\ \varphi \in \Gamma \end{array} \right\} \quad \Rightarrow \quad \mathcal{P}(\lfloor \Pi \rfloor ; \Gamma \backslash \{\varphi\} : \neg \varphi)$$

$$(\neg :) \qquad \mathcal{P}(\Pi ; \Gamma : \varphi) \quad \Rightarrow \quad \mathcal{P}(\lfloor \Pi \rfloor ; \neg \varphi, \Gamma : \bot)$$

$$(: \wedge) \qquad \left. \begin{array}{c} \mathcal{P}(\Pi_1 ; \Delta_1 : \varphi_1) \\ \mathcal{P}(\Pi_2 ; \Delta_2 : \varphi_2) \end{array} \right\} \Rightarrow \mathcal{P}(\lfloor \Pi_1, \Pi_2 \rfloor ; \Delta_1, \Delta_2 : \varphi_1 \wedge \varphi_2)$$

$$(\wedge :) \qquad \left. \begin{array}{c} \mathcal{P}(\Pi ; \Delta : \theta) \\ \varphi_1 \in \Delta \text{ or } \varphi_2 \in \Delta \end{array} \right\} \Rightarrow \mathcal{P}(\lfloor \Pi \rfloor ; \varphi_1 \wedge \varphi_2, \Delta \backslash \{\varphi_1, \varphi_2\} : \theta)$$

$(:\vee)$
$$\mathcal{P}(\Pi \; ; \; \Delta \!:\! \varphi_1) \Rightarrow \mathcal{P}(\lfloor \Pi \rfloor \; ; \; \Delta \!:\! \varphi_1 \vee \varphi_2)$$
$$\mathcal{P}(\Pi \; ; \; \Delta \!:\! \varphi_2) \Rightarrow \mathcal{P}(\lfloor \Pi \rfloor \; ; \; \Delta \!:\! \varphi_1 \vee \varphi_2)$$

$(\vee :)$

$$\left.\begin{array}{c} \mathcal{P}(\Pi_1 \; ; \; \Delta_1 \!:\! \theta) \\ \mathcal{P}(\Pi_2 \; ; \; \Delta_2 \!:\! \theta) \\ \varphi_1 \in \Delta_1 \\ \varphi_2 \in \Delta_2 \end{array}\right\} \Rightarrow \mathcal{P}(\lfloor \Pi_1, \Pi_2 \rfloor \; ; \; \varphi_1 \vee \varphi_2, \Delta_1 \backslash \{\varphi_1\}, \Delta_2 \backslash \{\varphi_2\} \!:\! \theta)$$

$$\left.\begin{array}{c} \mathcal{P}(\Pi_1 \; ; \; \Delta_1 \!:\! \bot) \\ \mathcal{P}(\Pi_2 \; ; \; \Delta_2 \!:\! \theta) \\ \varphi_1 \in \Delta_1 \\ \varphi_2 \in \Delta_2 \end{array}\right\} \Rightarrow \mathcal{P}(\lfloor \Pi_1, \Pi_2 \rfloor \; ; \; \varphi_1 \vee \varphi_2, \Delta_1 \backslash \{\varphi_1\}, \Delta_2 \backslash \{\varphi_2\} \!:\! \theta)$$

$$\left.\begin{array}{c} \mathcal{P}(\Pi_1 \; ; \; \Delta_1 \!:\! \theta) \\ \mathcal{P}(\Pi_2 \; ; \; \Delta_2 \!:\! \bot) \\ \varphi_1 \in \Delta_1 \\ \varphi_2 \in \Delta_2 \end{array}\right\} \Rightarrow \mathcal{P}(\lfloor \Pi_1, \Pi_2 \rfloor \; ; \; \varphi_1 \vee \varphi_2, \Delta_1 \backslash \{\varphi_1\}, \Delta_2 \backslash \{\varphi_2\} \!:\! \theta)$$

$(:\!\rightarrow)(a)$
$$\mathcal{P}(\Pi \; ; \; \Delta \!:\! \varphi_2) \Rightarrow \mathcal{P}(\lfloor \Pi \rfloor \; ; \; \Delta \backslash \{\varphi_1\} \!:\! \varphi_1 \rightarrow \varphi_2)$$

$(:\!\rightarrow)(b)$
$$\left.\begin{array}{c} \mathcal{P}(\Pi \; ; \; \Delta \!:\! \bot) \\ \varphi_1 \in \Delta \end{array}\right\} \Rightarrow \mathcal{P}(\lfloor \Pi \rfloor \; ; \; \Delta \backslash \{\varphi_1\} \!:\! \varphi_1 \rightarrow \varphi_2)$$

$(\rightarrow :)$
$$\left.\begin{array}{c} \mathcal{P}(\Pi_1 \; ; \; \Delta_1 \!:\! \varphi_1) \\ \mathcal{P}(\Pi_2 \; ; \; \Delta_2 \!:\! \theta) \\ \varphi_2 \in \Delta_2 \end{array}\right\} \Rightarrow \mathcal{P}(\lfloor \Pi_1, \Pi_2 \rfloor \; ; \; \varphi_1 \rightarrow \varphi_2, \Delta_1, \Delta_2 \backslash \{\varphi_2\} \!:\! \theta)$$

$(:\forall)$
$$\left.\begin{array}{c} \mathcal{P}(\Pi \; ; \; \Delta \!:\! \varphi_a^x) \\ a \text{ does not occur in any member of } \Delta \end{array}\right\} \Rightarrow \mathcal{P}(\lfloor \Pi \rfloor \; ; \; \Delta \!:\! \forall x \varphi)$$

$(\forall :)$
$$\left.\begin{array}{c} \mathcal{P}(\Pi \; ; \; \Delta \!:\! \theta) \\ \varphi_{t_1}^x, \ldots, \varphi_{t_n}^x \in \Delta \end{array}\right\} \Rightarrow \mathcal{P}(\lfloor \Pi \rfloor \; ; \; \forall x \varphi, \Delta \backslash \{\varphi_{t_1}^x, \ldots, \varphi_{t_n}^x\} \!:\! \theta)$$

$(:\exists)$
$$\mathcal{P}(\Pi \; ; \; \Delta \!:\! \varphi_t^x) \Rightarrow \mathcal{P}(\lfloor \Pi \rfloor \; ; \; \Delta \!:\! \exists x \varphi)$$

$$(\exists:) \quad \left.\begin{array}{c} \mathcal{P}(\Pi\;;\;\Delta, \varphi^x_a\!:\!\theta) \\ a \text{ does not occur in } \varphi \\ a \text{ does not occur in } \theta \\ a \text{ does not occur in any member of } \Delta \end{array}\right\} \;\Rightarrow\; \mathcal{P}(\lfloor\Pi\rfloor\;;\;\exists x\varphi, \Delta\!:\!\theta)$$

$$(:=) \quad \mathcal{P}(\emptyset\!:\!t\!=\!t\;;\;\emptyset\!:\!t\!=\!t)$$

$$(=:) \quad \left.\begin{array}{c} \mathcal{P}(\Pi\;;\;\Delta\!:\!\varphi) \\ \varphi^t_u = \psi^t_u \end{array}\right\} \;\Rightarrow\; \mathcal{P}(\lfloor\Pi\rfloor\;;\;t\!=\!u, \Delta\!:\!\psi)$$

Closure clause for system of core proof: if $\mathcal{P}(\Pi\;;\;\Delta\!:\!\varphi)$ in Core Logic, then this can be shown by appeal to the foregoing basis clause and inductive clauses.

Chapter 4 set out the rules of natural deduction for Core Logic. Each logical operator @ received an Introduction rule (@-I) and an Elimination rule (@-E). This chapter has set out the rules of the sequent calculus for Core Logic. Each logical operator @ has received a Right rule (:@) and a Left rule (@:).

5.4 The isomorphism between natural deductions and sequent proofs

Inspection reveals that

1. (@-I) translates directly into (:@), and vice versa;

2. (@-E) translates directly into (@:), and vice versa.

This is especially easy to see, given our formulation of the natural deduction rules as inductive clauses using the notation

$$\mathcal{P}_{\text{ND}}([\varphi, \{\Sigma, \}\Pi_1, \ldots, \Pi_n], \varphi, \Delta),$$

(where Σ is only ever a major premise for an elimination) and our formulation of the sequent rules as inductive clauses using the notation

$$\mathcal{P}_{\text{SC}}(\lfloor\Pi_1, \ldots, \Pi_n\rfloor, \Delta : \varphi).$$

Here we add the subscripts ND and SC for 'natural deduction' and 'sequent calculus' respectively, to keep matters clear.

Metatheorem 2. *There is an effective translation f of natural deductions into sequent proofs, and an effective translation g of sequent proofs into natural deductions, such that*

$$\forall \text{ natural deductions } \Pi \quad gf\Pi = \Pi$$

and

$$\forall \text{ sequent proofs } \Pi \quad fg\Pi = \Pi$$

and

$$\forall \Pi(\mathcal{P}_{ND}(\Pi, \varphi, \Delta) \Rightarrow \mathcal{P}_{SC}(f\Pi, \Delta : \varphi))$$

and

$$\forall \Pi(\mathcal{P}_{SC}(\Pi, \Delta : \varphi) \Rightarrow \mathcal{P}_{ND}(g\Pi, \varphi, \Delta))$$

Proof. Obvious from the definitional homology (created and) noted.   □

The signal advantage of Metatheorem 2 is that it ensures that many results (concerning Core Logic) that are proved at the meta-level for the system of natural deduction hold also for the sequent calculus. This is especially true of the main result of Chapter 6, which guarantees an adequate amount of *transitivity of deducibility* in Core Logic. Note that Metatheorem 2 generalizes to any system satisfying the intrinsic normality requirement, even if its ND and SC formulations respectively contain EFQ and Thinning on the Right.

5.5   On maintaining concentration

We have seen that the only structural rule in Core Logic is Reflexivity, which is usually stated in the form

$$\overline{\varphi : \varphi}$$

Reflexivity 'is' the basis clause in the inductive definition of the notion

$$\Pi \text{ is a proof of the sequent } \mathcal{S}$$

—in the notation used above,

$$\mathcal{P}_{SC}(\Pi; \mathcal{S}).$$

That basis clause says, simply,

$$\mathcal{P}_{SC}(\varphi\!:\!\varphi;\ \varphi\!:\!\varphi).$$

That is to say: to write down a sequent of the form $\varphi\!:\!\varphi$ is to prove it.[1]

In general a sequent rule of the form

$$\frac{\mathcal{S}_1 \dots \mathcal{S}_n}{\mathcal{S}}$$

'is' an inductive clause of the form

$$\left.\begin{array}{c} \mathcal{P}_{SC}(\Pi_1;\mathcal{S}_1) \\ \vdots \\ \mathcal{P}_{SC}(\Pi_n;\mathcal{S}_n) \end{array}\right\} \Rightarrow \mathcal{P}_{SC}\left(\frac{\Pi_1 \dots \Pi_n}{\mathcal{S}};\ \mathcal{S}\right)$$

in the inductive definition of the notion $\mathcal{P}(\Pi;\mathcal{S})$. For each rule, the task is to say how the concluding sequent $\mathcal{S}$ is to be determined from the premise sequents $\mathcal{S}_1, \dots, \mathcal{S}_n$.

For a given rule ρ, the proof formed by applying it to appropriate immediate subproofs $\Pi_1, \dots, \Pi_n$ can be denoted $\rho(\Pi_1, \dots, \Pi_n)$.

For any proof Π, we denote by $s\Pi$ the sequent that Π proves. It follows that for all proofs Π (natural deductions or sequent proofs) we have

$$\mathcal{P}(\Pi;s\Pi).$$

5.5.1   Dilution v. concentration

In the sequent calculus as traditionally presented—but *not* in Core Logic— there is a structural rule called Dilution, or Thinning, or Weakening.[2] It comes in two forms—one for dilution on the left, the other for dilution on the right:

$$\frac{\Delta:\Gamma}{\Delta,\varphi:\Gamma} \ (\text{where } \varphi \notin \Delta) \qquad\qquad \frac{\Delta:}{\Delta:\varphi}$$

Definition 6. $\Delta:\Gamma$ *is a* subsequent *of* $\Delta':\Gamma'$ *(abbreviated $\Delta:\Gamma \sqsubseteq \Delta':\Gamma'$)* $\equiv_{df} \Delta \subseteq \Delta'$ *and* $\Gamma \subseteq \Gamma'$.

---

[1] Note that in this context of talk about sequents, $\mathcal{S}$ is now a metavariable over *sequents*, not (as previously) over *systems* of rules of inference.

[2] 'Dilution' or 'thinning' in English strike me as better translations of Gentzen's *Verdünnung*. But the term 'weakening' is more current nowadays among logicians. This is a pity, for *concentration* is the natural converse of *dilution*, and some dilutions or thinnings are not genuine weakenings in the logical sense.

Definition 7. $\Delta':\Gamma'$ *is a* dilution *of* $\Delta:\Gamma$ *(abbreviated* $\Delta:\Gamma \sqsubset \Delta':\Gamma') \equiv_{df}$ $\Delta:\Gamma \sqsubseteq \Delta':\Gamma'$ *and either* $\Delta \subset \Delta'$ *or* $\Gamma \subset \Gamma'$. *Equivalently, we may say that* $\Delta:\Gamma$ concentrates, *or is a* concentration *of,* $\Delta':\Gamma'$.

5.5.2   Strengthening

Definition 8. Δ *is at least as strong as* Δ' *(abbreviated* $\Delta \vdash \Delta') \equiv_{df} \Delta$ *logically implies every sentence in* Δ'.

Definition 9. Δ *is stronger than* Δ' *(abbreviated* $\Delta \rhd \Delta') \equiv_{df} \Delta$ *is at least as strong as* Δ' *and* Δ' *is not at least as strong as* Δ; *that is* $\Delta \vdash \Delta'$ *and* $\Delta' \not\vdash \Delta$.

Definition 10. *(i) The sequent* $\Delta':\Phi$ *is stronger than the sequent* $\Delta:\Phi \equiv_{df}$ $\Delta \rhd \Delta'$. *(Note that* Φ *is empty or a singleton.)*
(ii) The sequent $\Delta':\emptyset$ *is stronger than the sequent* $\Delta':\varphi \equiv_{df} \varphi$ *is satisfiable. When sequent* $\mathcal{S}_1$ *is stronger than sequent* $\mathcal{S}_2$ *we write* $\mathcal{S}_1 \succ \mathcal{S}_2$.

Observation 1. $\Delta':\Gamma' \sqsubset \Delta:\Gamma$ *does not imply that* $\Delta':\Gamma' \succ \Delta:\Gamma$. *The relation of dilution, and its converse, concentration, is a matter only of inclusions, not of relative logical strengths.*

That being said, nevertheless:

Observation 2. *Some concentrations are strengthenings.*

Example: the sequent
$$A, \neg A :$$

is both a concentration and a strengthening of the sequent

$$A, \neg A : B.$$

5.5.3   Core logic is free of dilution

Consider the first part of the rule $(:\rightarrow)$, namely

$$\frac{\Delta, \varphi :}{\Delta : \varphi \rightarrow \psi}$$

and the second and third parts of the rule $(\vee :)$, namely

$$\frac{\Delta_1, \varphi_1 : \psi \quad \Delta_2, \varphi_2 :}{\Delta_1, \Delta_2, \varphi_1 \vee \varphi_2 : \psi} \qquad \frac{\Delta_1, \varphi_1 : \quad \Delta_2, \varphi_2 : \psi}{\Delta_1, \Delta_2, \varphi_1 \vee \varphi_2 : \psi}$$

The suspicion may arise, from inspection of these, that Core Logic does after all permit dilution (on the right)—that we are somehow incorporating dilution (on the right) into the rules for $\rightarrow$ and for $\vee$.

But this is in error. Consider the following

Global restriction on proof formation, Version I.
Never apply a rule if the conclusion sequent of the proof Π that would thereby result would be a dilution of any other sequent in Π.

The Core logician could be bound by this restriction, without loss; for, every valid sequent would still have a Core-provable subsequent.

The proposed restriction definitely prevents the formation of certain proofs that would otherwise be licit core proofs. Consider, for example, the following core proof:

$$\Omega \quad \cfrac{\cfrac{\cfrac{A:A \quad\quad B:B}{A, A \rightarrow B \;:\; B}}{\cfrac{A, A \rightarrow B \;:\; B \vee C}{A \wedge D, A \rightarrow B \;:\; B \vee C}} \quad\quad \cfrac{\cfrac{\cfrac{A:A \quad\quad B:B}{A, A \rightarrow B \;:\; B} \quad C:C}{A, A \rightarrow B, B \rightarrow C \;:\; C}}{A, A \rightarrow B, B \rightarrow C \;:\; B \vee C}}{(A \wedge D) \vee (B \rightarrow C), A \rightarrow B, A \;:\; B \vee C}$$

Note how the sequent $A, A \rightarrow B \;:\; B \vee C$ (three lines down on the left) is diluted by the conclusion sequent of (the final step of) Ω. *If* the proposed global restriction (Version I) were in effect, one would not be able to take that final step; the rule $(\vee :)$ would be inapplicable in such a context. Nevertheless, as already intimated, there is no deductive loss here. For, although deprived (by the would-be restriction) of the proof just displayed, we nevertheless already have the subproof whose conclusion is the sequent $A, A \rightarrow B \;:\; B \vee C$. And this more than suffices.

For consider: if one were set the problem of finding a proof of the sequent

$$(A \wedge D) \vee (B \rightarrow C), A \rightarrow B, A \;:\; B \vee C$$

it would certainly be acceptable to prove, in its stead, its proper subsequent

$$A \rightarrow B, A \;:\; B \vee C.$$

What is more, this is a *logically stronger* result. As the reader can quickly check, its premises $A \rightarrow B$, A fail to jointly imply the superfluous premise

$$(A \wedge D) \vee (B \rightarrow C).$$

The theme illustrated by this example extends to a much more general one: *opting to prove a proper subsequent of a given sequent never leads to loss.* In the search for a proof of any given sequent $\mathcal{S}$, one should *always* be able to rest content with a proof (if one is found) of a *proper subsequent* of $\mathcal{S}$. And this should be the case *iteratively*:

> If the call for a proof of $\mathcal{S}$ arises as a *sub*problem during a search for a proof of some other sequent $\mathcal{S}'$, say, then any proof of a *proper subsequent* of $\mathcal{S}$ should be just as valuable as a proof of $\mathcal{S}$ itself, insofar as the search for a proof of (a subsequent of) $\mathcal{S}'$ is concerned.

Call this the *Iterative Requirement that Proofs of Concentrations be Deployable.* We explicate it further in §5.5.5.

5.5.4   Diluters and non-diluters

Definition 11. *A proof Π is a diluter just in case $s\Pi$ dilutes some (other) sequent in Π; and Π is a non-diluter otherwise.*

The global restriction on proof formation (Version I) bruited in §5.5.3 can now be re-phrased as follows:

Global restriction on proof formation, Version I.
No rule application may produce a diluter.

We know from our earlier example Ω:

$$\frac{\dfrac{\dfrac{\dfrac{A:A \quad\quad B:B}{A,\,A\to B \ : \ B}}{A,\,A\to B \ : \ B\vee C}}{A\wedge D,\,A\to B \ : \ B\vee C} \quad\quad \frac{\dfrac{\dfrac{A:A \quad\quad B:B}{A,\,A\to B \ : \ B}\quad C:C}{A,\,A\to B,\,B\to C \ : \ C}}{A,\,A\to B,\,B\to C \ : \ B\vee C}}{(A\wedge D)\vee(B\to C),\,A\to B,\,A \ : \ B\vee C}$$

that some core proofs are diluters. Note, however, that Ω is a (proper) *substitution instance* (via a non-permutative relettering) of the following *non-diluter*:

$$\Xi \quad\quad \frac{\dfrac{\dfrac{\dfrac{A:A \quad\quad B:B}{A,\,A\to B \ : \ B}}{A,\,A\to B \ : \ B\vee C}}{A\wedge D,\,A\to B \ : \ B\vee C} \quad\quad \frac{\dfrac{\dfrac{E:E \quad\quad B:B}{E,\,E\to B \ : \ B}\quad C:C}{E,\,E\to B,\,B\to C \ : \ C}}{E,\,E\to B,\,B\to C \ : \ B\vee C}}{(A\wedge D)\vee(B\to C),\,A\to B,\,E,\,E\to B \ : \ B\vee C}$$

This example teaches us that if Version I of the global restriction on proof formation is imposed, then not every substitution instance of a proof is a proof. The good intention behind the global restriction (namely, to avoid forming diluters) will be thwarted by the familiar process of taking substitution instances.

One response might then be to make the global restriction even more exigent.

Global restriction on proof formation, Version II.
No rule application may produce a proof any of whose substitution instances obtained by a non-permutative relettering is a diluter.

Checking all non-permutative reletterings of primitive extralogical expressions occurring in a proof, and verifying that none of the resulting proofs is a diluter, is an effective matter, even in the context of an undecidable logic such as first-order Classical Logic with at least one two-place predicate, or first-order Core Logic with at least two monadic predicates. Thus there can be no principled objection to imposing Version II of the global restriction. We would still be able to determine, effectively, whether a would-be proof is indeed a proof.

Version II of the global restriction would have the effect that Ξ cannot count as a proof. For, as we have seen, the non-permutative relettering of Ξ that maps E to A (and leaves the other atoms untouched) produces the diluter Ω.

Intuitively, however, Ξ strikes one as a perfectly good propositional proof. Ξ is, itself, a non-diluter; it just happens to yield a diluter *via* a non-permutative relettering. Surely, the thought must go, we can just learn to live with that? Surely we should refrain from imposing Version II of the proposed global restriction? After all, a non-permutative relettering *enriches* the logical structure of the sequent to be proved. One can therefore expect that certain such reletterings will produce sequents that do not need (for their validity or provability) every one of their member sentences. And that sort of 'logical overkill' can extend even to certain non-diluting proofs (such as Ξ) which, upon the relettering in question, become proofs (such as Ω) that are diluters.

We are seeking to show that Core Logic has no need for dilution (in *any* form) when proving any of the sequents that it can prove. The reason why we stress '*any* form' here is that Core Logic already eschews applications of any *rule of dilution*. So core proofs are free of any (single-premise) step whose conclusion sequent dilutes its immediate premise sequent. But some core proofs, as we have seen, are diluters, in the sense defined above. Such a proof contains a sequent $\mathcal{S}$ that is diluted by the proof's conclusion sequent standing *more than one step down* from $\mathcal{S}$. So the 'dilution' within a diluter is not a matter of applying a rule of dilution. Rather, it arises from more global and 'distributed' features of the proof.

5.5.5 A new property for rules to preserve

Proofs are constructed by means of rules of inference. Typically the main concern, with any system of rules of inference, is to preserve *validity of argument*. Here we propose an additional desideratum:

rules should maintain concentrations of (sub)arguments.

The formal statement of this requirement is given below. Informal explanation will follow immediately afterwards.

Definition 12. *If we replace an immediate subproof* Π_i *of*

$$\Pi = \rho(\Pi_1, \ldots, \Pi_i, \ldots, \Pi_n)$$

by another proof Σ *and the result of doing so is a proof, then we shall denote that proof by*

$$\rho(\Pi_1, \ldots, \Sigma, \ldots, \Pi_n).$$

The requirement that any sequent rule ρ *maintain concentration*.

$$\left. \begin{array}{c} \mathcal{P}(\rho(\Pi_1, \ldots, \Pi_i, \ldots, \Pi_n), \mathcal{S}) \\ \mathcal{P}(\Pi_i, \mathcal{S}_i) \\ s\Sigma \sqsubset \mathcal{S}_i \\ s\Sigma \not\sqsubseteq \mathcal{S} \end{array} \right\} \;\Rightarrow\; s(\rho(\Pi_1, \ldots, \Sigma, \ldots, \Pi_n)) \sqsubseteq \mathcal{S}$$

We turn now to the promised informal explanation of this proposed requirement on sequent rules. We are attending to one of the immediate subproofs Π_i in a proof Π, say, of the form

$$\frac{\begin{array}{ccccc} \Pi_1 & & \Pi_i & & \Pi_n \\ \mathcal{S}_1 & \cdots & \mathcal{S}_i & \cdots & \mathcal{S}_n \end{array}}{\mathcal{S}}(\rho)$$

Moreover, we are in possession of a proof Σ of some *concentration* of the concluding sequent $\mathcal{S}_i$ of the proof Π_i. That concentration, in our notation, is $s\Sigma$. So $s\Sigma \sqsubset \mathcal{S}_i$. But $s\Sigma$, we are to suppose, is *not* a subsequent of $\mathcal{S}$. The question naturally arises:

> Can we deploy the proof Σ in place of the proof Π_i, leaving all the other proof work in Π untouched, so as to obtain, by so doing, a proof of the concluding sequent $\mathcal{S}$ of Π (or perhaps even of some *proper subsequent* of $\mathcal{S}$)? That is, can we effectively find some $\mathcal{S}' \sqsubseteq \mathcal{S}$ so that the following is a proof:

$$\cfrac{\begin{matrix} \Pi_1 & & \Sigma & & \Pi_n \\ \mathcal{S}_1 & \cdots & s\Sigma & \cdots & \mathcal{S}_n \end{matrix}}{\mathcal{S}'}{\scriptstyle (\rho)} \quad ?$$

Our proposed requirement is that the answer to this question must always be affirmative. This ends our informal explanation of the proposed requirement.

Observation 3. *By Observation 2, concentrations effected in conformity with our proposed requirement (that sequent rules maintain concentration) can lead to strengthenings of the overall results proved—or, as one might say, to 'epistemic gain'.*

The reader will have noted that all the sequent rules stated in §5.2 involve just one or two premise sequents. So they have one or other of the gross forms

$$\frac{\mathcal{S}_1}{\mathcal{S}} \qquad\qquad \frac{\mathcal{S}_1 \quad \mathcal{S}_2}{\mathcal{S}}$$

Moreover one can easily see that concentrations are indeed maintained by the rules of Core Logic:

Metatheorem 3. *The sequent rules of Core Logic maintain concentration. That is, for every sequent rule ρ of Core Logic,*

$$\left.\begin{matrix} \mathcal{P}(\rho(\Pi_1,\ldots,\Pi_i,\ldots,\Pi_n),\mathcal{S}) \\ \mathcal{P}(\Pi_i,\mathcal{S}_i) \\ s\Sigma \sqsubset \mathcal{S}_i \\ s\Sigma \not\sqsubseteq \mathcal{S} \end{matrix}\right\} \;\Rightarrow\; s(\rho(\Pi_1,\ldots,\Sigma,\ldots,\Pi_n)) \sqsubseteq \mathcal{S}.$$

Proof. By inspection. □

Note that Metatheorem 3 does *not* generalize to all systems enjoying formulations that satisfy the intrinsic normality requirement. The standard systems **M**, **I**, and **C**, for example, even in their formulations satisfying the intrinsic normality requirement, do not have their ND introduction and elimination rules (nor their corresponding SC left- and right-logical rules) formulated in a way that maintains concentration. This further entails that Corollary 2 below will not hold for the standard systems either.

Some of the rules of Core Logic come in more than one 'part'. The same is true of Intuitionistic Logic, for example, concerning the sequent rule $(: \vee)$ for introducing $\vee$ on the right (see the list of rules stated in §5.2.)

It is less well known that certain other sequent rules—see §5.2 again—can be stated in more than one part: for example, Core Logic's two-part rule $(:\rightarrow)$ for introducing $\rightarrow$ on the right, and its three-part rule $(\vee:)$ for introducing $\vee$ on the left. The proof of Metatheorem 3 makes use of the fact that one can have recourse to a *different part* of the rule $(:\rightarrow)$, or of the rule $(\vee:)$, when dealing with certain concentrations of the premise sequents.

Corollary 2. *Core Logic meets the Iterative Requirement that Proofs of Concentrations be Deployable.*

Proof. By Metatheorem 3 we meet the requirement in the case where the concentrating (sub)proof Σ is such that $s\Sigma \sqsubset S_i$ (the condition for concentration) and $s\Sigma \not\sqsubseteq S$. In that case we take the proof $\rho(\Pi_1, \ldots, \Sigma, \ldots \Pi_n)$ in place of the 'overall proof' $\rho(\Pi_1, \ldots, \Pi_i, \ldots \Pi_n)$.

If, however, it happens to be the case that $s\Sigma \sqsubseteq S$, then *there is no need to apply the rule ρ with Σ deployed in place of Π_i*. Instead, we can take, for the overall proof, the proof Σ itself, since it establishes a *subsequent* of the target sequent S. ∎

5.5.6   Two uniqueness conjectures

There are certain systemic properties enjoyed by Core Logic, whose conjunction we shall call Ω. Call any subsystem of Intuitionistic Logic that satisfies Ω an Ω-*system*; and call any subsystem of Classical Logic that satisfies Ω a *classical* Ω-*system*.

Conjecture 1. *Core Logic is the only Ω-system whose rules maintain concentration.*

Conjecture 2. *Classical Core Logic is the only classical Ω-system whose rules maintain concentration.*

Chapter 6

Transitivity of Deducibility

Abstract

§6.1 discusses some examples of nontransitive logical consequence relations in the recent literature on languages extended to contain vague predicates and/or a transparent truth predicate. We explain how Core Logic's eschewal of unrestricted transitivity in the *un*extended, *non-vague* language of first-order logic is an altogether more radical departure from orthodoxy.

§6.2 looks at sequent calculi for multiple-conclusion arguments, both to note their existence and their historical importance, and to set them aside in order to concentrate on the ubiquitous and much more natural case of single-conclusion arguments.

§6.3 discusses the transitivity 'problem' for Core Logic (where the scare quotes are intended to diminish its importance altogether). We make the reader aware of the main result to be proved in this chapter, which is the Cut-Elimination Theorem for Core Logic.

§6.4 explains the important distinction between having a rule of Cut as a primitive rule *of* one's system, as opposed to its being 'merely' *admissible for* the system. We explain a suite of syntactic notions that are needed for the eventual proof of the Cut-Elimination Theorem for Core Logic: two proofs *connecting*; the *target sequent* resulting from such connection; and the *subsequent* relation. The subsequent relation for the orthodox system involves subsetting down only on the left; whereas for the Core systems it involves subsetting down not only on the left, but also on the right.

§6.5 stresses once more how admissible cuts can provide epistemic gain.

§6.6 restates the graphic rules for the orthodox systems **M**, **I**, and **C** *in our preferred format*. This is the one in which *all* elimination rules are *parallelized*, and major premises of their applications *stand proud*, with no proof work above them.

§6.7 restates the rules for Core Logic in graphic form, in the same format; and explains how to extend Core Logic to Classical Core Logic by adopting either the rule of Classical Reductio (with non-vacuous discharge) or the rule of Dilemma (allowing the positive horn to prove absurdity rather than the sought conclusion).

§6.8 introduces some compact Polish notation—allowing linear designations of complicated proofs with almost no use of parentheses. This will afford a much more succinct statement of the various transformations that will be involved in the recursive definition of the binary operation $[\Pi, \Sigma]$ on proofs.

§6.9 and §6.10 take care of the classification of all the various proof pairs $\langle \Pi, \Sigma \rangle$ that can arise, and the appropriate kind of transformation—distribution conversion, permutative conversion, or reduction—to apply to them in the first step taken to determine the reduct $[\Pi, \Sigma]$.

§6.11 organizes the proof pairs into natural classes, based on their respective structures, and tells one what kind of transformation will be appropriate for each such pair.

§6.12 deals in detail with distribution conversions and permutative conversions; while §6.13 does the same for reductions.

§6.14 summarizes in a comprehensive table the transformations that are invoked for each of the systems $\mathbf{M}$, $\mathbf{I}$, $\mathbf{C}$, $\mathbb{C}$, and $\mathbb{C}^+$ in order to establish Cut-Elimination for them. This table is based on nomenclature from natural deduction.

§6.15 provides the same summary, but using nomenclature for sequent calculi.

§6.16 proves all the lemmas needed for, and then establishes, the main result of Cut-Elimination for each of the systems $\mathbf{M}$, $\mathbf{I}$, $\mathbf{C}$, $\mathbb{C}$, and $\mathbb{C}^+$. The common form of this result for these systems is: if Π uses premises Δ to prove A and Σ uses A with other premises Γ to prove φ, then $[\Pi, \Sigma]$ uses premises in $\Delta \cup \Gamma$ to prove [either $\bot$ or] φ. Cut-Elimination for $\mathbf{M}$, $\mathbf{I}$, and $\mathbf{C}$ omits the last bit of bracketed material; while Cut-Elimination for the two Core systems includes it. So the Core systems can enjoy epistemic gains that the orthodox systems cannot.

§6.17 presents *extraction theorems* about how to extract from any intuitionistic proof a core proof of a result at least as strong; and how to do the same with any classical proof.

§6.18 takes stock.

6.1   Nontransitive consequence relations

First, a few words are in order as to what our treatment of deducibility is *not*.

The enterprise of the Core logician needs to be distinguished clearly from two apparently related, but on closer inspection fundamentally distinct, inventions or definitions of unorthodox logical consequence relations. These have been furnished not for the language of mathematics, but rather for languages either containing a new 'logical' expression in the form of a transparent truth predicate, or containing such new extra-logical expressions as vague, or 'tolerant' predicates (such as '... is bald').

First, in Zardini [2008], for example, the focus is entirely on the formal-semantic side of things, not on rules of inference and the sorts of proofs that can be constructed in accordance with them. The phenomena that Zardini is examining are nowhere to be found in mathematics; their examination lies outside the scope of our more narrowly Fregean enterprise of giving the best possible formalizing account of mathematical proofs. Moreover, even though Zardini explores *tableaux* as the finitary syntactic objects generating deducibility relations to be matched with his respective logical consequence relations, it has to be borne in mind that *mathematical proofs are impossible to formalize faithfully as tableaux*. By 'faithfully' here we mean: in a way that corresponds, homologously, to the passages of reasoning actually engaged in by the mathematician. This notion of homology, although not precisely explicated, is nevertheless precise enough for our claim about faithful formalizability to be immediately plausible to anyone well-versed both in writing out mathematical proofs and in constructing tableau-style derivations of the arguments proved.

Zardini's logical consequence relations, like the deducibility relations of the Core systems, are not unrestrictedly transitive. That is about the only general point of contact that one can try to make between his approach and mine. Zardini's semantics also makes for some surprising sequent invalidities. One of these, as noted in Cobreros et al. [2012] at p. 374, is

$$\varphi, \varphi \to \psi, \psi \to \chi \not\models \chi.$$

So it would appear that the contact between Zardini's $\models$ and Core $\vdash$ is tenuous at best.

Second: another recent departure from transitivity is to be found in Ripley [2013]. He advocates *nontransitive* extensions of Classical Logic when dealing with a transparent truth predicate and/or tolerant (vague) predicates. It is only upon such extension of the language that he considers giving up unrestricted transitivity; he maintains it, as usual, for Classical Logic itself insofar as it deals only with the usual connectives and quantifiers. In each of his three systems (one with just a truth predicate; one with just vague predicates; and one with both) there are

> ... instances ... where $A \vdash B$ and $B \vdash C$ are both derivable, but $A \vdash C$ is not derivable

> ... there is no problem maintaining full Classical Logic, ... and conservatively extending it to accommodate transparent truth and tolerant vagueness. One simply needs to recognize that full Classical Logic does not, on its own, guarantee the transitivity of all its extensions. Of course *Classical Logic itself remains tran-*

sitive—nothing here should be construed as denying this well-known fact. But a transitive logic can have nontransitive extensions, and it's this possibility that's been exploited here ....
(pp. 7–8; emphasis added)

The Core logician's eschewal of unrestricted transitivity of $\vdash$ is much more radical than this. It goes to the very marrow of classical (and intuitionistic) logic, restricting it in fundamental and theoretically well-motivated ways that *do not await extensions* of the language with the likes of a transparent truth predicate, or vague predicates. Unrestricted transitivity *for both Classical Logic and Intuitionistic Logic themselves* is a systematic dogma that the Core logician challenges. The challenge is to reconsider Gentzenian Cut even for the *unextended* language of just connectives and quantifiers—the language used for the formalization of mathematics.

By contrast with the work just discussed on formal-semantical definitions of logical consequence, it needs to be emphasized that this chapter is concerned solely with matters syntactic—that is, with *deducibility*. We shall be examining systems of proof (sequent calculus and natural deduction) *in our new style* for Minimal, Intuitionistic and Classical Logic, and for Core Logic and Classical Core Logic. We aim to show how closely the three orthodox systems can be made to resemble the Core systems as far as transitivity is concerned. By re-casting the former systems in the mold of the latter, we can simultaneously make welcome innovations in the traditional proof theory of the orthodox systems, and motivate the application of those innovations to the Core systems. This approach enables one to appreciate how *innocuously*—but also how *profoundly*—Core Logic differs from Intuitionistic Logic, and Classical Core Logic from Classical Logic likewise.

6.2   Single- v. multiple-conclusion sequent calculi

In Gentzen's original formulation of sequent calculi, the sequent calculus for Intuitionistic Logic was a 'single-conclusion calculus'. That is, it dealt only with sequents whose succedents were at most singletons—sequents of one or other of the two forms

$$\{\varphi_1, \ldots, \varphi_n\} : \{\psi\} \qquad\qquad \{\varphi_1, \ldots, \varphi_n\} : \emptyset\,,$$

which we are writing, respectively, as

$$\varphi_1, \ldots, \varphi_n : \psi \qquad\qquad \varphi_1, \ldots, \varphi_n : \bot\,.$$

By contrast, Gentzen's peculiar way of obtaining Classical Logic was to make it a 'multiple-conclusion calculus'. That is, Gentzen's classical sequent calculus allows succedents to contain any (finite) number of sentences. So sequents of Gentzen's classical system can be of the form

$$\{\varphi_1, \ldots, \varphi_n\} : \{\psi_1 \ldots, \psi_m\}$$

—or, more economically,

$$\varphi_1, \ldots, \varphi_n : \psi_1 \ldots, \psi_m.$$

This trick afforded Gentzen the opportunity to prove the Law of Excluded Middle, and Double Negation Elimination, as already pointed out in §1.5.3:

$$\cfrac{\cfrac{\cfrac{A : A}{: A, \neg A}(:\neg)}{: A \vee \neg A}(:\vee)} \qquad (\neg:)\cfrac{\cfrac{A : A}{: A, \neg A}(:\neg)}{\neg\neg A : A}$$

Note how the second lines of these proofs exploit the permission to have 'multiple conclusions' (i.e. succedents containing more than one sentence).

We, however, shall be 'classicizing' by adopting either a rule of Dilemma, or a rule of Classical Reductio (see §6.6.4 and §6.7.2). Both these classical rules can be formulated in single-conclusion fashion. So the reader is forewarned that, even when we are considering Classical Logic itself, and Classical Core Logic, we shall be dealing with *single*-conclusion calculi. This feature of our sequent calculi is also partly responsible for the isomorphism, already discussed, between sequent proofs and their corresponding natural deductions in the Core systems. In the Core systems, the formulation of the rules governing logical operators—Right and Left rules in their sequent calculi; Introduction and Elimination rules in their systems of natural deduction—also contribute to the isomorphism mentioned.

6.3 The transitivity 'problem' for Core Logic

We shall be seeing that the quotes in this section heading are definitely scare quotes. In core proofs in the system of natural deduction, every major premise for an elimination (MPE) stands proud. So *all core proofs are in strong normal form*. This is because no MPE stands as the conclusion of an application of the corresponding Introduction rule. But by having MPEs stand proud, we are *also* preventing MPEs from standing as the conclusion of an application of any *Elimination* rule.

This raises the question: how can one ensure transitivity of core proof? To explain how the misgiving might arise that we *cannot* ensure it, we now digress to explain the (merely) *apparent* predicament.

Suppose one has core (hence normal) proofs

$$\begin{array}{ccc} \Delta & & A,\Gamma \\ \Pi & \text{and} & \Sigma \\ A & & \theta \end{array}$$

where, by virtue of A's being displayed separately, it is to be assumed that $A \notin \Gamma$. (Note that A is for present purposes an arbitrary sentence, not necessarily an atomic one.) In systems of proof *allowing* accumulation of proof trees (such as **M**, **I**, and **C**), one would be able to form a proof by grafting a copy of Π onto every undischarged assumption occurrence of A within Σ:

$$\begin{array}{c} \Delta \\ \Pi \\ (A),\Gamma \\ \Sigma \\ \theta \end{array}$$

and one would have the automatic assurance that the resulting construction *would* count as a proof of the overall conclusion θ from the set $\Delta \cup \Gamma$ of combined assumptions. Finite repetitions of the accumulation operation would also of course be countenanced:

$$\begin{array}{ccccc} & & & \Delta_1 & \Delta_n \\ \Delta_1 & \Delta_n & A_1,\ldots,A_n,\Gamma & \Pi_1 & \Pi_n \\ \Pi_1 \;,\ldots,\; \Pi_n & \text{and} & \Sigma & \text{together yield} & (A_1),\ldots,(A_n),\Gamma \\ A_1 & A_n & \theta & & \Sigma \\ & & & & \theta \end{array}$$

where $A_1,\ldots,A_n \notin \Gamma$. This is what is commonly understood as constituting the *transitivity of proof*—within a system allowing the formation of *ab*normal proofs.

Why the stress on '*abnormal*'? To answer this, we direct attention to the last display but one: in the core (hence normal) proof Σ, the 'cut sentence' A might stand at one of its undischarged assumption-occurrences as the major premise of an elimination. In a system of *normal* proof, the 'proof accumulation' given above of Π on top of Σ would therefore not always count as a proof (of θ from $\Delta \cup \Gamma$). *A fortiori*, in the n-fold case, in a system of *normal* proof, the 'proof accumulation' just indicated of $\Pi_1,\ldots,\Pi_n$ on top of Σ would not always count as a proof (of θ from $\Delta_1 \cup \ldots \cup \Delta_n \cup \Gamma$).

This brings to a close our explanatory digression about the source of the possible worry as to whether core deducibility—which is based entirely on *normal* proofs—will be transitive.

This apparent absence of unrestricted transitivity, however, (for the system of core, hence *normal*, proof) imposes no limitation in principle. This perhaps surprising, but very welcome, result is secured by the metatheorem below called **Cut-Elimination for Core Proof**. The terminology might strike the proof-theoretically informed reader as a little odd. Why speak of Cut-elimination when one is in the setting of natural deduction? The answer is really quite simple, on brief reflection: it is because of the thoroughgoing isomorphism between natural deductions and sequent proofs (in systems set up in the manner of Core Logic and Classical Core Logic). This makes the grafting of proof trees in the system of natural deduction the same operation, essentially, as a would-be application of Cut to the corresponding sequent proofs. The interesting wrinkle, of course, is that Cut *is not a rule of our sequent system for either Core Logic or Classical Core Logic!* Still, one can 'feign' cuts in a most satisfying way, as will be revealed by Metatheorem 4 below (and by its extension to deal with Classical Core Logic). Also, the natural-deduction setting makes it easier to follow what is going on in the rearrangements of bits of proof that are called for by each of the many steps of proof transformation below.

The reader should bear in mind that this result is at root very simple. It is about ways in which one can shunt around tree-like arrays of sentences, re-arranging twigs, sprigs, and sub-trees in order to achieve certain desired patterns. The mathematics is really very simple indeed, even if the patterns are numerous and at times a little complex. All that is needed is a little combinatorial imagination, and some intellectual stamina. Moreover, all the meta-reasoning is constructive and relevant. It is carried out in Core Logic at the metalevel.

Note that the title chosen for Metatheorem 4 ('Cut-Elimination for Core Proof') is, strictly speaking, a misnomer. For no applications of Cut are ever allowed within core proofs. In a traditional Gentzenian sequent system, to be sure, formal applications of the Cut rule are allowed; and the whole point of his *Hauptsatz*, the Cut-Elimination Theorem, is to show that one can avoid having recourse to such cuts, by eliminating them from sequent proofs that happen to contain them.

From our point of view, however, which is that of Core Logic, there simply are no cuts to get rid of. There is, however, the task of showing how Core Logic nevertheless *gives one the desired effect* of having Cut as a rule in the system. From this point of view, perhaps a more appropriate label for Metatheorem 4 would be **Cut-Eschewal for Core Proof**. But the terminology of Cut-elimination is now so firmly established that we use it here, albeit with these cautionary preparatory remarks.

Metatheorem 4 (Cut-Elimination for Core Proof).
There is an effective method [,] that transforms any two core proofs

$$
\begin{array}{ll}
\Delta & A, \Gamma \\
\Pi & \Sigma \qquad \text{(where } A \notin \Gamma \text{ and } \Gamma \text{ may be empty)} \\
A & \theta
\end{array}
$$

into a core proof $[\Pi, \Sigma]$ *of* θ *or of* $\perp$ *from (some subset of)* $\Delta \cup \Gamma$.

Corollary 3 (Multiple Cut-Elimination for Core Proof).
One can effectively transform the core proofs

$$
\begin{array}{llc}
\Delta_1 & \Delta_n & \overbrace{A_1, \ldots, A_n, \Gamma} \\
\Pi_1, \ldots, \Pi_n, & & \Sigma \qquad \text{(where } A_1, \ldots, A_n \notin \Gamma \text{ and } \Gamma \text{ may be empty)} \\
A_1 & A_n & \theta
\end{array}
$$

into a core proof $[\Pi_1, \ \ldots \ [\Pi_n, \Sigma] \ldots]$ *of* θ *or of* $\perp$ *from (some subset of)* $\Delta_1 \cup \ldots \Delta_n \cup \Gamma$.

Comment. In general the core proof $[\Pi_1, \ \ldots \ [\Pi_n, \Sigma] \ldots]$ will depend on the order of the Π_i. So, for example, $[\Pi_1, [\Pi_2, \Sigma]]$ need not be identical to $[\Pi_2, [\Pi_1, \Sigma]]$.

Thus far our square-bracket notation for the recursive function $[\Pi, \Sigma]$ that we intend to define has been used only in connection with Core Logic $\mathbb{C}$. But the leading idea and the detailed methods involved generalize or extend to the other main systems under discussion, namely Classical Core Logic $\mathbb{C}^+$, Minimal Logic **M**, Intuitionistic Logic **I**, and Classical Logic **C**.[1] So, taking any of these systems as system $\mathcal{S}$, one can inductively define the corresponding *recursive* 'reduction' or 'Cut-elimination' function $[\Pi, \Sigma]_{\mathcal{S}}$.

We shall now work towards a suitably general form of proof for Metatheorem 4 for any one of the five systems under consideration. The discussion will show how even the usual trio of systems **M**, **I**, and **C** can be shown to enjoy cut-admissibility in this fashion, provided only that they are formulated in a similar way to the Core systems $\mathbb{C}$ and $\mathbb{C}^+$ in one crucial

---

[1] It is in this regard that the treatment here marks an advance on both Tennant [2012a] and Tennant [2015b]. The reader who does not wish to get bogged down in any of the technicalities that follow may proceed immediately to §6.16. (One might, however, need to page back for the occasional definition of important terms used in stating results.)

regard. This is that *all their elimination rules should be parallelized, and MPEs should be required to stand proud, with no proof work above them.* That signal requirement—as we have repeatedly emphasized—ensures also that sequent proofs in these systems 'are' (i.e. are isomorphic to) the corresponding natural deductions. This observation will allow us to conduct much of the following discussion as though it is sequent proofs that are under consideration. The reader should bear in mind, though, that we can make the segue back and forth between a system's sequent formulation and its natural-deduction formulation, depending on expository need. What we say about a system in either one of its formulations will hold for it also in the other formulation.

It is worth pointing out that it is *not* the case that the restriction of Cut that Core Logic occasions is one that would affect any system of natural deduction as soon as one were to impose upon it the (strong) requirement that *all its elimination rules should be parallelized, and MPEs should be required to stand proud, with no proof work above them*—let alone just the weaker requirement that all its proofs be in normal form.[2] For, as we shall be seeing in Metatheorem 5, one can formulate the standard systems **M**, **I**, and **C** as systems meeting the stronger requirement, yet *still* prove that *unrestricted* Cut is admissible for them. This shows that the real source of Cut's (epistemically gainful) restriction in the Core systems is to be found, rather, in our careful modifications of rules of inference such as (→I) and (∨E), and our refusal to have EFQ (or Thinning) as a rule of the system.

6.3.1 Terminology

Given a proof Π,

> *its set of premises will be denoted* $p\Pi$;
> *its conclusion will be denoted* $c\Pi$; *and, as already mentioned:*
> *the sequent that it establishes will be denoted* $s\Pi$.

Note that $s\Pi$ is the sequent $p\Pi : c\Pi$.

---

[2] I am grateful to an anonymous referee for making me aware of the need to explain this consideration.

6.4   Cut

6.4.1   Cut as a structural rule

Cut is usually thought of as a so-called structural rule *of* or *in* a sequent calculus. In a single-conclusion sequent calculus, Cut takes the form

$$\frac{\Delta : \varphi \quad \varphi, \Gamma : \psi}{\Delta, \Gamma : \psi} .$$

φ is the *cut sentence*. Think of φ as a lemma in some branch of mathematics. The premise sets Δ and Γ might be different selections from the set of mathematical axioms for that branch; and the overall conclusion ψ might be a deep and interesting mathematical theorem. The lemma φ is a deductive 'halfway house' or 'stepping stone'—standing as the conclusion of the first proof Π, but serving as a premise for the second proof Σ.

   Being able to break down the deductive passage from one's mathematical axioms ($\Delta \cup \Gamma$) to a desired mathematical theorem (ψ) by interpolating a judiciously chosen lemma (φ) makes for greater economy of proof. Typically, the sum of the lengths of the (cut-free) sequent proofs Π and Σ:

$$\frac{\overset{\Pi}{\Delta : \varphi} \quad \overset{\Sigma}{\varphi, \Gamma : \psi}}{\Delta, \Gamma : \psi}$$

is much less—hyperlogarithmically so—than the length of any cut-free sequent proof

$$\overset{\Xi}{\Delta, \Gamma : \psi}$$

not proceeding via φ.

6.4.2   Must the system contain a rule of cut?

It is important, methodologically, for deducibility in one's system of logical proof in mathematics to be transitive. The rule of Cut in the sequent calculus gives direct expression to the requirement of transitivity. But *nota bene*: this familiar rule of Cut is a rule of *unrestricted* Cut. If we keep the adjective 'unrestricted' in mind, it will help the ensuing discussion.

   Compatibly with securing the required measure of transitivity, do we really *need* to express the requirement of (unrestricted) transitivity of deducibility as a rule of (unrestricted) Cut *in* the system? Might there not be some alternative way to know that one is entitled to enjoy the fruits of transitivity within legitimate reach without having to build (unrestricted) Cut *into* the system itself?

Note that if one's sequent calculus does not contain (unrestricted) Cut as a structural rule, then, trivially, all its proofs are cut-free, or *normal.* (This remark applies to **M, I,** and **C,** just as much as to $\mathbb{C}$ and $\mathbb{C}^+$.) The worry will then be whether cut-free proofs prove enough. In particular, can one make the deductive progress that is required by 'chaining together' (in some way to be specified) various cut-free proofs? Will there always be a cut-free proof that establishes the overall expected result?

6.4.3 Cut-admissibility

This is where the notion of *Cut-admissibility* gains purchase. Let us develop it first in the case of the three traditional systems **M, I,** and **C.** We shall say that Cut is *admissible for the sequent system* $\mathcal{S}$ (= **M, I,** or **C**) just in case the following (metalinguistic) rule holds:

$$\frac{\Delta \vdash_{\mathcal{S}} \varphi \quad \varphi, \Gamma \vdash_{\mathcal{S}} \psi}{\Delta, \Gamma \vdash_{\mathcal{S}} \psi}.$$

When we write $\Omega \vdash_{\mathcal{S}} \theta$ we have the orthodox meaning in mind: for some $\Omega' \subseteq \Omega$, there is a proof, in the system $\mathcal{S}$, of the sequent $\Omega' : \theta$. Formally:

$$\Omega \vdash_{\mathcal{S}} \theta \equiv_{df} \exists \Omega' \subseteq \Omega \, \exists \Pi \, \mathcal{P}_{\mathcal{S}}(\Pi, \Omega' : \theta).$$

So Cut-admissibility in the foregoing form is really the inference

$$\frac{\exists \Delta' \subseteq \Delta \, \exists \Pi \, \mathcal{P}_{\mathcal{S}}(\Pi, \Delta' : \varphi) \quad \exists \Gamma' \subseteq \Gamma \cup \{\varphi\} \, \exists \Sigma \, \mathcal{P}_{\mathcal{S}}(\Sigma, \Gamma' : \psi)}{\exists \Omega \subseteq \Delta \cup \Gamma \, \exists \Xi \, \mathcal{P}_{\mathcal{S}}(\Xi, \Omega : \psi)}.$$

This fails to impose any requirement to the effect that any of the proofs Ξ whose existence justifies the deducibility claim $\Delta, \Gamma \vdash_{\mathcal{S}} \psi$ bears an intelligible relationship to such proofs Π and Σ as might respectively justify the two deducibility claims $\Delta \vdash_{\mathcal{S}} \varphi$ and $\varphi, \Gamma \vdash_{\mathcal{S}} \psi$. Thus, some proof Π might establish the sequent $\Delta : \varphi$, and another proof Σ might establish the sequent $\{\varphi\} \cup \Gamma : \psi$, without Π and Σ bearing any intelligible relationship to any proof Ξ establishing the 'target sequent' $\Delta \cup \Gamma : \psi$.

Far better would be a result like this (for the sequent system $\mathcal{S}$):

$$\frac{\mathcal{P}_{\mathcal{S}}(\Pi, \Delta : \varphi) \quad \mathcal{P}_{\mathcal{S}}(\Sigma, \{\varphi\} \cup \Gamma : \psi)}{\exists \Delta' \subseteq \Delta \, \exists \Gamma' \subseteq \Gamma \, \mathcal{P}_{\mathcal{S}}([\Pi, \Sigma]_{\mathcal{S}}, \Delta' \cup \Gamma' : \psi)}.$$

where $[\Pi, \Sigma]_\mathcal{S}$ is a proof *effectively determinable from* the $\mathcal{S}$-proofs Π and Σ.

In a system with the rule Cut, such a result is immediate: for $[\Pi, \Sigma]_\mathcal{S}$ take the result of applying Cut to (the concluding sequents of) Π and Σ. That secures, in the time-honored words of Russell, all the advantages of theft over honest toil. Here we undertake some honest toil.

Let $\mathcal{S}$ be a system of *cut-free*, or *normal*, proof for any one of **C**, **I**, or **M**. We seek to establish, for $\mathcal{S}$, a general result of the form

There is an effective method $[\ ,\ ]_\mathcal{S}$ that transforms any two $\mathcal{S}$-proofs

$$\begin{array}{cc} \Pi & \Sigma \\ \Delta : \varphi & \varphi, \Gamma : \theta \end{array}$$

(where $\varphi \notin \Gamma$ and Γ may be empty) into an $\mathcal{S}$-proof

$$\begin{array}{c} [\Pi, \Sigma]_\mathcal{S} \\ \Omega : \theta \end{array}$$

for some $\Omega \subseteq \Delta \cup \Gamma$.

We can indeed establish such a result for $\mathcal{S} = \mathbf{C}$, the system of Classical Logic. We shall establish it relatively easily by choosing a set $\mathcal{C}$ of *optimally formulated primitive rules of inference* for **C**. Relative to this choice, the systems **M** and **I** will be obtained as subsystems, $\mathbf{M} \subset \mathbf{I} \subset \mathbf{C}$, by choosing appropriate sets $\mathcal{M} \subset \mathcal{I} \subset \mathcal{C}$, respectively, of these optimally formulated primitive rules of inference. It will follow from our suitably 'modular' method of proof of the foregoing Cut-admissibility result for **C**-proofs that the result holds for **I**-proofs and for **M**-proofs.

6.4.4 Cut-admissibility strengthened for the Core systems

We shall then see that this Cut-admissibility result for **C** (hence also for **I** and **M**) can be *strengthened* for the sequent systems for $\mathbb{C}$ and $\mathbb{C}^+$ (taken as $\mathcal{S}$ in the following strengthening):

There is an effective method $[\ ,\ ]$ that transforms any two $\mathcal{S}$-proofs

$$\begin{array}{cc} \Pi & \Sigma \\ \Delta : \varphi & \varphi, \Gamma : \{\theta\} \end{array}$$

(where $\varphi \notin \Gamma$ and Γ may be empty) into an $\mathcal{S}$-proof

$$[\Pi, \Sigma]$$
$$\Omega : \Xi$$

for some $\Omega \subseteq \Delta \cup \Gamma$ *and* $\Xi \subseteq \{\theta\}$.

Note the potential 'subsetting down' not only on the left (as before, for the three traditional systems) but now also on the *right* (for the Core systems).

   We established Cut-admissibility in this form for Core Logic $\mathbb{C}$, and for Classical Core Logic $\mathbb{C}^+$, in Tennant [2012a] and Tennant [2015b] respectively. The method of proof generalizes to systems of *cut-free*, or *normal*, proof for **C**, **I** and **M**, when their elimination rules are framed in the way described above.

6.4.5 Cut-admissibility for Core systems of Natural Deduction

Both of the papers just cited dealt with certain *natural deduction* formulations of the Core systems, for which Cut-admissibility is better expressed as follows.

Cut-admissibility for Core Systems of Natural Deduction

There is an effective method [,] that transforms any two core natural deductions

$$
\begin{array}{cc}
\Delta & \varphi, \Gamma \\
\Pi & \Sigma \\
\varphi & \theta
\end{array}
\qquad \text{(where } \varphi \notin \Gamma \text{ and } \Gamma \text{ may be empty)}
$$

into a core natural deduction $[\Pi, \Sigma]$ of some subsequent of $\Delta \cup \Gamma : \{\theta\}$.

The difference, however, between treating of sequent proofs and treating of natural deductions is absolutely inessential. This is because of the isomorphism, already remarked, between any sequent proof and the corresponding natural deduction in a system whose elimination rules are suitably framed.

6.4.6 On two proofs connecting (in *any* system $\mathcal{S}$)

Let us think first in terms of sequent proofs. We shall say that Π *connects with* Σ just in case $c\Pi \in p\Sigma$, i.e. just in case they are of the forms

$$
\begin{array}{cc}
\Pi & \Sigma \\
\Delta : A & A, \Gamma : \theta
\end{array} .
$$

Here A (which is obviously not $\bot$) is called the *cut sentence*. A may be atomic or complex. By virtue of A's being displayed separately in the antecedent of $s\Sigma$, it is to be assumed that $A \notin \Gamma$. Of course, Γ itself may be empty.

Now let us think in terms of natural deductions. We shall say that Π *connects with* Σ just in case they are of the forms

$$
\begin{array}{cc}
\Delta & A,\Gamma \\
\Pi & \Sigma \\
A & \theta
\end{array}\ .
$$

Here A (which is obviously not $\bot$) will still be called the *cut sentence*, even though we are working with natural deductions. By virtue of A's being displayed separately as a premise of Σ, it is to be assumed that $A \notin \Gamma$.

We shall refer to the cut sentence A often, even in the absence of the proof schemata above. If A is compound, its dominant operator is α.

It is clear that the isomorphism between sequent proofs and natural deductions allows us to define further notions involving proofs, without specifying whether we are thinking specifically of proofs in a sequent calculus, or of proofs in a system of natural deduction.

Definition 13. *Take Π and Σ of the forms above, i.e. with Π connecting with Σ. $\Pi*\Sigma$ is defined to be $\Delta,\Gamma : \theta$, and is called the* target sequent.

Definition 14. $\sqsubseteq_{\mathcal{S}}$ *is the subsequent relation appropriate for the system $\mathcal{S}$.*

What is crucial is how this notion of 'subsequent' is defined. For **M**, **I**, and **C** one obtains a subsequent by subsetting down *on the left only*. For $\mathbb{C}$ and $\mathbb{C}^+$ one can *also* subset down *on the right*.

6.4.7 Cut-admissibility for systems $\mathcal{S}$ in general

Cut-admissibility for systems $\mathcal{S}$ in general can now be stated in the following form:

> There is an effective method $[\ ,\ ]_{\mathcal{S}}$ that transforms any two $\mathcal{S}$-proofs Π and Σ (where Π connects with Σ) into an $\mathcal{S}$-proof $[\Pi, \Sigma]_{\mathcal{S}}$ such that $s[\Pi, \Sigma]_{\mathcal{S}} \sqsubseteq_{\mathcal{S}} \Pi*\Sigma$.

Formally:

$\exists\,\mathrm{eff}.f\,(\forall\Pi \text{ in } \mathcal{S})\,(\forall\Sigma \text{ in } \mathcal{S})(\Pi \text{ connects with } \Sigma \to s(f(\Pi, \Sigma)) \sqsubseteq_{\mathcal{S}} \Pi*\Sigma)$

So in general we have $s[\Pi, \Sigma]_{\mathcal{S}} \sqsubseteq_{\mathcal{S}} \Pi * \Sigma$. That is, the $[\ ,\ ]_{\mathcal{S}}$-transform of two connecting $\mathcal{S}$-proofs Π and Σ proves *some subsequent* (in the sense of $\mathcal{S}$) of their target sequent.

6.5 How admissible cuts can achieve epistemic gain

Consider the obvious proof Π of $A, B : A \wedge B$, and the obvious proof Σ of $A \wedge B : A$.

$$\Pi : \frac{A : A \qquad B : B}{A, B : A \wedge B} \qquad\qquad \Sigma : \frac{A : A}{A \wedge B : A} \ .$$

Then $[\Pi, \Sigma]$ is the trivial proof of the sequent $A : A$. It is *not* a proof of the 'thinned' sequent $A, B : A$.

That is how it should be. When Π proves $\Delta : \varphi$ and Σ proves $\varphi, \Gamma : \psi$, the sequent $\Omega : \psi$ established by the proof $[\Pi, \Sigma]$ might have $\Omega \subset \Delta \cup \Gamma$ (i.e. Ω a *proper* subset of $\Delta \cup \Gamma$). That is, we can expect on (frequent) occasion to 'subset down on the left' when applying the operation $[\ ,\ ]$. By doing so, we can often achieve a logically stronger result. This is to be welcomed as *epistemic gain*.

There is no principled reason to confine the potential 'subsetting down' (and the epistemic gain that it provides) to the *left* of the target sequent $\Delta \cup \Gamma : \{\psi\}$ when determining the normal proof

$$\left[\begin{array}{cc} \Pi & \Sigma \\ \Delta : \varphi & , & \varphi, \Gamma : \{\psi\} \end{array} \right]$$

After all, if we can have $\dfrac{[\Pi, \Sigma]}{\Omega : \{\psi\}}$ for some $\Omega \subseteq \Delta \cup \Gamma$, then why should we

not also be allowed to have $\dfrac{[\Pi, \Sigma]}{\Omega : \Psi}$ for some $\Omega \subseteq \Delta \cup \Gamma$ and some $\Psi \subseteq \{\psi\}$?

This would allow the case where strengthening occurs on *both* sides of the sequent:

$$\frac{[\Pi, \Sigma]}{\Omega : \emptyset} \quad \text{for some } \Omega \subset \Delta \cup \Gamma.$$

Definition 15. *An* explosive *logical system $\mathcal{S}$ is one in which there is an $\mathcal{S}$-proof of either or both of the following sequents:*

$$\neg\varphi, \varphi : \psi \qquad\qquad \neg\varphi, \varphi : \neg\psi \ .$$

These are called the positive and negative forms, respectively, of Lewis's First Paradox.

The three explosive systems of Classical, Intuitionistic, and Minimal Logic are very well known through their classification by modern proof theorists in terms of their rules of natural deduction, *in a particular formulation*—essentially that of Gentzen (1934–5), reprised by Prawitz (1965).

Minimal Logic **M** (hence also Intuitionistic Logic **I** and Classical Logic **C**) proves the *negative* form of Lewis's First Paradox:

$$\neg\varphi, \varphi : \neg\psi$$

Intuitionistic Logic (hence also Classical Logic) also proves the *positive* form of Lewis's First Paradox:

$$\neg\varphi, \varphi : \psi$$

Non-explosive logical systems are called *paraconsistent*.[3] Core Logic is paraconsistent, as is its classicized extension. One of the main motivating ideas behind the adoption of a paraconsistent logic is that it permits there to be distinct non-trivial inconsistent theories. It affords logical closure of a set of beliefs about a wide variety of different domains and topics, with any inconsistencies that might crop up being 'local', and not infectious. With Intuitionistic Logic (hence also with Classical Logic) there is only one inconsistent theory in a given language: the set of *all* its sentences, regardless of the domains or topics they concern. And closure of an inconsistent theory in Minimal Logic is almost as drastic, since it contains *all negations*.

6.6   Rules of Minimal, Intuitionistic, and Classical Logic

The traditional natural-deduction rules for Minimal Logic **M** are just the introduction and elimination rules for the logical operators, *in the particular formulation due to Gentzen and Prawitz.*

As a reminder: to obtain *Intuitionistic* Logic **I**, one adds the Absurdity Rule (also known as *Ex Falso Quodlibet* (EFQ)). *Classical* Logic **C** then results by adding any one of four classical rules of negation: the Law of Excluded Middle; Classical Dilemma; Double Negation Elimination; or Classical *Reductio ad Absurdum*. (See Tennant [1978], §4.5.)

---

[3] See, for example, the definition of paraconsistent logics provided by Priest [2002], at pp. 288–9.

6.6.1 Graphic conventions for stating rules

We are going to reframe some of the traditional rules, but in such a way that we still obtain **M**, **I**, and **C** in the usual fashion described above. We shall use the same labels for the rules, even though some of them have been subtly changed. I beg the reader's indulgence for any element of reprise in what follows; but it is meet and right to emphasize that this all applies to **M**, **I**, and **C** when they are formulated in our preferred way.

In the detailed statements that follow of all the aforementioned natural-deduction rules, we observe the following graphic conventions. They have already been explained or employed in §§1.5 and 2.3.4, but we restate them here in the interests of having a self-contained exposition within this chapter, and in order to underscore their uniform applicability in characterizing how the various rules differ as between the standard trio **M**, **I**, and **C**, and the two Core systems $\mathbb{C}$ and $\mathbb{C}^+$.

1. A *box* next to a discharge stroke placed over a dischargeable assumption indicates that vacuous discharge is not allowed. There must be an assumption of the indicated form available for discharge.[4]

2. A *diamond* next to a discharge stroke indicates that it is not required that the assumption in question should have been used and be available for discharge. But if it is available, then it is discharged.

3. The *absence of vertical dots* above major premises for eliminations indicates that those MPEs stand proud, with no proof work above them.

(3) is the main innovation to highlight, especially in the context of our formulations of **M**, **I**, and **C**. It ensures that all natural deductions in these systems (as in the Core systems) are already in normal form, and so never have to be normalized.

The rules of natural deduction that are stated graphically below are reformulations, subject to the constraints already mentioned concerning elimination rules, of the rules originally framed for **I** and **C** by Gentzen and reprised by Prawitz. (Note that Johansson's invention, in 1936, of the system **M** postdates the invention of natural deduction in Gentzen [1934, 1935].) The use of the boxes and diamonds (explained above) is fully in accord with both the spirit and the letter of the formulations of natural deduction to be found in both Gentzen and Prawitz.

---

[4] With ($\wedge$E) and ($\forall$E) we require only that *at least one* of the indicated assumptions should have been used, and be available for discharge.

Next to each natural-deduction rule we state the corresponding rule of the sequent calculus. Because of the way we choose to formulate the natural-deduction rules, they correspond *directly* to the sequent rules. This is an essential advance on Gentzen. Any natural deduction is *isomorphic* (*qua* proof tree of rule applications) to its corresponding sequent proof. No longer needed are the complicated transformations that Gentzen provided, to turn a natural deduction into a sequent proof or vice versa.

In rendering the Gentzen–Prawitz rules for the three standard systems **M**, **I**, and **C** in our preferred way (parallelized elimination rules, with MPEs standing proud) we do not otherwise alter the effect of their original rules. Thus, for example, (¬I) is still stated (for all three of these systems) as a 'vacuous discharge' rule—meaning that it can be applied even if there is no assumption of the indicated form to be discharged. The reader can easily check that our 'parallelized', normal proofs for the respective systems **M**, **I**, and **C** generate all their deducibilities.

Nota bene:

- **Introduction** rules in natural deduction correspond to **Right** logical rules in the sequent calculus.

- **Elimination** rules in natural deduction correspond to **Left** logical rules in the sequent calculus.

- **Ex Falso Quodlibet** in natural deduction corresponds to **Thinning on the Right** in the sequent calculus.

- **Classical Dilemma** in natural deduction translates directly into **Classical Dilemma** in the *single-conclusion* sequent calculus.

We stress once again, as we did in §6.2: *our formulation of the sequent calculus involves sequents with at most singleton succedents*. We do *not* 'go classical' by exploiting Gentzen's unnatural trick, in his sequent calculus for **C**, of allowing succedents to contain more than one sentence.

What follows are the rules of both natural deduction and the sequent calculus for **M**, **I**, and **C**, in our preferred format. As pointed out in §2.3.8: had Gentzen but followed through on a result he published in his very first paper (Gentzen [1932]), he might have arrived at the rules for **I** and **C** in these forms himself. For, working only with structural rules at the outset, Gentzen established a normalization theorem according to which every proof could be brought into a normal form with only one application of Thinning, which moreover would be terminal. See Tennant [2015c] for details.

6.6.2 Rules of Minimal Logic (in our preferred format)

$(\neg I)$

$$
\begin{array}{c}
\diamond\!\!-\!\!(i) \\
\varphi \\
\vdots \\
\dfrac{\bot}{\neg\varphi}\,(i)
\end{array}
$$

$(\neg R)$

$$
\dfrac{\Delta :}{\Delta\setminus\{\varphi\} : \neg\varphi}
$$

$(\neg E)$

$$
\dfrac{\neg\varphi \quad \overset{\vdots}{\varphi}}{\bot}
$$

$(\neg L)$

$$
\dfrac{\Delta : \varphi}{\neg\varphi, \Delta :}
$$

$(\wedge I)$

$$
\dfrac{\overset{\vdots}{\varphi} \quad \overset{\vdots}{\psi}}{\varphi\wedge\psi}
$$

$(\wedge R)$

$$
\dfrac{\Delta : \varphi \qquad \Gamma : \psi}{\Delta, \Gamma : \varphi\wedge\psi}
$$

$(\wedge E)$

$$
\begin{array}{c}
(i)\!\!-\!\!\square\!\!-\!\!(i) \\
\overbrace{\varphi , \psi} \\
\vdots \\
\dfrac{\varphi\wedge\psi \qquad \theta}{\theta}\,(i)
\end{array}
$$

$(\wedge L)$

$$
\dfrac{\Delta : \theta}{\varphi\wedge\psi, \Delta\setminus\{\varphi,\psi\} : \theta}
$$
$$
\text{where } \Delta\cap\{\varphi,\psi\}\neq\emptyset
$$

$(\vee I)$

$$
\dfrac{\overset{\vdots}{\varphi} \quad \overset{\vdots}{\psi}}{\varphi\vee\psi \quad \varphi\vee\psi}
$$

$(\vee R)$

$$
\dfrac{\Delta : \varphi}{\Delta : \varphi\vee\psi} \qquad \dfrac{\Delta : \psi}{\Delta : \varphi\vee\psi}
$$

$(\vee E)$

$$
\begin{array}{c}
\square\!\!-\!\!(i) \quad \square\!\!-\!\!(i) \\
\varphi \qquad \psi \\
\vdots \qquad \vdots \\
\dfrac{\varphi\vee\psi \quad \theta \quad \theta}{\theta}\,(i)
\end{array}
$$

$(\vee L)$

$$
\dfrac{\varphi, \Delta : \theta \qquad \psi, \Gamma : \theta}{\varphi\vee\psi, \Delta, \Gamma : \theta}
$$

$(\rightarrow I)$

$$\begin{array}{c} \diamond\!\!-\!\!(i) \\ \varphi \\ \vdots \\ \dfrac{\psi}{\varphi \rightarrow \psi}(i) \end{array}$$

$(\rightarrow R)$
$$\dfrac{\Delta : \psi}{\Delta \setminus \{\varphi\} : \varphi \rightarrow \psi}$$

$(\rightarrow E)$

$$\dfrac{\varphi \rightarrow \psi \quad \varphi \quad \begin{array}{c}\square\!\!-\!\!(i)\\ \psi\\ \vdots\\ \theta\end{array}}{\theta}(i)$$

$(\rightarrow L)$
$$\dfrac{\Delta : \varphi \quad \psi, \Gamma : \theta}{\varphi \rightarrow \psi, \Delta, \Gamma : \theta}$$

$(\exists I)$

$$\begin{array}{c} \vdots \\ \dfrac{\varphi_t^x}{\exists x \varphi} \end{array}$$

$(\exists R)$
$$\dfrac{\Delta : \varphi_t^x}{\Delta : \exists x \varphi}$$

$(\exists E)$

$$\dfrac{\exists x \varphi \,^{\textcircled{a}} \qquad \underbrace{\textcircled{a} \ldots \varphi_a^x \ldots \textcircled{a}}_{\begin{array}{c}\square\!\!-\!\!(i)\\ \vdots\\ \psi\,^{\textcircled{a}}\end{array}}}{\psi}(i)$$

$(\exists L)$
$$\dfrac{\varphi_a^x, \Delta : \psi}{\exists x \varphi, \Delta : \psi} \;\textcircled{a}$$

$(\forall I)$

$$\begin{array}{c} \textcircled{a} \\ \vdots \\ \dfrac{\varphi}{\forall x \varphi_x^a} \end{array}$$

$(\forall R)$
$$\dfrac{\Delta : \varphi}{\Delta : \forall x \varphi_x^a} \;\textcircled{a}$$

$(\forall E)$

$$\dfrac{\forall x \varphi \qquad \begin{array}{c}(i)\underbrace{\rule{1em}{0.4pt} \cdots \square \cdots \rule{1em}{0.4pt}}(i)\\ \varphi_{t_1}^x , \;\ldots\; , \varphi_{t_n}^x\\ \vdots\\ \theta\end{array}}{\theta}(i)$$

$(\forall L)$
$$\dfrac{\varphi_{t_1}^x , \;\ldots\; , \varphi_{t_n}^x, \Delta : \theta}{\forall x \varphi, \Delta : \theta}$$

6.6.3 Rules of Intuitionistic Logic (in our preferred format)

The rules of Minimal Logic, plus

$$\text{(EFQ)} \quad \frac{\bot}{\varphi} \qquad\qquad \text{(T}_R\text{)} \quad \frac{\Delta:}{\Delta:\varphi} \quad \text{`Thinning on the Right'}$$

When EFQ is applied to obtain a conclusion with α dominant, we shall call it an application of αEFQ. Note, however, that both EFQ and Thinning on the Right are 'operatorless' rules. *They are both stated without any explicit mention of any particular logical operator.*

6.6.4 Rules of Classical Logic (in our preferred format)

The rules of Intuitionistic Logic, plus

$$\text{(Dil)} \quad \begin{array}{cc} \square\!\!-\!\!(i) & \square\!\!-\!\!-\!\!(i) \\ \varphi & \neg\varphi \\ \vdots & \vdots \\ \psi & \psi \\ \hline \multicolumn{2}{c}{\rule{0pt}{1em}} \end{array}\!\!(i) \qquad \text{(Dil)} \quad \frac{\varphi,\Delta:\psi \quad \neg\varphi,\Gamma:\psi}{\Delta,\Gamma:\psi}$$

$$\psi$$

So much for the rules (in our preferred format) of the orthodox Gentzen–Prawitz systems **M**, **I**, and **C**. We now address the Core Systems, whose sets of rules are more austere, in that they contain no 'operatorless' rules at all.

6.7 Rules of Inference for the Core Systems

6.7.1 Rules of Core Logic

$$\text{(}\neg\text{I)} \quad \begin{array}{c} \square\!\!-\!\!(i) \\ \varphi \\ \vdots \\ \bot \\ \hline \neg\varphi \end{array}\!\!(i) \qquad \text{(}\neg R\text{)} \quad \frac{\varphi,\Delta:}{\Delta:\neg\varphi}$$

$(\neg E)$

$$\frac{\neg\varphi \quad \varphi}{\bot}$$

$(\neg L)$

$$\frac{\Delta : \varphi}{\neg\varphi, \Delta :}$$

$(\wedge I)$

$$\frac{\varphi \quad \psi}{\varphi \wedge \psi}$$

$(\wedge R)$

$$\frac{\Delta : \varphi \qquad \Gamma : \psi}{\Delta, \Gamma : \varphi\wedge\psi}$$

$(\wedge E)$

$$\overbrace{\varphi, \psi}^{(i)-\square-(i)}$$

$$\frac{\varphi \wedge \psi \qquad \theta}{\theta}{}_{(i)}$$

$(\wedge L)$

$$\frac{\Delta : \theta}{\varphi\wedge\psi, \Delta \setminus \{\varphi, \psi\} : \theta}$$

$$\text{where } \Delta \cap \{\varphi, \psi\} \neq \emptyset$$

$(\vee I)$

$$\frac{\varphi}{\varphi \vee \psi} \quad \frac{\psi}{\varphi \vee \psi}$$

$(\vee R)$

$$\frac{\Delta : \varphi}{\Delta : \varphi\vee\psi} \quad \frac{\Delta : \psi}{\Delta : \varphi\vee\psi}$$

$(\vee E)$

$$\square\!-\!(i) \quad \square\!-\!(i)$$
$$\varphi \qquad \psi$$

$$\frac{\varphi \vee \psi \quad \theta/\bot \quad \theta/\bot}{\theta/\bot}{}_{(i)}$$

$(\vee L)$

$$\frac{\varphi, \Delta : \theta/\bot \qquad \psi, \Gamma : \theta/\bot}{\varphi\vee\psi, \Delta, \Gamma : \theta/\bot}$$

$(\to I)(a)$

$$\begin{array}{c}\square\!-\!(i)\\ \varphi\\ \vdots\\ \bot\end{array}$$
$$\frac{}{\varphi\to\psi}{}_{(i)}$$

$(\to I)(b)$

$$\begin{array}{c}\lozenge\!-\!(i)\\ \varphi\\ \vdots\\ \psi\end{array}$$
$$\frac{}{\varphi\to\psi}{}_{(i)}$$

$(\to R)(a)$

$$\frac{\Delta, \varphi :}{\Delta : \varphi\to\psi}$$

$(\to R)(b)$

$$\frac{\Delta : \psi}{\Delta \setminus \{\varphi\} : \varphi\to\psi}$$

$(\to E)$

$$\frac{\varphi\to\psi \qquad \varphi \qquad \theta}{\theta}\,{\scriptstyle(i)}$$

with discharge
$$\overset{\square\!-\!(i)}{\underset{\vdots}{\psi}}$$
above θ

$(\to_L)$
$$\frac{\Delta:\varphi \qquad \psi,\Gamma:\theta}{\varphi\to\psi,\Delta,\Gamma:\theta}$$

$(\exists I)$
$$\frac{\overset{\vdots}{\varphi^x_t}}{\exists x\varphi}$$

$(\exists_R)$
$$\frac{\Delta:\varphi^x_t}{\Delta:\exists x\varphi}$$

$(\exists E)$
$$\frac{\exists x\varphi\,^{\textcircled{a}} \qquad \psi^{\textcircled{a}}}{\psi}\,{\scriptstyle(i)}$$
with
$$\underbrace{\textcircled{a}\ldots\varphi^x_a\ldots\textcircled{a}}_{}\,{\scriptstyle\square\!-\!(i)}$$

$(\exists_L)$
$$\frac{\varphi^x_a,\Delta:\psi}{\exists x\varphi,\Delta:\psi}\ \textcircled{a}$$

$(\forall I)$
$$\frac{\overset{\textcircled{a}}{\underset{\vdots}{\varphi}}}{\forall x\varphi^a_x}$$

$(\forall_R)$
$$\frac{\Delta:\varphi}{\Delta:\forall x\varphi^a_x}\ \textcircled{a}$$

$(\forall E)$
$$\frac{\forall x\varphi \qquad \theta}{\theta}\,{\scriptstyle(i)}$$
with
$$\underbrace{\varphi^x_{t_1},\ \ldots\ ,\ \varphi^x_{t_n}}_{}\,{\scriptstyle(i)\!-\!\cdots\square\cdots\!-\!(i)}$$

$(\forall_L)$
$$\frac{\varphi^x_{t_1},\ \ldots\ ,\ \varphi^x_{t_n},\Delta:\theta}{\forall x\varphi,\Delta:\theta}$$

6.7.2 Rules of Classical Core Logic $\mathbb{C}^+$

In order to obtain Classical Core Logic from Core Logic, it suffices to add either Classical Reductio or Dilemma. (These rules, like all the rules of Core Logic, are subject to the Global Anti-Dilution Condition on Proof Formation.) These two strictly classical rules are interderivable in Core Logic. In each of these rules, it is the sentence φ that is its 'classical focus'. This is because the reasoner who applies the rule is presuming that φ is determinately true, or false, even though it might not be known (or indeed, even knowable) which is the case. (See Tennant [1997] for extended philosophical discussion of this point.)

(CR)
$$\begin{array}{c} \square\!\!-\!\!-(i) \\ \neg\varphi \\ \vdots \\ \dfrac{\bot}{\varphi}(i) \end{array}$$

(CR)
$$\dfrac{\neg\varphi, \Delta : \emptyset}{\Delta : \varphi}$$

(Dil)
$$\begin{array}{cc} \square\!\!-\!\!-(i) & \square\!\!-\!\!-(i) \\ \varphi & \neg\varphi \\ \vdots & \vdots \\ \psi & \psi \\ \hline & \psi \end{array}(i)$$

(Dil)
$$\dfrac{\varphi, \Delta : \psi \quad \neg\varphi, \Gamma : \psi}{\Delta, \Gamma : \psi}$$

$$\begin{array}{cc} \square\!\!-\!\!-(i) & \square\!\!-\!\!-(i) \\ \varphi & \neg\varphi \\ \vdots & \vdots \\ \psi & \bot \\ \hline & \psi \end{array}(i)$$

$$\dfrac{\varphi, \Delta : \psi \quad \neg\varphi, \Gamma : \emptyset}{\Delta, \Gamma : \psi}$$

6.7.3   Remarks on avoiding irrelevance

Note the special extra form allowed for Dilemma in Classical Core Logic— where the negative horn proves absurd.

There is a striking contrast between the form of $(\neg I)$ in Minimal Logic (which is inherited by both Intuitionistic Logic and Classical Logic) and the form of the 'same' rule (which it is not, really!) in Core Logic (and in Classical Core Logic). The former form (in the standard systems) permits *vacuous* discharge of the *reductio* assumption; the latter form (in the Core systems) does not. Thus the standard systems contain the negative form of Lewis's First Paradox, $\neg A, A : \neg B$. The proof is

$$\frac{\dfrac{\neg A \quad A}{\bot}}{\neg B} \quad ,$$

where the final step counts as an application of $(\neg I)$, with vacuous discharge of (the non-existent) assumption B. The Core systems do not contain this proof, since its final step violates their requirement of non-vacuous discharge of the assumption B.

Of course, both Intuitionistic Logic and Classical Logic would be committed to this last result anyway, since they both contain the *positive* form of Lewis's First Paradox, courtesy of *Ex Falso Quodlibet*. So they can regard the final step of the last proof as an application of EFQ rather than of $(\neg I)$.

What is all too infrequently realized is that even if *Ex Falso Quodlibet* were not a permitted rule, Classical Logic as based on a rule of Classical Reductio permitting vacuous discharge would *still* contain Lewis's First Paradox. This is because the proof

$$\frac{\dfrac{\neg A \quad A}{\bot}}{B}(CR)$$

would officially be allowed as correct, despite the fact that the final step, masquerading (but legitimately so!) as an application of Classical Reductio, is *really*—so one wants to maintain—an application of *Ex Falso Quodlibet*.

The relevantizer must therefore be extremely careful about not permitting vacuous discharge of assumptions. But that means also that she has to be vigilant about nefarious ways in which certain proofs might be constructed—if such means were permitted—so as to make an 'assumption for the sake of argument' *spuriously* appear as one on which the conclusion (of a subordinate) proof officially depends. This disastrous possibility is always rearing its ugly head in a natural deduction system that allows the formation of abnormal proofs that have conclusions of introductions standing as major premises of the corresponding elimination. For then one can make *any* 'assumption' θ spuriously relevant to a conclusion ψ below. Here is how one can do so, regardless of whether one is using the serial form of ($\wedge$E), or its parallelized form, respectively:

$$\frac{\dfrac{\varphi \quad \theta}{\varphi \wedge \theta}(\wedge \text{I})}{\varphi}(\wedge \text{E})$$

$$\vdots$$

$$\psi$$

$$\frac{\varphi \quad \theta}{\varphi \wedge \theta}(\wedge \text{I})$$

$$\frac{\overline{\varphi}^{(1)} \qquad \vdots \qquad \psi}{\psi}(1)\,(\wedge \text{I})$$

This immediately puts the relevantizer in mind of necessary measures to be taken, to prevent this kind of spurious 'dependency' (of ψ on θ) from arising within properly constructed proofs. The carefully considered result is Core Logic's insistence, with its *parallelized* elimination rules, that their major premises should *stand proud*.

6.8 Compact Polish notation for proofs

We introduce here a compact notation that will be extremely useful in due course, when stating numerous transformations of proofs as succinctly as possible. These are the transformations that will be involved in providing a recursive definition of the effective operation $[\Pi, \Sigma]$ on pairs of proofs, which plays such a crucial role in establishing the epistemically gainful transitivity of (classical) core deducibility.

Recall that a sequent rule ρ, applied to subproofs $\Sigma_1, \ldots, \Sigma_n$, produces a proof that will be denoted

$$\rho\Sigma_1, \ldots, \Sigma_n .$$

One can also write

$$\rho\Sigma_1, \ldots, \Sigma_n\theta$$

when it is important, in context, to know or to be reminded that the conclusion is θ. *The very same notation* will also describe the corresponding *natural deductions* establishing the same results as the sequent proofs. This convenience is owed, once again, to the isomorphism between sequent proofs and natural deductions that we have already (frequently!) remarked upon. The only wrinkle is that with an Elimination rule ρ (or, in the sequent case, a Left logical rule) there is tacit dependency on a particular major premise, which we shall assume is known from the context. From this knowledge flows also knowledge of the set of undischarged assumptions on which the conclusion θ depends.

6.8.1 On the $\mathcal{S}$-relativity of our notations

'One and the same' inference rule can have importantly different formulations, depending on whether one is dealing with one of the Core systems ($\mathbb{C}$ or $\mathbb{C}^+$) or one of the standard trio (**M** or **I** or **C**). The rules falling in the scope of this observation are ($\neg$I), ($\to$I), and ($\vee$E), and, in the classical case, both (Dil) and (CR). With the foregoing ρ-notation, therefore, one should be mindful of implicit subscripts on the designations of these five rules. For a particular system $\mathcal{S}$, one is really dealing with proofs of the form

$$\rho_{\mathcal{S}} \Sigma_1, \ldots, \Sigma_n$$

(or of the form

$$\rho_{\mathcal{S}} \Sigma_1, \ldots, \Sigma_n \theta$$

if one needs to emphasize the conclusion).

The same relativity to system affects our notation for transforms. One is really dealing with $[\Pi, \Sigma]_{\mathcal{S}}$, rather than just $[\Pi, \Sigma]$. We shall, however, suppress system subscripts whenever possible. We shall write $[\Pi, \Sigma]$ rather than $[\Pi, \Sigma]_{\mathcal{S}}$ when it is not necessary to specify any particular system or systems $\mathcal{S}$. The same holds for our designations of inference rules within our compact notations for proofs. The following notational conventions, however, for the four quantifier rules, happen to be the same for each of the Core systems $\mathbb{C}$ and $\mathbb{C}^+$ and each of the standard systems **M**, **I**, and **C**. So there is no need for any explicit subscripts $\mathcal{S}$.

A word is in order here about the level of abstraction in our Polish notation for proofs. By abstracting to this level we are able to use the same notation whether we are talking about natural deductions or sequent proofs. All we need to remember is the following correspondence between rules of natural deduction and those of sequent calculus. Note that we now write $@_L$ for ($@$:), and $@_R$ for (:$@$).

ND	$\neg$I	$\wedge$I	$\vee$I	$\to$I	$\exists$I	$\forall$I	$\neg$E	$\wedge$E	$\vee$E	$\to$E	$\exists$E	$\forall$E	EFQ	Dil
SC	$\neg_R$	$\wedge_R$	$\vee_R$	$\to_R$	$\exists_R$	$\forall_R$	$\neg_L$	$\wedge_L$	$\vee_L$	$\to_L$	$\exists_L$	$\forall_L$	T_R	Dil

6.8.2 Notation for proofs of existentials

The notation

$$\exists_R \Xi(t)$$

will denote a proof of a sequent $\Delta : \exists x \varphi$, for some formula φ with just x free. The proof is formed by an application of $\exists_R$—or, in the natural-deduction case, ($\exists$I)—to the immediate subproof Ξ, whose conclusion is the sequent $\Delta : \varphi_t^x$.

6.8.3   Notation for proofs of universals

The notation

$$\forall_R \Xi(a)$$

will denote a proof of a sequent of the form $\Delta : \forall x \varphi_x^a$, for some sentence φ involving a parameter a. The proof is formed by an application of $\forall_R$—or, in the natural-deduction case, ($\forall$I)—to the immediate subproof Ξ, whose conclusion is the sequent $\Delta : \varphi$, with no sentence in Δ involving a.

6.8.4   Notation for proofs from existentials

The notation

$$\exists_L \Xi(a)$$

will denote a proof of a sequent of the form $\exists x \varphi, \Delta : \psi$, containing no occurrences of the parameter a. The proof is formed by an application of $\exists_L$—or, in the natural-deduction case, ($\exists$E)—to the immediate subproof Ξ of the sequent $\varphi_a^x, \Delta : \psi$.

6.8.5   Notation for proofs from universals

Finally, the notation

$$\forall_L \Xi(t_1, \ldots, t_n)$$

will denote a proof of a sequent of the form $\forall x \varphi, \Delta : \psi$, for some formula φ with just x free. The proof is formed by an application of $\forall_L$—or, in the natural-deduction case, ($\forall$E)—to the immediate subproof Ξ, whose conclusion is the sequent $\varphi_{t_1}^x, \ldots, \varphi_{t_n}^x, \Delta : \psi$.

6.9   Grounding cases

Grounding cases for our inductive definition of transforms $[\Pi, \Sigma]$ are Π–Σ proof pairs whose transforms can be defined outright, without the need for any further application of the binary transformation $[\ , \ ]$.

6.9.1 Grounding cases for the three standard systems

The following table characterizes the grounding cases for proof pairs Π–Σ in any one of the three standard systems **M**, **I** or **C** (denoted in the table by $\mathcal{S}$).

Cases	Sub-cases	in which	$[\Pi,\Sigma]_\mathcal{S}$	$p[\Pi,\Sigma]_\mathcal{S}$	$c[\Pi,\Sigma]_\mathcal{S}$
$c\Pi \in p\Sigma$	Π trivial and Σ trivial	$p\Pi = \{c\Pi\}$ $= p\Sigma = \{c\Sigma\}$ $= \{A\}$	$=_{df} A$	$= \{A\} = \{A\}\cup\emptyset$ $= \{A\}\cup(\{A\}\setminus\{A\})$ $= p\Pi\cup(p\Sigma\setminus\{c\Pi\})$	$= A$ $= c\Sigma$
	Π trivial and Σ non-trivial	$p\Pi = \{c\Pi\}$	$=_{df} \Sigma$	$= p\Sigma$ $= \{c\Pi\}\cup(p\Sigma\setminus\{c\Pi\})$ $= p\Pi\cup(p\Sigma\setminus\{c\Pi\})$	$= c\Sigma$
	Π non-trivial and Σ trivial	$\{c\Pi\}=p\Sigma$ $=\{c\Sigma\}$; and $c\Pi = c\Sigma$	$=_{df} \Pi$	$= p\Pi$ $= p\Pi\cup\emptyset$ $= p\Pi\cup(p\Sigma\setminus\{c\Pi\})$	$= c\Pi$ $= c\Sigma$
$c\Pi \notin p\Sigma$	$\bot = c\Pi$	N/A	N/A	N/A	N/A
	$\bot \neq c\Pi$		$=_{df} \Sigma$	$= p\Sigma$ $= p\Sigma\setminus\{c\Pi\}$ $\subseteq p\Pi\cup(p\Sigma\setminus\{c\Pi\})$	$= c\Sigma$

We have the following transformations in grounding cases for the standard systems.

1. Π and Σ are trivial, so $\Pi = \Sigma = A$. Here $[\Pi,\Sigma] =_{df} A$.

2. Π is trivial, Σ is non-trivial, and $c\Pi \in p\Sigma$. Here $[\Pi,\Sigma] =_{df} \Sigma$.

3. Π is non-trivial, Σ is trivial, and $c\Pi \in p\Sigma$. Here $[\Pi,\Sigma] =_{df} \Pi$.

4. $\bot \neq c\Pi \notin p\Sigma$. Here $[\Pi,\Sigma] =_{df} \Sigma$.

Lemma 1. *In each grounding case for the standard systems,*

(i) $p[\Pi,\Sigma] \subseteq p\Pi \cup (p\Sigma\setminus\{c\Pi\})$; and
(ii) $c[\Pi,\Sigma] = c\Sigma$.

Proof. Clear by inspection of the table above. ☐

For the systems **M**, **I**, and **C** we have the following:

If Π connects with Σ, then in the recursive unwinding of $[\Pi,\Sigma]$ one will never generate any $[\Pi',\Sigma']$ where (Π' fails to connect with Σ' because) $c\Pi' = \bot$.

This is immediate by inspection of the forms of the rules of inference in the systems **M**, **I**, and **C**, and of the transformations for non-grounding cases to be stated below. The entries 'N/A' in the preceding table reflect the lack of any felt need, on the part of advocates of the standard systems **M**, **I**, and **C**, to consider the possibility (when eliminating cuts) of an argument-strengthening that takes the form of showing that the premises of the argument are jointly inconsistent.

6.9.2   Grounding cases for the two Core systems

Matters are different in this regard, however, with the two Core systems. This is because of their liberalized rules of $(\to I)$ and $(\vee E)$. These allow subordinate proofs for an application of one of these rules to have $\bot$ as conclusion while the main conclusion of such application is distinct from $\bot$:

$$
\begin{array}{ccc}
\dfrac{\quad}{\varphi}{\scriptstyle(i)} & \dfrac{\quad}{\varphi}{\scriptstyle(i)}\ \dfrac{\quad}{\psi}{\scriptstyle(i)} & \dfrac{\quad}{\varphi}{\scriptstyle(i)}\ \dfrac{\quad}{\psi}{\scriptstyle(i)}
\end{array}
$$

$$
\begin{array}{ccc}
\vdots & \vdots\quad\vdots & \vdots\quad\vdots \\
\dfrac{\bot}{\varphi\to\psi}{\scriptstyle(i)} & \dfrac{\varphi\vee\psi\quad\bot\quad\theta}{\theta}{\scriptstyle(i)} & \dfrac{\varphi\vee\psi\quad\theta\quad\bot}{\theta}{\scriptstyle(i)}\ .
\end{array}
$$

Hence the need to fill that 'hole in logical space' in the first table above, which **M**, **I**, and **C** studiously avoid. The Core logician provides a grounding transform for possible cases where $(c\Pi \not\ni p\Sigma$ and) $c\Pi = \bot$. This is because the Core logician *welcomes* the prospect of an argument-strengthening that takes the form of showing that the premises of the argument are jointly inconsistent.

The table below explains the grounding cases for the Core systems. The subscript $\mathbb{C}$ can also be read as $\mathbb{C}^+$.

Cases	Sub-cases	in which	$[\Pi,\Sigma]_{\mathbb{C}}$	$p[\Pi,\Sigma]_{\mathbb{C}}$	$c[\Pi,\Sigma]_{\mathbb{C}}$
$c\Pi \in p\Sigma$	Π trivial and Σ trivial	$p\Pi = \{c\Pi\}$ $= p\Sigma = \{c\Sigma\}$ $= \{A\}$	$=_{df} A$	$= \{A\} = \{A\}\cup\emptyset$ $= \{A\}\cup(\{A\}\backslash\{A\})$ $= p\Pi\cup(p\Sigma\backslash\{c\Pi\})$	$= A$ $= c\Sigma$
	Π trivial and Σ non-trivial	$p\Pi = \{c\Pi\}$	$=_{df} \Sigma$	$= p\Sigma$ $= \{c\Pi\}\cup(p\Sigma\backslash\{c\Pi\})$ $= p\Pi\cup(p\Sigma\backslash\{c\Pi\})$	$= c\Sigma$
	Π non-trivial and Σ trivial	$\{c\Pi\} = p\Sigma$ $= \{c\Sigma\}$; and $c\Pi = c\Sigma$	$=_{df} \Pi$	$= p\Pi$ $= p\Pi\cup\emptyset$ $= p\Pi\cup(p\Sigma\backslash\{c\Pi\})$	$= c\Pi$ $= c\Sigma$
$c\Pi \notin p\Sigma$	$\bot = c\Pi$		$=_{df} \Pi$	$= p\Pi$ $\subseteq p\Pi\cup(p\Sigma\backslash\{c\Pi\})$	$= c\Pi$ $= \bot$
	$\bot \neq c\Pi$		$=_{df} \Sigma$	$= p\Sigma$ $= p\Sigma\backslash\{c\Pi\}$ $\subseteq p\Pi\cup(p\Sigma\backslash\{c\Pi\})$	$= c\Sigma$

As the table shows, we have the following transformations in grounding cases for the Core systems.[5]

1. Π and Σ are trivial, so $\Pi = \Sigma = A$. Here $[\Pi,\Sigma] =_{df} A$.

2. Π is trivial, Σ is non-trivial, and $c\Pi \in p\Sigma$. Here $[\Pi,\Sigma] =_{df} \Sigma$.

3. Π is non-trivial, Σ is trivial, and $c\Pi \in p\Sigma$. Here $[\Pi,\Sigma] =_{df} \Pi$.

4. $\bot = c\Pi \notin p\Sigma$. Here $[\Pi,\Sigma] =_{df} \Pi$.

5. $\bot \neq c\Pi \notin p\Sigma$. Here $[\Pi,\Sigma] =_{df} \Sigma$.

Lemma 2. *In each grounding case for the Core systems,*

(i) $p[\Pi,\Sigma] \subseteq p\Pi \cup (p\Sigma\backslash\{c\Pi\})$; and
(ii) either $c[\Pi,\Sigma] = c\Sigma$ or $c[\Pi,\Sigma] = \bot$.

Proof. Clear by inspection of the table above. □

---

[5] Grounding transformation (4) was inadvertently omitted from Tennant [2012a], but its omission was made good in Tennant [2015b], by clause (5) on p. 248. That (4) is needed is immediate upon considering conversions (see below) in cases where Σ has two immediate subproofs, one of which (say Σ') lacks the cut sentence as a premise. In such a case, the part $[\Pi,\Sigma']$ within the converted result needs to be set equal to Σ', in order to achieve the aim behind the operation of conversion, which is (metaphorically) to 'place Π on top of' each premise occurrence of the cut sentence in Σ.

6.10 Non-grounding Π–Σ proof pairs

All non-grounding Π–Σ proof pairs satisfy the following conditions.

(i) Π is non-trivial;

(ii) Σ is non-trivial; and

(iii) Π connects with Σ, i.e. $c\Pi \in p\Sigma$

Recall that α is the dominant operator of A.

In a non-grounding case the arguments Π, Σ for the operation $[\Pi, \Sigma]$ can be represented graphically by the annotated *natural-deduction* schemata

$$
\begin{array}{cc}
\begin{array}{l} \Pi \text{ non-trivial} \\ \pi \neq \neg\text{-}E \end{array} \longrightarrow
\begin{array}{c} \Delta \\ \Pi_{\,\pi} \\ A \end{array}
& \qquad
\Sigma \text{ non-trivial} \longrightarrow
\begin{array}{c} A\,,\Gamma \\ \Sigma_{\,\sigma} \\ \theta \end{array}
\end{array}
$$

Note: the last step of Π is called π; the last step of Σ is called σ.

Equivalently, we have the *sequent-proof* schemata

$$
\begin{array}{cc}
\begin{array}{c} \Pi_{\,\pi} \\ \Delta : A \end{array}
& \qquad\qquad
\begin{array}{c} \Sigma_{\,\sigma} \\ A,\Gamma : \theta \end{array}
\end{array}
$$

where Π and Σ are non-trivial, with last steps π and σ respectively; and π is not an application of $\neg_L$.

6.11 Classification of transformations for non-grounding cases

Remember, σ is the final step of Σ, and π is the final step of Π. There is an eightfold partition of the 'non-grounding' possibilities for the ordered pair $\langle \Pi, \Sigma \rangle$. The partition is given below. A word is in order as to how it is generated. We start with the major dichotomy—whether the cut sentence A stands as the MPE of the final step σ of the proof Σ. If, on one hand, A does *not* so stand (because σ is not an elimination, or, if it is one, has as its MPE a sentence other than A), then we look at a fourfold partition of possibilities, depending on whether σ is an application of an I-rule (i.e. a Right rule), an E-rule (i.e. a Left rule), EFQ (i.e. Thinning on the Right), or Dilemma. If, on the other hand, A *does* so stand, then we look at a similar fourfold partition of possibilities, determined this time by the final step π of the proof Π. One just needs to bear in mind that even if this step is an elimination, it cannot be $\neg$E (i.e. $\neg_R$), because $c\Pi = A \neq \bot$.

6.11.1 σ does *not* have the cut sentence A as its MPE

σ is an elimination	[E-Distribution Conversions]
σ is an introduction	[I-Distribution Conversions]
σ is an application of EFQ	[EFQ-Distribution Conversions]
σ is an application of Dilemma	[Dil-Distribution Conversions]

6.11.2 σ *does* have the cut sentence A as its MPE

π is βE, $\beta \neq \neg$	[Permutative Conversions]
π is αI	[I-reductions]
π is an application of EFQ	[EFQ-reductions]
π is an application of Dilemma	[Dil-reductions]

6.12   Terminology for conversions

For all Conversions (other than EFQ-Distribution Conversions), no matter what rule $\rho_{\mathcal{S}}$ of system $\mathcal{S}$ is applied at the final step σ of Σ, we use the following definitions.

When $\rho_{\mathcal{S}}$ involves just one immediate subproof Σ_1 we define

$$[\Pi, \rho_{\mathcal{S}}\Sigma_1]_{\mathcal{S}} =_{df} \quad [\Pi, \Sigma_1]_{\mathcal{S}} \quad \text{if } s[\Pi, \Sigma_1]_{\mathcal{S}} \sqsubseteq_{\mathcal{S}} \Pi *\Sigma \;;$$
$$=_{df} \quad \rho_{\mathcal{S}}[\Pi, \Sigma_1]_{\mathcal{S}} \quad \text{if } s[\Pi, \Sigma_1]_{\mathcal{S}} \not\sqsubseteq_{\mathcal{S}} \Pi *\Sigma .$$

We abbreviate this definition by writing

$$[\Pi, \rho_{\mathcal{S}}\Sigma_1]_{\mathcal{S}} =_{df} \{\rho_{\mathcal{S}}\}[\Pi, \Sigma_1]_{\mathcal{S}} .$$

When $\rho_{\mathcal{S}}$ involves two immediate subproofs Σ_1 and Σ_2, we define

$$[\Pi, \rho_{\mathcal{S}}\Sigma_1\Sigma_2]_{\mathcal{S}} =_{df} \quad [\Pi, \Sigma_1]_{\mathcal{S}} \text{ if } s[\Pi, \Sigma_1]_{\mathcal{S}} \sqsubseteq_{\mathcal{S}} \Pi *\Sigma \;;$$
$$\text{otherwise,}$$
$$=_{df} \quad [\Pi, \Sigma_2]_{\mathcal{S}} \text{ if } s[\Pi, \Sigma_2]_{\mathcal{S}} \sqsubseteq_{\mathcal{S}} \Pi *\Sigma \;;$$
$$\text{otherwise,}$$
$$=_{df} \quad \rho_{\mathcal{S}}[\Pi, \Sigma_1]_{\mathcal{S}}[\Pi, \Sigma_2]_{\mathcal{S}} .$$

We abbreviate this definition by writing

$$[\Pi, \rho_S \Sigma_1 \Sigma_2]_S =_{df} \{\rho_S\}[\Pi, \Sigma_1]_S [\Pi, \Sigma_2]_S .$$

Note that with these definitions, it is crucial to know the exact nature of the subsequent relation $\sqsubseteq_S$ for the system S in question.

Even though it is important, for a proper understanding of the transformations to follow, to bear in mind the dependency of both rules and operations on the system S in question, we shall nevertheless suppress the subscripts S wherever possible, in the interests of a cleaner notation.

6.12.1   Distribution conversions

$$
\begin{aligned}
[\Pi, \beta_E \Sigma_1] &=_{df} & \{\beta_E\}[\Pi, \Sigma_1] \\
[\Pi, \beta_E \Sigma_1 \Sigma_2] &=_{df} & \{\beta_E\}[\Pi, \Sigma_1][\Pi, \Sigma_2] \\
[\Pi, \beta_I \Sigma_1] &=_{df} & \{\beta_I\}[\Pi, \Sigma_1] \\
[\Pi, \beta_I \Sigma_1 \Sigma_2] &=_{df} & \{\beta_I\}[\Pi, \Sigma_1][\Pi, \Sigma_2] \\
[\Pi, \text{EFQ}\, \Sigma_1] &=_{df} & \text{EFQ}\,[\Pi, \Sigma_1] \text{ (only for } \mathbf{I} \text{ and } \mathbf{C}) \\
[\Pi, Dil\, \Sigma_1 \Sigma_2] &=_{df} & \{Dil\}[\Pi, \Sigma_1][\Pi, \Sigma_2] \text{ (only for } \mathbf{C} \text{ and } \mathbb{C}^+)
\end{aligned}
$$

6.12.2   Permutative conversions

With a permutative conversion, the basic aim is to get the terminal elimination of Π (equivalently, the terminal application of a left rule in Π) to be terminal in the transform $[\Pi, \Sigma]$, so that the cut sentence A, standing as it does as the MPE of the final step of Σ, is not left as an undischarged assumption in the reduct. In order to achieve this basic aim, one has to take into account the internal structure of Π.

The effect of every permutative conversion is to reduce the combined complexity of the proofs Π and Σ to which the operation $[\,,\,]$ needs to be applied.

$$
\begin{aligned}
[\beta_E \Pi_1, \Sigma] &=_{df} & \{\beta_E\}[\Pi_1, \Sigma] \text{ if } \beta \text{ is } \wedge, \exists \text{ or } \forall ; \\
[\to_E \Pi_1 \Pi_2, \Sigma] &=_{df} & \{\to_E\}\Pi_1[\Pi_2, \Sigma] ; \\
[\vee_E \Pi_1 \Pi_2, \Sigma] &=_{df} & \{\vee_E\}[\Pi_1, \Sigma][\Pi_2, \Sigma] .
\end{aligned}
$$

Every sequence of Distribution and Permutative Conversions is clearly at most finitely long.

6.13 Reductions

6.13.1 I-reductions

I-reductions $[\Pi, \Sigma]$ are called for when:

1. π is an α-Introduction (= application of the Right rule for α) with conclusion A; and

2. σ is an α-Elimination (= application of the Left rule for α) with MPE A

Remember that both Π *and* Σ *are normal,* even when the system we are dealing with is **C**, **I**, or **M**. I-reductions are stated by revealing a certain level of structure within each of the proofs Π and Σ. So it must be borne in mind that references to Π in the definientia on the right-hand sides below are to the first argument of the $[\ ,\ ]$-operation on the left-hand side.

Note that the embedded terms of the form $[\Pi, \Sigma_i]$ serve to take care, recursively, of other possible assumption occurrences, within Σ, of the cut sentence A.

$$
\begin{aligned}
[\neg_I \Pi_1, \neg_E \Sigma_1] &=_{df} [[\Pi, \Sigma_1], \Pi_1] \\
[\wedge_I \Pi_1 \Pi_2, \wedge_E \Sigma_1] &=_{df} [\Pi_1, [\Pi_2, [\Pi, \Sigma_1]]] \\
[\vee_I \Pi_1, \vee_E \Sigma_1 \Sigma_2] &=_{df} [\Pi_1, [\Pi, \Sigma_i]], (i = 1, 2) \\
[\rightarrow_{I(a)} \Pi_1, \rightarrow_E \Sigma_1 \Sigma_2] &=_{df} [[\Pi, \Sigma_1], \Pi_1] \\
[\rightarrow_{I(b)} \Pi_1, \rightarrow_E \Sigma_1 \Sigma_2] &=_{df} [[\Pi, \Sigma_1], [\Pi_1, [\Pi, \Sigma_2]]] \\
[\exists_I \Pi_1(t), \exists_E \Sigma_1(a)] &=_{df} [\Pi_1, [\Pi, \Sigma_{1 t}^a]] \\
[\forall_I \Pi_1(a), \forall_E \Sigma_1(t_1, \ldots, t_n)] &=_{df} [\Pi_{1 t_1}^a, \ldots, [\Pi_{1 t_n}^a, [\Pi, \Sigma_1]] \ldots]
\end{aligned}
$$

6.13.2 EFQ-reductions (only for I and C)

EFQ-reductions $[\Pi, \Sigma]$ are called for when:

1. π is an application of αEFQ (= application of Thinning on the Right) with conclusion A (whose dominant operator, recall, is α);

2. σ is an α-Elimination (= application of the Left rule for α) with MPE A

Note that the conclusion of each immediate subproof Π_1 of Π below is $\perp$; and the conclusion of Σ is θ (if not $\perp$).

The obvious uniformity of output with EFQ-reductions underscores the idiosyncratic nature of EFQ. With every EFQ-reduction, including the one for negation, the internal details of Σ are irrelevant in determining $[\Pi, \Sigma]$.

Note that for **M**, $\mathbb{C}$, and $\mathbb{C}^+$ *there are no EFQ reductions.*

$$
\begin{array}{lll}
[\neg_{\mathrm{EFQ}}\Pi_1, \neg_E\Sigma_1] & =_{df} & \Pi_1 \\
[\wedge_{\mathrm{EFQ}}\Pi_1, \wedge_E\Sigma_1] & =_{df} & {\mathrm{EFQ}}\,\Pi_1\theta \\
[\vee_{\mathrm{EFQ}}\Pi_1, \vee_E\Sigma_1\Sigma_2] & =_{df} & {\mathrm{EFQ}}\,\Pi_1\theta \\
[\rightarrow_{\mathrm{EFQ}}\Pi_1, \rightarrow_E\Sigma_1\Sigma_2] & =_{df} & {\mathrm{EFQ}}\,\Pi_1\theta \\
[\exists_{\mathrm{EFQ}}\Pi_1, \exists_E\Sigma_1] & =_{df} & {\mathrm{EFQ}}\,\Pi_1\theta \\
[\forall_{\mathrm{EFQ}}\Pi_1, \forall_E\Sigma_1] & =_{df} & {\mathrm{EFQ}}\,\Pi_1\theta
\end{array}
$$

6.13.3   Dil-reductions (only for C and $\mathbb{C}^+$)

Dil-reductions $[\Pi, \Sigma]$ are called for when:

1. π is an application of Dilemma with conclusion A;

2. σ is an α-Elimination (= application of the Left rule for α) with MPE A

Note that for **M**, **I**, and $\mathbb{C}$ *there are no Dil-Reductions.*

First we need to define the following operation.

$$
\begin{array}{lll}
\langle Dil^{\varphi} \rangle \Xi\Theta & =_{df} & \Xi \qquad \text{if } \varphi \notin p\Xi; \text{ otherwise} \\
& =_{df} & \Theta \qquad \text{if } \neg\varphi \notin p\Theta; \text{ otherwise} \\
& =_{df} & Dil^{\varphi}\Xi\Theta.
\end{array}
$$

We can then perform a *Dil*-Reduction in accordance with the following definition:

$$
[Dil^A\,\Pi_1\Pi_2, \Sigma] =_{df} \langle Dil^A \rangle [\Pi_1, \Sigma]\,[\Pi_2, \Sigma].
$$

Note that in the case where both Π_1 and Π_2 prove the cut sentence A, the terms of the form $[\Pi_i, \Sigma]$ on the right-hand side already take care of all potential assumption occurrences of A in Σ. Thus there is no need for any embeddings of Π-transforms on the right-hand side.

We choose to use Dilemma as our strictly classical rule of negation for a rather interesting reason. With Dilemma, but not with CR (as sole classical rule), there is yet further epistemic gain to be had in forming reducts: one can often reduce the complexity of 'classical foci' (see §6.7.2), thereby making the sought proof 'more constructive'. (All the relevant considerations are set out in great detail in Tennant [2015b].)

6.14 Table of transformations for each system, in the terminology of natural deduction

Possible forms for $[\Pi, \Sigma]$		M	I	C	$\mathbb{C}$	$\mathbb{C}^+$

Distribution conversions

		M	I	C	$\mathbb{C}$	$\mathbb{C}^+$
$[\Pi, \neg_E \Sigma_1]$	$\{\neg_E\}[\Pi, \Sigma_1]$	•	•	•	•	•
$[\Pi, \wedge_E \Sigma_1]$	$\{\wedge_E\}[\Pi, \Sigma_1]$	•	•	•	•	•
$[\Pi, \exists_E \Sigma_1]$	$\{\exists_E\}[\Pi, \Sigma_1]$	•	•	•	•	•
$[\Pi, \forall_E \Sigma_1]$	$\{\forall_E\}[\Pi, \Sigma_1]$	•	•	•	•	•
$[\Pi, \to_E \Sigma_1 \Sigma_2]$	$\{\to_E\}[\Pi, \Sigma_1][\Pi, \Sigma_2]$	•	•	•	•	•
$[\Pi, \vee_E \Sigma_1 \Sigma_2]$	$\{\vee_E\}[\Pi, \Sigma_1][\Pi, \Sigma_2]$	•	•	•	•	•
$[\Pi, \neg_I \Sigma_1]$	$\{\neg_I\}[\Pi, \Sigma_1]$	•	•	•	•	•
$[\Pi, \vee_I \Sigma_1]$	$\{\vee_I\}[\Pi, \Sigma_1]$	•	•	•	•	•
$[\Pi, \to_I \Sigma_1]$	$\{\to_I\}[\Pi, \Sigma_1]$	•	•	•	•	•
$[\Pi, \exists_I \Sigma_1]$	$\{\exists_I\}[\Pi, \Sigma_1]$	•	•	•	•	•
$[\Pi, \forall_I \Sigma_1]$	$\{\forall_I\}[\Pi, \Sigma_1]$	•	•	•	•	•
$[\Pi, \wedge_I \Sigma_1 \Sigma_2]$	$\{\wedge_I\}[\Pi, \Sigma_1][\Pi, \Sigma_2]$	•	•	•	•	•
$[\Pi, \text{EFQ}\, \Sigma_1]$	$\text{EFQ}\,[\Pi, \Sigma_1]$	•	•			
$[\Pi, Dil\, \Sigma_1 \Sigma_2]$	$\{Dil\}[\Pi, \Sigma_1][\Pi, \Sigma_2]$	•				•

Permutative conversions

		M	I	C	$\mathbb{C}$	$\mathbb{C}^+$
$[\wedge_E \Pi_1, \Sigma]$	$\{\wedge_E\}[\Pi_1, \Sigma]$	•	•	•	•	•
$[\exists_E \Pi_1, \Sigma]$	$\{\exists_E\}[\Pi_1, \Sigma]$	•	•	•	•	•
$[\vee_E \Pi_1, \Sigma]$	$\{\forall_E\}[\Pi_1, \Sigma]$	•	•	•	•	•
$[\to_E \Pi_1 \Pi_2, \Sigma]$	$\{\to_E\}\Pi_1[\Pi_2, \Sigma]$	•	•	•	•	•
$[\vee_E \Pi_1 \Pi_2, \Sigma]$	$\{\vee_E\}[\Pi_1, \Sigma][\Pi_2, \Sigma]$	•	•	•	•	•

I-reductions

		M	I	C	$\mathbb{C}$	$\mathbb{C}^+$
$[\neg_I \Pi_1, \neg_E \Sigma_1]$	$[[\Pi, \Sigma_1], \Pi_1]$	•	•	•	•	•
$[\wedge_I \Pi_1 \Pi_2, \wedge_E \Sigma_1]$	$[\Pi_1, [\Pi_2, [\Pi, \Sigma_1]]]$	•	•	•	•	•
$[\vee_I \Pi_1, \vee_E \Sigma_1 \Sigma_2]$	$[\Pi_1, [\Pi, \Sigma_i]], (i = 1, 2)$	•	•	•	•	•
$[\to_{I(a)} \Pi_1, \to_E \Sigma_1 \Sigma_2]$	$[[\Pi, \Sigma_1], \Pi_1]$	•	•	•	•	•
$[\to_{I(b)} \Pi_1, \to_E \Sigma_1 \Sigma_2]$	$[[\Pi, \Sigma_1], [\Pi_1, [\Pi, \Sigma_2]]]$	•	•	•	•	•
$[\exists_I \Pi_1(t), \exists_E \Sigma_1(a)]$	$[\Pi_1, [\Pi, \Sigma_{1\,t}^{a}]]$	•	•	•	•	•
$[\forall_I \Pi_1(a), \forall_E \Sigma_1(t_1,...,t_n)]$	$[\Pi_{1\,t_1}^{a}, \ldots, [\Pi_{1\,t_n}^{a}, [\Pi, \Sigma_1]]\ldots]$	•	•	•	•	•

EFQ-reductions

		M	I	C	$\mathbb{C}$	$\mathbb{C}^+$
$[\neg_{\text{EFQ}} \Pi_1, \neg_E \Sigma_1]$	Π_1	•	•			
$[\wedge_{\text{EFQ}} \Pi_1, \wedge_E \Sigma_1]$	$\text{EFQ}\, \Pi_1 \theta$	•	•			
$[\vee_{\text{EFQ}} \Pi_1, \vee_E \Sigma_1 \Sigma_2]$	$\text{EFQ}\, \Pi_1 \theta$	•	•			
$[\to_{\text{EFQ}} \Pi_1, \to_E \Sigma_1 \Sigma_2]$	$\text{EFQ}\, \Pi_1 \theta$	•	•			
$[\exists_{\text{EFQ}} \Pi_1, \exists_E \Sigma_1]$	$\text{EFQ}\, \Pi_1 \theta$	•	•			
$[\forall_{\text{EFQ}} \Pi_1, \forall_E \Sigma_1]$	$\text{EFQ}\, \Pi_1 \theta$	•	•			

Dil-reductions

		M	I	C	$\mathbb{C}$	$\mathbb{C}^+$
$[Dil^A\, \Pi_1 \Pi_2, \Sigma]$	$\langle Dil^A \rangle[\Pi_1, \Sigma]\, [\Pi_2, \Sigma]$				•	•

6.15 Table of transformations for each system, in the terminology of sequent calculus

Possible forms for $[\Pi, \Sigma]$		**M**	**I**	**C**	**ℂ**	**ℂ⁺**
Distribution conversions						
$[\Pi, \neg_L \Sigma_1]$	$\{\neg_L\}[\Pi, \Sigma_1]$	•	•	•	•	•
$[\Pi, \wedge_L \Sigma_1]$	$\{\wedge_L\}[\Pi, \Sigma_1]$	•	•	•	•	•
$[\Pi, \exists_L \Sigma_1]$	$\{\exists_L\}[\Pi, \Sigma_1]$	•	•	•	•	•
$[\Pi, \forall_L \Sigma_1]$	$\{\forall_L\}[\Pi, \Sigma_1]$	•	•	•	•	•
$[\Pi, \rightarrow_L \Sigma_1\Sigma_2]$	$\{\rightarrow_L\}[\Pi, \Sigma_1][\Pi, \Sigma_2]$	•	•	•	•	•
$[\Pi, \vee_L \Sigma_1\Sigma_2]$	$\{\vee_L\}[\Pi, \Sigma_1][\Pi, \Sigma_2]$	•	•	•	•	•
$[\Pi, \neg_R \Sigma_1]$	$\{\neg_R\}[\Pi, \Sigma_1]$	•	•	•	•	•
$[\Pi, \vee_R \Sigma_1]$	$\{\vee_R\}[\Pi, \Sigma_1]$	•	•	•	•	•
$[\Pi, \rightarrow_R \Sigma_1]$	$\{\rightarrow_R\}[\Pi, \Sigma_1]$	•	•	•	•	•
$[\Pi, \exists_R \Sigma_1]$	$\{\exists_R\}[\Pi, \Sigma_1]$	•	•	•	•	•
$[\Pi, \forall_R \Sigma_1]$	$\{\forall_R\}[\Pi, \Sigma_1]$	•	•	•	•	•
$[\Pi, \wedge_R \Sigma_1\Sigma_2]$	$\{\wedge_R\}[\Pi, \Sigma_1][\Pi, \Sigma_2]$	•	•	•	•	•
$[\Pi, \text{EFQ}\, \Sigma_1]$	$\text{EFQ}\,[\Pi, \Sigma_1]$	•	•			
$[\Pi, Dil\, \Sigma_1\Sigma_2]$	$\{Dil\}[\Pi, \Sigma_1][\Pi, \Sigma_2]$			•		•
Permutative conversions						
$[\wedge_L \Pi_1, \Sigma]$	$\{\wedge_L\}[\Pi_1, \Sigma]$	•	•	•	•	•
$[\exists_L \Pi_1, \Sigma]$	$\{\exists_L\}[\Pi_1, \Sigma]$	•	•	•	•	•
$[\forall_L \Pi_1, \Sigma]$	$\{\forall_L\}[\Pi_1, \Sigma]$	•	•	•	•	•
$[\rightarrow_L \Pi_1\Pi_2, \Sigma]$	$\{\rightarrow_L\}\Pi_1[\Pi_2, \Sigma]$	•	•	•	•	•
$[\vee_L \Pi_1\Pi_2, \Sigma]$	$\{\vee_L\}[\Pi_1, \Sigma][\Pi_2, \Sigma]$	•	•	•	•	•
I-reductions						
$[\neg_R \Pi_1, \neg_L \Sigma_1]$	$[[\Pi, \Sigma_1], \Pi_1]$	•	•	•	•	•
$[\wedge_R \Pi_1\Pi_2, \wedge_L \Sigma_1]$	$[\Pi_1, [\Pi_2, [\Pi, \Sigma_1]]]$	•	•	•	•	•
$[\vee_R \Pi_1, \vee_L \Sigma_1\Sigma_2]$	$[\Pi_1, [\Pi, \Sigma_R]], (i=1,2)$	•	•	•	•	•
$[\rightarrow_{I(a)} \Pi_1, \rightarrow_L \Sigma_1\Sigma_2]$	$[[\Pi, \Sigma_1], \Pi_1]$	•	•	•	•	•
$[\rightarrow_{I(b)} \Pi_1, \rightarrow_L \Sigma_1\Sigma_2]$	$[[\Pi, \Sigma_1], [\Pi_1, [\Pi, \Sigma_2]]]$	•	•	•	•	•
$[\exists_R \Pi_1(t), \exists_L \Sigma_1(a)]$	$[\Pi_1, [\Pi, \Sigma_{1_t}^a]]$	•	•	•	•	•
$[\forall_R \Pi_1(a), \forall_L \Sigma_1(t_1,...,t_n)]$	$[\Pi_{1_{t_1}}^a, \ldots, [\Pi_{1_{t_n}}^a, [\Pi, \Sigma_1]]\ldots]$	•	•	•	•	•
T_R-Reductions						
$[\neg_{T_R}\Pi_1, \neg_L \Sigma_1]$	Π_1			•	•	
$[\wedge_{T_R}\Pi_1, \wedge_L \Sigma_1]$	$T_R \Pi_1\theta$			•	•	
$[\vee_{T_R}\Pi_1, \vee_L \Sigma_1\Sigma_2]$	$T_R \Pi_1\theta$			•	•	
$[\rightarrow_{T_R}\Pi_1, \rightarrow_L \Sigma_1\Sigma_2]$	$T_R \Pi_1\theta$			•	•	
$[\exists_{T_R}\Pi_1, \exists_L \Sigma_1]$	$T_R \Pi_1\theta$			•	•	
$[\forall_{T_R}\Pi_1, \forall_L \Sigma_1]$	$T_R \Pi_1\theta$			•	•	
Dil-reductions						
$[Dil^A \Pi_1\Pi_2, \Sigma]$	$\langle Dil^A\rangle[\Pi_1, \Sigma][\Pi_2, \Sigma]$			•		•

Observation 4. *In the foregoing tables for the ND and SC renderings of the same transformations of non-grounding cases of proof pairs Π–Σ, the definiendum reduct $[\Pi, \Sigma]$ —whose non-grounding cases are listed in the first column—is defined in terms of other reducts in the respective ways listed in the second column. Each of the latter reducts is of one of the forms $[\Pi', \Sigma']$, $[\Pi, \Sigma']$, $[\Pi', \Sigma]$, or $T_R \Pi'$, where Π' is an immediate subproof of Π, and Σ' is an immediate subproof of Σ. Thus the proof pairs involved in reducts within the definiens are of lower complexity than the definiendum reduct.*

6.16 Main results

Lemma 3. *In each non-grounding case for the standard systems, the property*

$p[\Pi, \Sigma] \subseteq p\Pi \cup (p\Sigma \setminus \{c\Pi\})$ *and* $c[\Pi, \Sigma] = c\Sigma$
(*whence* $s[\Pi, \Sigma] \sqsubseteq s(\Pi * \Sigma)$)

is preserved from the reducts in the definiens to the reduct being defined.

Proof. Clear by inspection of the definitions of the reducts in non-grounding cases. It is a straightforward exercise for the reader to supply enough detail to carry out the argument for each of the remaining definitional clauses dealing with non-grounding cases. $\square$

Now suppose that one is dealing with a Core system (either $\mathbb{C}$ or $\mathbb{C}^+$).

Lemma 4. *In each non-grounding case for the Core systems, the property*

$p[\Pi, \Sigma] \subseteq p\Pi \cup (p\Sigma \setminus \{c\Pi\})$ *and* $c[\Pi, \Sigma] = c\Sigma$ *or* $c[\Pi, \Sigma] = \bot$
(*whence* $s[\Pi, \Sigma] \sqsubseteq s(\Pi * \Sigma)$)

is preserved from the reducts in the definiens to the reduct being defined.

Proof. Clear by inspection of the definitions of the reducts in non-grounding cases. It is a straightforward exercise for the reader to supply enough detail to carry out the argument for each of the remaining definitional clauses dealing with non-grounding cases. $\square$

We may summarize as follows.

Metatheorem 5. *For $S = \mathbf{M}, \mathbf{I}$ or $\mathbf{C}$, we have*

$p[\Pi, \Sigma] \subseteq p(\Pi * \Sigma)$ *and* $c[\Pi, \Sigma] = c\Sigma$

whence also

$$s[\Pi, \Sigma] \sqsubseteq_{\mathcal{S}} s(\Pi * \Sigma).$$

Proof. By induction on the definition of $[\Pi, \Sigma]$ for the system in question. The basis step (dealing with grounding cases) is accomplished by Lemma 1. The inductive step, by Observation 4, is accomplished by Lemma 3. □

Metatheorem 6. *For $\mathcal{S} = \mathbb{C}$ or $\mathbb{C}^+$, we have*

$$p[\Pi, \Sigma] \subseteq p(\Pi * \Sigma) \text{ and either } c[\Pi, \Sigma] = c\Sigma \text{ or } c[\Pi, \Sigma] = \bot$$

whence also

$$s[\Pi, \Sigma] \sqsubseteq_{\mathcal{S}} s(\Pi * \Sigma).$$

Proof. By induction on the definition of $[\Pi, \Sigma]$ for the system in question. The basis step (dealing with grounding cases) is accomplished by Lemma 7. The inductive step, by Observation 4, is accomplished by Lemma 4. □

Observation 5. *Our proofs of Metatheorems 5 and 6 are core meta-proofs, in a theory of iterated inductive definitions. They appeal respectively to Lemmas 3 and 4. Thus they appear to involve treating the latter as cut sentences (in the meta-reasoning). So the reader might worry that upon eliminating these cuts, the resulting metaproofs might be infeasibly long and/or infeasibly time-consuming to produce, as outputs of the normalization function. But this worry is misplaced. The burden of the two Lemmas is that of verifying that in each of the cases in which a transformation applies, the appropriate subsequent relation obtains between the conclusion of the transform and the target sequent. It is easy to see that this burden of proof can be distributed across the various cases for the inductive step (as determined by the forms of Π and Σ) for the inductive proofs of the Metatheorems. This ensures that our overall proofs of the Metatheorems are suitably* direct, *by induction on the complexity of these Π–Σ proof pairs. This point about the structure of our meta-reasoning is important in meeting a certain criticism by Burgess, to be discussed in Chapter 12.*

6.17   Extraction theorems

The rules of inference stated above for the systems **I**, **C**, $\mathbb{C}$, and $\mathbb{C}^+$ permit only the formation of proofs in which major premises of eliminations stand proud. This is by now a familiar refrain. This feature has allowed us to establish Cut-admissibility for all these systems (and for **M**), in a satisfyingly uniform and purely syntactic way.

In this section we continue our exploration of what our chosen formulation of the rules affords us by way of systemic (and still syntactic) insights. We focus now on the relationship between the standard system **I** and its relevantized counterpart $\mathbb{C}$, and between the standard system **C** and *its* relevantized counterpart $\mathbb{C}^+$. There are two homologous Metatheorems to be proved, which we shall call *Extraction* Theorems.

Metatheorem 7 (Extraction of Core Proofs from Intuitionistic Proofs). *There is an effective procedure τ such that for any (normal) intuitionistic proof Π of $\Delta : \psi$, $\tau\Pi$ is a core proof of some subsequent of $\Delta : \psi$.*[6]

Proof. Straightforward by induction on the length of (normal) intuitionistic proofs Π. The procedure τ can be described as 'eliminating thinnings' from Π. We shall also say that the core proof $\tau\Pi$ has been *extracted* from the intuitionistic proof Π.

The basis step is obvious. The inductive hypothesis is to the effect that the result holds for all intuitionistic proofs less complex than Π. The inductive step is by cases according to the final step of Π. Let the final step of Π be an application of the intuitionistic rule ρ.

(i) Suppose ρ is T_R. Then the immediate subproof Π' of Π proves the sequent $p\Pi : \bot$, which by definition is a subsequent of $s\Pi$; and by IH the core proof $\tau\Pi'$ proves a subsequent of $p\Pi : \bot$. For $\tau\Pi$ take $\tau\Pi'$.

(ii) Suppose ρ is an introduction or elimination rule.

> (a) Suppose on one hand that Π' is the leftmost immediate subproof of Π for which the core proof $\tau\Pi'$ proves some subsequent of $s\Pi$. Then for $\tau\Pi$ take $\tau\Pi'$.

> (b) Suppose on the other hand that none of the immediate subproofs Π_i of Π ($1 \leq i \leq n$) is such that the core proof $\tau(\Pi'_i)$ proves a subsequent of $s\Pi$. Then one can apply the obvious *core* rule corresponding to ρ to the core proofs $\tau\Pi_1, \ldots, \tau\Pi_n$ to achieve the desired result—a core proof of some subsequent of $s\Pi$.

$\square$

Examination of (ii)(b) in all the cases involved is left to the reader. Note how directly and effortlessly our chosen forms of rules for Core Logic—in

---

[6] This result was proved in Tennant [1994b].

particular, the liberalized versions of $\to$I and $\lor$E—afford a proof of this important metatheorem.

Observation 6. *The procedure τ of Metatheorem 7 is not only effective; it is a* linear-time *algorithm.*

Metatheorem 8 (Extraction of Classical Core Proofs from Classical Proofs). *There is an effective procedure τ such that for any (normal) classical proof Π of $\Delta : \psi$, $\tau\Pi$ is a classical core proof of some subsequent of $\Delta : \psi$.[7]*

Proof. Let us assume that the system of Classical Core Logic that we are dealing with is obtained by adding the rule of Dilemma to Core Logic.

We proceed as in the proof of Metatheorem 7, needing only to examine, in the inductive step, the extra case where the final step of Π is an application of Dilemma. Suppose the classical proof Π is of the form

$$
\cfrac{
\cfrac{\overset{(i)}{\underline{\qquad}}}{\varphi, \Delta_1} \quad \begin{matrix}\Pi_1 \\ \psi\end{matrix}
\qquad
\cfrac{\overset{(i)}{\underline{\qquad}}}{\neg\varphi, \Delta_2} \quad \begin{matrix}\Pi_2 \\ \psi\end{matrix}
}{\psi}{}_{(i)}
$$

If $\varphi \notin p\tau\Pi_1$, take $\tau\Pi_1$ for $\tau\Pi$. Otherwise, if $\neg\varphi \notin p\tau\Pi_2$, take $\tau\Pi_2$ for $\tau\Pi$. It remains to consider the four cases where both $\varphi \in p\tau\Pi_1$ and $\neg\varphi \in p\tau\Pi_2$, and where, by IH, $p\tau\Pi_1 = \Delta_1', \varphi$ (for some $\Delta_1' \subseteq \Delta_1$) and $p\tau\Pi_2 = \Delta_2', \neg\varphi$ (for some $\Delta_2' \subseteq \Delta_2$). These four cases are determined by whether $\tau\Pi_i$ $(i = 1, 2)$ has ψ or $\bot$ as its conclusion.

1. If $c\tau\Pi_1 = \psi$ and $c\tau\Pi_2 = \psi$, then for $\tau\Pi$ we take the classical core proof

$$
\cfrac{
\cfrac{\overset{(i)}{\underline{\qquad}}}{\varphi, \Delta_1'} \quad \begin{matrix}\tau\Pi_1 \\ \psi\end{matrix}
\qquad
\cfrac{\overset{(i)}{\underline{\qquad}}}{\neg\varphi, \Delta_2'} \quad \begin{matrix}\tau\Pi_2 \\ \psi\end{matrix}
}{\psi}{}_{(i)}
$$

2. If $c\tau\Pi_1 = \psi$ and $c\tau\Pi_2 = \bot$, then for $\tau\Pi$ we take the classical core proof

$$
\cfrac{
\cfrac{\overset{(i)}{\underline{\qquad}}}{\varphi, \Delta_1'} \quad \begin{matrix}\tau\Pi_1 \\ \psi\end{matrix}
\qquad
\cfrac{\overset{(i)}{\underline{\qquad}}}{\neg\varphi, \Delta_2'} \quad \begin{matrix}\tau\Pi_2 \\ \bot\end{matrix}
}{\psi}{}_{(i)}
$$

---

[7] This result was proved in Tennant [1984].

Here we exploit the special extra form allowed for Dilemma in Classical Core Logic—where the negative horn proves absurd.

3. If $c\tau\Pi_1 = \bot$ and $c\tau\Pi_2 = \psi$, then for $\tau\Pi$ we take the following reduct, which is a classical core proof of some subsequent of $\Delta'_1, \Delta'_2 : \psi$, hence also of $\Delta_1, \Delta_2 : \psi$.

$$
\left[
\begin{array}{cc}
\underline{}^{(i)} & \\
\varphi, \Delta'_1 & \neg\varphi, \Delta'_2 \\
\tau\Pi_1 & , \quad \tau\Pi_2 \\
\underline{}\bot_{(i)} & \psi \\
\neg\varphi &
\end{array}
\right]
$$

4. If $c\tau\Pi_1 = \bot$ and $c\tau\Pi_2 = \bot$, then for $\tau\Pi$ we likewise take the classical core proof

$$
\left[
\begin{array}{cc}
\underline{}^{(i)} & \\
\varphi, \Delta'_1 & \neg\varphi, \Delta'_2 \\
\tau\Pi_1 & , \quad \tau\Pi_2 \\
\underline{}\bot_{(i)} & \bot \\
\neg\varphi &
\end{array}
\right]
$$

$\square$

Note that in cases (3) and (4) we shall actually have eliminated an application of Dilemma, thereby helping to make the final sought proof more constructive. There is no guarantee, however, that we shall thereby end up finding a core proof rather than a classical core proof. For either $\tau\Pi_1$ or $\tau\Pi_2$ could contain applications of Dilemma that do not get eliminated in the same way in the course of the extraction process. Moreover, as one can see in cases (3) and (4), the step of extraction with an application of Dilemma can involve recourse to the Cut-elimination operation [,], which will make the extraction algorithm worse than polynomial time, let alone linear time. Thus we cannot enter here, for the classical case, an analog of Observation 6—at least, so long as the rule we choose in order to classicize Core Logic is Dilemma.

If we choose Classical Reductio instead, matters are different. For then the outstanding case to dispose of is that in which the classical proof Π has the form

$$
\begin{array}{c}
\underline{}^{(i)} \\
\Delta, \neg\varphi \\
\Pi' \\
\underline{}\bot_{(i)} \\
\varphi
\end{array}
$$

If $\neg\varphi \notin p\tau\Pi'$, then take the classical core proof $\tau\Pi'$ for $\tau\Pi$. If, however, $\neg\varphi \in p\tau\Pi'$, then for $\tau\Pi$ take the classical core proof $\text{CR}\tau\Pi'$, discharging $\neg\varphi$.

Now we have earned the right to the analogous observation mentioned above.

Observation 7. *The procedure τ of Metatheorem 8 (for Classical Core Logic based on the rule of Classical Reductio) is not only effective; it is a* linear-time *algorithm.*

Observation 8. *Metatheorem 7 assures the orthodox intuitionistic logician that we do not need to go to the trouble of establishing the completeness of Core Logic with respect to the usual Kripkean model-theoretic semantics. Given the completeness of ordinary Intuitionistic Logic, if φ is an intuitionistic logical consequence of Δ, then there is a core proof whose premises are in Δ and whose conclusion is either φ or $\bot$. Moreover, any metaproof (and there is at least one), using Intuitionistic Logic in the metalanguage, of the completeness of Intuitionistic Logic can itself be turned into a proof in Core Logic. The latter's conclusion will be the statement of completeness, rather than $\bot$, on the reasonable consensual assumption that the axioms of the (meta)mathematics used are jointly consistent.*

Observation 9. *Metatheorem 8 assures the orthodox Classical logician that we do not need to go to the trouble of establishing the completeness of Classical Core Logic with respect to the usual Tarskian model-theoretic semantics. Given the completeness of ordinary Classical Logic, if φ is a classical logical consequence of Δ, then there is a classical core proof whose premises are in Δ and whose conclusion is either φ or $\bot$. Moreover, the metaproof, using Classical Logic in the metalanguage, of the completeness of Classical Logic can itself be turned into a proof in Classical Core Logic. The latter's conclusion will be the statement of completeness, rather than $\bot$, on the reasonable consensual assumption that the axioms of the (meta)mathematics used are jointly consistent.*[8]

6.18   Taking stock

We have offered a new way to formulate the well-known trio of Minimal Logic **M**, Intuitionistic Logic **I**, and Classical Logic **C**. We have made natural deductions isomorphic to their corresponding (single-conclusion) sequent proofs by parallelizing all elimination rules, and requiring that their major

---

[8] In the classical first-order case, the axioms of WKL$_0$ suffice. See Simpson [1999].

premises stand proud. This ensures that all proofs are *normal*. It also ensures that the sequent calculus, even in the classical case, deals only with
single-conclusion sequents.

In response to ensuing qualms about transitivity of deduction, we have
shown how to establish the Admissibility of Cut, by recursively defining an
operation $[\Pi, \Sigma]$ on normal proofs, with an eye to the special case where the
conclusion of Π is a premise of Σ. All it takes is a simple, systematic, and
decidable partition-classification of proof pairs. (This decidability keeps all
the meta-reasoning constructive.) Admissibility of Cut for both **M** and **I**
follows obviously from Admissibility of Cut for **C**. We find that in general
(when the system in question is **M**, **I**, or **C**) the reduct $[\Pi, \Sigma]$ can involve
subsetting down on the left of the overall sequent proved.

An obvious question to raise is: *Why stop there?*—Why not see if one
can enjoy even further epistemic gain, by allowing for the possibility that
reducts can involve *subsetting down on the right* in addition to subsetting
down on the left? This is exactly how one arrives at Core Logic $\mathbb{C}$ and its
classical extension $\mathbb{C}^+$.

One revisits the natural-deduction rules of **M**, and imposes, where needed,
requirements of *non-vacuous discharge* on rules that would otherwise produce fallacies of relevance. One insists also on vindicating the truth tables
for the connectives. Upon making these reforms, $\mathbb{C}$ is what results, and it
takes over from **M** as the properly relevantized version of **I**.

Thereafter, one can 'classicize' as usual, by adopting, say, Classical Reductio or Dilemma (in suitably relevantized forms). Upon doing so, we
obtain $\mathbb{C}^+$, the properly relevantized version of **C**.

It is tempting to summarize the foregoing findings as follows.

Constructively minded Johansson was off-target with **M**; he should have
hit $\mathbb{C}$. Classically minded Anderson and Belnap were off-target with **R**; they
should have hit $\mathbb{C}^+$. **I** is relevantized by $\mathbb{C}$, and **C** is relevantized by $\mathbb{C}^+$, in
exactly the same way. Cut-admissibility for $\mathbb{C}^+$ (hence also for $\mathbb{C}$) follows
effortlessly by employing a less extensive classification of proof pairs than
one had to for **C**. (This is because the Core systems eschew EFQ.) The
logical rules of the Core systems, however, are formulated in such a way
that the epistemic gains to be had can be enjoyed *both* on the left *and* on
the right.

Chapter 7

Epistemic Gain

Abstract

§7.1 explains how Core Logic avoids Lewis's First Paradox, even though it contains (∨I) and a form of (∨E) that permits core proof of Disjunctive Syllogism. The reason why Core Logic avoids Lewis is that the method of cut elimination will unearth the fact that the newly combined premises form an inconsistent set. Hence transitivity of deducibility 'fails', only to provide us with a more valuable logical insight concerning the 'target' sequent.

§7.2 explores an alternative form of definition of the formal-semantical relation ⊨ of logical consequence, according to which $A, \neg A \not\models B$. The alternative definiens captures the notion of logical consequence in a way that is entirely in keeping with the original spirit of the Bolzano–Tarski analysis. It is available as an alternative to the conventionally defined relation ⊨, should we wish to resort to it; but we stress that we can make do with the conventional definition, yet still show that (Classical) Core Logic is adequate unto *it*.

§7.3 explains once again how Core Logic eschews unrestricted Cut, and proceeds to argue that despite this eschewal (i) Core Logic is adequate for all intuitionistic mathematical deduction; (ii) Classical Core Logic is adequate for all classical mathematical deduction; and (iii) Core Logic is adequate for all the deduction involved in the empirical testing of scientific theories.

§7.4 offers a summation, and heralds the replies to critics that will be given in Chapter 12.

7.1 How Core Logic avoids Lewis's First Paradox

Lewis's First Paradox is the sequent, or argument,

$$A, \neg A : B.$$

It counts as a 'paradox' for a logical system only if it is provable in that system. The paradoxicality is obvious, in the eyes of anyone uncorrupted by a training in orthodox logic: there need not be any connection in meaning between the propositions A and B. So how could the contradiction engendered by A and $\neg A$ succeed in *provably implying* a completely unrelated proposition B?

The usual intuitionistic proof of Lewis's First Paradox amounts to simply claiming that it *does* so succeed! The proof uses the so-called Absurdity Rule, or *Ex Falso Quodlibet*, at its final step:

$$\frac{\neg A \quad A}{\dfrac{\bot}{B}} \quad .$$

In Core Logic, the first step

$$\frac{\neg A \quad A}{\bot}$$

is of course perfectly in order, being an application of $(\neg E)$. But Core Logic does not allow EFQ at all. It does not recognize the final step of the intuitionistic proof of Lewis's First Paradox as correct or permissible.

In Core Logic one *can*, however, construct the following proof:

$$\frac{\neg A \quad \dfrac{}{A}{\scriptstyle(1)}}{\dfrac{\bot}{A \to B}{\scriptstyle(1)}}$$

whose final step is an application of the 'first half' of the rule of $(\to I)$ stated above. The assumption A was actually used in the subordinate proof of $\bot$. Thus one may infer $A \to B$, and thereby discharge the assumption A. The result is a proof telling us that $A \to B$ follows logically from $\neg A$. And this is exactly what we are told also by the last two lines of the truth table for $\to$. If the antecedent φ is false, then the conditional $\varphi \to \psi$ is true:[1]

φ	ψ	$\varphi \to \psi$
T	T	T
T	F	F
F	T	T
F	F	T

---

[1] The third row, of course, overdetermines the value T for the conditional, since the truth of the consequent is also by itself sufficient for the truth of the conditional. But this does not impugn the point being made here, about the validity of $\neg A : A \to B$.

It might be thought that the First Lewis Paradox could now be proved in a slightly more roundabout way, exploiting this last proof of $\neg A : A \to B$. The suggestion from an objector might be that we should be allowed to 'assume A again', and then detach in order to obtain B. The result of doing so is displayed below. Note that the imagined objector is cooperating by using our parallelized form of $(\to E)$, which involves, alongside the major premise $A \to B$, a minor subproof of its antecedent A, and a major subproof that uses its consequent B as an *assumption* (to be discharged), and whose conclusion is then brought down as the main conclusion. In this application, the minor subproof is degenerate—a proof of A from A itself. Likewise, the major subproof is degenerate—a proof of B from B itself. But the imagined objector is *not* cooperating with us when making so bold as to place some non-trivial proof work above the major premise $A \to B$:

$$
\cfrac{\cfrac{\neg A \quad \cfrac{}{A}{}^{(1)}}{\cfrac{\perp}{A \to B}{}^{(1)} \quad \cfrac{}{A} \quad \cfrac{}{B}{}^{(1)}}}{B}{}^{(1)}
$$

By thus 'accumulating proofs'—a proof of $A \to B$ from $\neg A$, 'on top of' a proof of B from $A \to B$ and A—the objector is violating the restriction that *every major premise for an elimination must stand proud*. The major premise $A \to B$ for $(\to E)$ is not allowed to have any 'proof work' placed above it! So this last 'proof' is ill-formed. It is not a proof in Core Logic.

There is, however, a proof in the intuitionistic natural deduction system of Gentzen and Prawitz (with parallelized elimination rules) that is very close in appearance to the above attempted 'proof'; but it too is not in normal form. That intuitionistic proof is

$$
\cfrac{\cfrac{\neg A \quad \cfrac{}{A}{}^{(1)}}{\cfrac{\cfrac{\perp}{B}}{A \to B}{}^{(1)} \quad \cfrac{}{A} \quad \cfrac{}{B}{}^{(1)}}}{B}{}^{(1)}
$$

and it involves a conspicuous application of EFQ. The abnormality of this last proof consists in its having $A \to B$ stand as the conclusion of an application of $(\to I)$ and as the major premise of $(\to E)$. In the Gentzen–Prawitz system, this proof really *is* a proof—it counts as well-formed according to the rules of proof formation, albeit with the maximal occurrence of $A \to B$ within it. If we apply the reduction procedure for $\to$ in order to get rid of this maximal occurrence of $A \to B$, then we obtain the reduct

$$\frac{\neg A \quad A}{\frac{\bot}{B}}$$

which is exactly what the intuitionist first offered by way of proof of the First Lewis Paradox. And it still contains the offending application of EFQ.

The situation for the Core logician is that the following two core proofs can be constructed:

$$\frac{\neg A \quad \overline{\quad}^{(1)} A}{\frac{\bot}{A \to B}^{(1)}} \qquad \frac{A \to B \quad A \quad \overline{\quad}^{(1)} B}{B}^{(1)}$$

They cannot, however, be 'accumulated' so as to produce a core proof of the First Lewis Paradox $A, \neg A : B$. But we can apply the 'Cut-elimination' procedure of Chapter 6 to these two core proofs, so as to produce a core proof of the sequent $A, \neg A : \bot$, which is a *proper subsequent* of $A, \neg A : B$. Formally expressed:

$$\left[\frac{\neg A \quad \overline{\quad}^{(1)} A}{\frac{\bot}{A \to B}^{(1)}} \ , \ \frac{A \to B \quad A \quad \overline{\quad}^{(1)} B}{B}^{(1)} \right] = \frac{\neg A \quad A}{\bot}$$

This means that we have the benefit, in this case, of *epistemic gain*. It is better to learn (from the combination of the two proofs) that their combined premises are inconsistent, than it is to 'learn' that an arbitrary proposition B 'follows' from those premises. For B *does not* thus follow.

The same lesson emerges from an examination of a slightly different argument that Lewis himself gave for his First Paradox. Lewis in effect pointed out that we have proofs of the two sequents

$$A : A \vee B \qquad A \vee B, \neg A : B$$

and he proposed that one should 'cut through' the disjunction $A \vee B$ in order to obtain, by accumulation, a proof of $A, \neg A : B$.

Let us consider the formal situation more closely from the standpoint of an intuitionistic natural-deduction theorist in the Gentzen–Prawitz tradition. We have the two intuitionistic proofs

$$
\frac{A}{A \vee B}
\qquad\qquad
\frac{A \vee B \qquad \dfrac{\neg A \quad \overset{\displaystyle\frac{}{A}^{(1)}}{}}{\dfrac{\perp}{B}} \qquad \overset{\displaystyle\frac{}{B}^{(1)}}{}}{B}{}^{(1)}
$$

If we follow the suggestion that we 'cut through' $A \vee B$, then we are in effect forming the proof

$$
\frac{\dfrac{A}{A \vee B} \qquad \dfrac{\neg A \quad \overset{\displaystyle\frac{}{A}^{(1)}}{}}{\dfrac{\perp}{B}} \qquad \overset{\displaystyle\frac{}{B}^{(1)}}{}}{B}{}^{(1)}
$$

But this proof is not in normal form, since $A \vee B$ stands as the conclusion of an application of $(\vee I)$ and as the major premise of an application of $(\vee E)$. Applying the reduction procedure for $\vee$, the Gentzen–Prawitz theorist would obtain the reduct

$$
\frac{\neg A \quad A}{\dfrac{\perp}{B}}
$$

just as she did before. Note that it contains a conspicuous application of EFQ, which is not allowed in Core Logic.

Let us now reconsider the same formal situation more closely from the standpoint of the Core logician—rather than that of the Gentzen–Prawitz intuitionist logician, for whom EFQ is permissible. Sure enough, the same two sequents

$$
A : A \vee B \qquad\qquad A \vee B, \neg A : B
$$

are provable in Core Logic. (This is in fact one of the desirable features of

Core Logic—it preserves Disjunctive Syllogism.[2]) The two core proofs are

$$\frac{A}{A \vee B} \qquad \frac{A \vee B \quad \dfrac{\neg A \quad \overline{A}^{(1)}}{\bot} \quad \overline{B}^{(1)}}{B}{}^{(1)}$$

Note that in the second of these proofs, the first case proof for $(\vee E)$ closes off with $\bot$, and the conclusion B of the other (degenerate!) case proof is thereby brought down as the main conclusion. This is an example of the 'liberalized' rule of $(\vee E)$ in Core Logic at work. If the truth does not lie in one of the two cases, then it must lie in the other case—that is an insight that does not need EFQ to support it.

We cannot, in Core Logic, try to 'accumulate' these two proofs in order to obtain a 'proof' of the First Lewis Paradox $A, \neg A : B$. For, once again, the major premise of the elimination—in this case $A \vee B$—must stand proud, with no proof work above it. We can, however, once again apply the 'Cut-elimination' procedure of Chapter 6 to the last two proofs, and we thereby obtain, in the now familiar epistemically gainful way, the core proof

$$\frac{\neg A \quad A}{\bot}$$

telling us once again that the result of 'accumulating' the two proofs would be to produce a *proof* that the overall set of newly combined premises is inconsistent. Formally expressed:

$$\left[\frac{A}{A \vee B} \;,\; \frac{A \vee B \quad \dfrac{\neg A \quad \overline{A}^{(1)}}{\bot} \quad \overline{B}^{(1)}}{B}{}^{(1)} \right] = \frac{\neg A \quad A}{\bot}$$

Via 'cut elimination' we *learn that fact*, which could in principle lurk undetected for the logician working in the laxer system of standard Intuitionistic Logic. Once again, we have to ask: Why should one seek to have a proof of the potentially irrelevant conclusion B from those inconsistent premises anyway? Surely it is better to have learned, instead, of that inconsistency? For *that* is a matter of unquestionable epistemic gain.

---

[2] In the main alternative relevant logics, due to Anderson and Belnap and their followers, Disjunctive Syllogism is a casualty of their reforms. See Anderson and Belnap [1975]. See also some further discussion of the contrast between our approach and that of Anderson and Belnap, at §10.1.

7.2   On whether $A, \neg A \models B$

Note the shift, now, in topic. We are moving from the previous section's considerations about *deducibility* ($\vdash$) to considerations about the formal-semantical notion of *logical consequence* ($\models$).

There is a familiar rearguard reaction on the part of adherents of Intu-itionistic or Classical Logic, to the suggestion that one should give up all methods of proof that can be exploited to 'deduce' an arbitrary conclusion B from the manifestly mutually inconsistent premises $A, \neg A$. It runs like this.

> Look, this may *appear* strange and counterintuitive to you, an unschooled beginner in logic; but on the best definition we have of the formal notion of logical consequence, an arbitrary con-clusion B simply *does* follow logically from (i.e. B is a logical consequence of) the jointly unsatisfiable premises $A, \neg A$. A sen-tence φ is *defined* to be a logical consequence of a set Δ if and only if any model interpreting the extralogical vocabulary involved that makes all the members of Δ true makes φ true also. Now, when Δ is an unsatisfiable set like $\{A, \neg A\}$, it natu-rally enjoys *no models at all* making all its members true. But that in turn means that *there is no counterexample* to the claim that every model that makes all of Δ true makes φ true also! In such circumstances—absence of any counterexample—the uni-versal claim in question must be allowed to be true. So, φ *is* (by definition) a logical consequence of any inconsistent set Δ. In particular, B is a logical consequence of $\{A, \neg A\}$. You just have to learn to live with that. Moreover, since we want deducibility to capture logical consequence—that is, we want
>
> $$\Delta \models \varphi \Rightarrow \exists \Delta' \subseteq \Delta \, \exists \Pi \, \mathcal{P}(\Pi, \varphi, \Delta')$$
>
> —we have to be able to deduce B from $\{A, \neg A\}$.

It is such orthodoxy that we wish to challenge, on behalf of both Core Logic and Classical Core Logic; for both these systems lack EFQ, and do not permit any proof of B from $\{A, \neg A\}$.

Our imagined orthodox interlocutor lays considerable stress on the *con-ventional* definition of logical consequence; or at least on the conventional *method of definition* of that notion. Let us pause to consider a neglected possibility in this regard.

Does our definition of $\models$ really have to be the conventional one? The simple answer to this question is negative. Recall that we use the abbreviation

$$M \Vdash \varphi$$

for 'the model M makes φ true'.

Observation 10. *For every model M we have $M \not\models \perp$. That is, in no model is $\perp$ true.*

We use the Frobenian abbreviation

$$M \Vdash \Delta$$

for 'the model M makes every member of Δ true'. In symbols:

$$\forall \theta \in \Delta \; M \Vdash \theta.$$

Let us now address the question of how best to define the notion 'Δ logically implies φ', which is to be abbreviated as

$$\Delta \models \varphi.$$

I contend that there is no compelling reason why $\Delta \models \varphi$ should be given the conventional (metalinguistic) definition

$$\forall M (M \Vdash \Delta \implies M \Vdash \varphi).$$

For *this* particular definiens, as we well recognize, simply builds into the definiendum notion $\models$ the untoward feature that for any Δ that *has no model* and for *any* sentence φ, we shall have $\Delta \models \varphi$. On *this* would-be (but now widely adopted) definition of logical consequence, one is 'Lewised' from the outset. One will have

$$A, \neg A \models B.$$

There is an altogether different and vastly preferable way to define $\models$. It respects the fundamental insight of Bolzano and Tarski, that logical consequence is a matter of the conclusion being true under *every* interpretation *that makes all the premises true*. The alternative definition is as follows.

Definition 16.
Suppose Δ has no model. Then $\Delta \not\models \perp$.
Suppose Δ has a model. Then $\forall \varphi (\Delta \models \varphi \Leftrightarrow \forall M (M \Vdash \Delta \implies M \Vdash \varphi))$.

Let us now very carefully draw out some consequences of Definition 16.

Lemma 5. *Suppose* $\Delta \models \perp$. *Then according to Definition 16,* Δ *has no model.*

Proof. Suppose $\Delta \models \perp$. Now suppose for *reductio* that Δ has a model. So the second clause in Definition 16 applies. We obtain

$$\forall \varphi (\Delta \models \varphi \Leftrightarrow \forall M(M \Vdash \Delta \Rightarrow M \Vdash \varphi)),$$

whence, taking $\perp$ for conclusion φ, we have

$$\Delta \models \perp \Leftrightarrow \forall M(M \Vdash \Delta \Rightarrow M \Vdash \perp).$$

It follows from this and our main supposition that

$$\forall M(M \Vdash \Delta \Rightarrow M \Vdash \perp).$$

We are supposing for *reductio* that Δ has a model. Call it N. We have

$$N \models \Delta.$$

It follows from the last two displayed sentences that

$$N \models \perp.$$

But this, by Observation 10, is impossible. Our *reductio* is now completed. We conclude that Δ has no model. □

Lemma 6. *According to Definition 16, we have* $\emptyset \models \theta$ *if and only if* θ *is a logical truth—that is, if and only if* θ *is true in all models.*

Proof. It is easy to furnish a model for the empty set ('of sentences'): simply take *any* model! Whatever one's choice of model, it will make *every* member of $\emptyset$ true—because there *aren't any* members of $\emptyset$! So $\emptyset$ has a model. Thus the second clause in Definition 16 applies. We obtain, taking $\emptyset$ for Δ,

$$\forall \varphi (\emptyset \models \varphi \Leftrightarrow \forall M(M \Vdash \emptyset \Rightarrow M \Vdash \varphi)).$$

Take now the arbitrary instance

$$\emptyset \models \theta \Leftrightarrow \forall M(M \Vdash \emptyset \Rightarrow M \Vdash \theta)$$

For every model M we have $M \Vdash \emptyset$ (for the reason just explained). Thus the right-hand side of the last biconditional amounts to

$$\forall M \; M \Vdash \theta$$

It follows that

$$\emptyset \models \theta \Leftrightarrow \forall M \; M \Vdash \theta,$$

for arbitrary θ—which is what had to be proved. □

7.2.1 An important dialectical qualification

We have just seen that we can isolate and define, or formally explicate, a formal relation $\models$ of logical consequence, fundamentally in the Bolzano–Tarski spirit, such that $\neg A, A \not\models B$. The reader may therefore expect this whole study to proceed on the governing assumption that *this* relevantized notion of logical consequence is the one that will have to be employed when we come to discuss such matters as the soundness and completeness of *systems of proof* with respect to 'the' formally defined notion of logical consequence.

This is a very understandable expectation; but we do not need to meet it here. We shall actually proceed on the *concessive* assumption that the *usual* (irrelevant) notion of logical consequence is the one to which our proof systems are beholden. We shall show that *even then* the Core systems are arguably both sound and complete. As far as completeness is concerned, the considerations are subtle; we shall deal with them in §12.5.2. *But*—we do not thereby give vulnerable hostage to argumentative fortune. For, suppose the orthodox skeptic about relevance were to find any insuperable objection to our meeting the *conventional* demand that one's proof system be adequate unto whatever formal-semantical relation of logical consequence might be taken as the 'gold standard'. In such a situation—which we consider remote, if not impossible—we would still have our *alternative* definition of $\models$ to fall back on. In keeping with the exclusively proof-theoretic focus of these investigations, we can keep that fall-back strategy in our back pockets. We can refrain from foisting on the reader a heavily revised notion of logical consequence to match our revised notion of deducibility. Instead, we can keep the conventional notion of logical consequence in place, but proceed to argue that our revised notion of deducibility is still adequate unto *it*.

7.3 The status of Cut, and the methodological adequacy of the Core systems

7.3.1 The status of Cut

The lesson pressed by the Core logician is that standard systems of logic—such as Intuitionistic and Classical Logic, which allow EFQ or its sequent-calculus equivalent of 'thinning on the right'—presume that the rule of Cut can be applied in an altogether cavalier way. Proponents of Intuitionistic and Classical Logic state Cut in the following completely unrestricted form:

$$\frac{\Delta : \varphi \qquad \Gamma, \varphi : \psi}{\Delta, \Gamma : \psi}$$

One spectacularly untoward instance of this unrestricted form of Cut is

$$\frac{A : A \vee B \qquad \neg A, A \vee B : B}{A, \neg A : B}$$

Another one is

$$\frac{\neg A : A \rightarrow B \qquad A, A \rightarrow B : B}{\neg A, A : B}$$

In *each* of these instances, as we have seen, the Core logician can prove *both* the premise sequents involved. So, in order to avoid the ministrations of unrestricted Cut in the form of the ensuing First Lewis Paradox, the Core logician must moderate Cut in an appropriate way. To this end he will state only the following:

> Given any core proof Π of $\Delta : \varphi$, and given any core proof Σ of $\Gamma, \varphi : \psi$, one can effectively determine a core proof $[\Pi, \Sigma]$ of *some subsequent of* $\Delta, \Gamma : \psi$.

This is indeed a metatheorem for Core Logic; the result was proved in Chapter 6. Let us pause to reflect on some of the consequences that it secures.

7.3.2 Core Logic is adequate for intuitionistic mathematics

Consider the concern of the intuitionistic mathematician to ensure deductive progress in his discipline. He proves a lemma φ from some selection Δ of his axioms. Then he uses another, possibly different, selection Γ of his axioms, along with the lemma φ, to prove the mathematical theorem ψ. He therefore has two proofs Π and Σ, say, as follows:

$$\begin{array}{ccc} \Delta & & \varphi, \Gamma \\ \Pi & \text{and} & \Sigma \\ \varphi & & \psi \end{array} .$$

Does this mean that ψ 'really follows' from his axioms, by the lights of the Core logician?

Yes, it does—on the presumption, shared with the intuitionistic mathematician, that the combined selection Δ, Γ of axioms is consistent. The Core logician can assure the mathematician that his proof work, divided as between his proof Π of the lemma φ and his subsequent proof Σ of the theorem ψ from the lemma φ, has not been in vain. The Cut-elimination result for Core Logic assures us that there will be a core proof $[\Pi, \Sigma]$ (effectively determinable from the two proofs[3] Π and Σ that the mathematician has already provided) whose conclusion is ψ and whose premises Ξ are drawn from Δ, Γ:[4]

$$\begin{array}{c} \Xi \;\subseteq (\Delta, \Gamma) \\ [\Pi, \Sigma] \\ \psi \end{array}$$

That the desired mathematical theorem ψ will feature as the conclusion is secured by the assumption that the set Δ, Γ of mathematical axioms is consistent. For, if ψ did not survive as the conclusion of the proof $[\Pi, \Sigma]$, we would have $\bot$ as the conclusion in its stead, thereby rendering Ξ, hence also Δ, Γ, *in*consistent.[5]

7.3.3 Classical Core Logic suffices for classical mathematics

The considerations of the previous section can simply be reprised for the classical case.

7.3.4 Core Logic suffices for the experimental testing of scientific hypotheses

Consider the concern of the methodologist or natural scientist who wishes to be reassured that any empirical refutation of a scientific theory will survive the strictures of the Core logician. The scientist wishes to apply a mathematical theorem φ, proved (in pure mathematics) from the mathematical axioms Δ, by means of a proof Π, say. So her methodological concern *might* be understood as that of needing to be reassured that the appropriate system of Core Logic ($\mathbb{C}$ or $\mathbb{C}^+$) should provide sufficient deductive power to deliver the mathematical results that she wishes to apply in her overall scientific theorizing. As a first step in allaying this concern, then, let us first separate out the 'mathematical proving' as embodied in the core proof Π (in $\mathbb{C}$ or in $\mathbb{C}^+$, depending on the mathematical theorem in question). Her scientific hypotheses, along with her statements of boundary and initial conditions, and her statements of the observational evidence concerning the outcome,

---

[3] Or, rather, from the full formalizations of these proofs!

[4] We are proceeding on the assumption, which will surely be granted for the sake of this discussion, that the sought mathematical theorem ψ is not itself a theorem of *logic*! But *if it is*, we could very well learn that that is so, by 'subsetting down' on the left all the way to the empty set. That is, the proof $[\Pi, \Sigma]$ of ψ might have *all* its assumptions discharged, thereby establishing ψ as a *logical theorem*, and not just as a theorem of the branch of mathematics in question.

[5] These considerations were advanced in Tennant [1994b].

form the set Γ. By means of a second core proof Σ, say, she has proved the
sequent Γ, φ : ⊥—that is, that the observational evidence contradicts the
predictions that she had extracted (deductively) from the scientific hypotheses, along with statements of boundary and initial conditions. The picture
is as follows. There are two core proofs (certainly in ℂ⁺, possibly both in ℂ)

$$
\begin{array}{cc}
\Delta & \varphi, \Gamma \\
\Pi & \Sigma \\
\varphi & \bot
\end{array}
$$

that can be represented a little more informatively as follows:

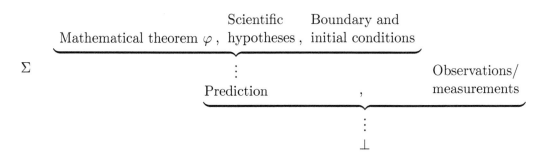

The summary logical picture is as follows:

$$
\begin{array}{ccc}
\Delta & & \varphi, \Gamma \\
\Pi & \text{and} & \Sigma \\
\varphi & & \bot
\end{array}
$$

where Γ, to repeat, contains the scientific hypotheses, the statements of
boundary and initial conditions, and the record of observations and measurements (which latter can also be called the *observational evidence*).

The proofs Π and Σ, if one were able to 'accumulate' them, would form
the overall *reductio*

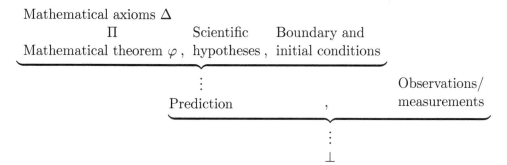

(In the proof Σ, the role played by φ would have been to furnish the right form of solution for a differential equation, say.) The Core logician, of course, cannot actually form this accumulation. So does her refutation of the scientific hypotheses really hold?

Of course it does. This is a case where the overall conclusion is already $\bot$. The Cut-elimination theorem for (Classical) Core Logic assures us that there is a proof $[\Pi, \Sigma]$ of some (possibly proper) subsequent of $\Delta, \Gamma : \bot$. Thus any 'subsetting down' of this sequent can occur only on the left. Recall that Δ is a set of mathematical axioms, which are presumably consistent. So, in the concluding sequent $\Delta', \Gamma' : \bot$ of the (possibly classical) core proof $[\Pi, \Sigma]$, the subset Γ' of Γ is non-empty. We can conclude that in (Classical) Core Logic we can construct a *reductio* of the following form:

$$\left. \begin{array}{r} \text{Mathematical axioms } \Delta' \\ \text{Scientific hypotheses} \\ \text{Statements of boundary and initial conditions} \\ \text{Observational evidence} \end{array} \right] \Gamma' \right\} \;\; \Longrightarrow \;\; \bot$$

The question now arises: what are the various possible make-ups of Γ', with regard to the three possible kinds of statement that Γ' might contain?

In a properly controlled experiment, the experimenter ensures the truth of the statements of boundary and initial conditions, and correctly records the observational evidence. So the union of these two kinds of statement is consistent—the world makes them true. Moreover, the mathematical axioms are consistent with these observational truths about the world. So the overall *reductio* $[\Pi, \Sigma]$ must have at least some of the scientific hypotheses among its premises.

What about the other two kinds of statements that might be found in Γ'—the statements of boundary and initial conditions, and of the observational evidence? If no such statements feature as premises of $[\Pi, \Sigma]$, then the scientific hypotheses have been shown to be internally inconsistent; so they would not even count as candidates for experimental testing. So we can provisionally conclude that $[\Pi, \Sigma]$ has, among its premises, at least one statement of boundary or initial conditions, and/or at least one statement of observational evidence.

Consider now the following two forms of 'reduced *reductio*'. By a 're-duced' *reductio* we mean one that produces $\perp$ as its conclusion even though it lacks premises of one of the two remaining kinds that can make up Γ'.

1. A reduced *reductio* lacking any statements of boundary or initial conditions among its premises, but having among its premises at least some statements of observational evidence, is of the following form:

$$\left. \begin{array}{c} \text{Mathematical axioms } \Delta' \\ \text{Scientific hypotheses} \\ \text{Observational evidence} \end{array} \right] \Gamma' \right\} \quad \Longrightarrow \quad \perp$$

If $[\Pi, \Sigma]$ is of this form, then it reveals that the scientific hypotheses are peculiarly strong, in predicting (wrongly) on the observational outcome, without even conditioning that prediction on any boundary or initial conditions characterizing the experimental set-up. Still, the scientific hypotheses will have been refuted in this situation.

2. A reduced *reductio* lacking any statements of observational evidence among its premises, but having among its premises at least some of the statements of boundary and initial conditions, is of the following form:

$$\left. \begin{array}{c} \text{Mathematical axioms } \Delta' \\ \text{Scientific hypotheses} \\ \text{Statements of boundary and initial conditions} \end{array} \right] \Gamma' \right\} \quad \Longrightarrow \quad \perp$$

If $[\Pi, \Sigma]$ is of this form, then it reveals that the experimental test was ill-conceived. For it is a requirement on experimental testing that the envisaged boundary and initial conditions be at least consistent with the scientific hypotheses. This is in order to ensure that the prediction is a genuine one made by the theory being tested, and not one that would 'follow' by EFQ![6]

From our consideration of these two forms of reduced *reductio* we may conclude that in the case of a well-conceived experimental test of a scientific theory, the set Γ' of premises for the *reductio* $[\Pi, \Sigma]$ must contain at least

---

[6] The reader should agree that this is an unexpected dialectical irony.

some of the scientific hypotheses, and at least some of the statements of observational evidence (records of observations and measurements). Thus the theory is, after all, refuted. Moreover, if Γ' does not contain *all* the scientific hypotheses originally at work, then we have obtained, by the process of 'cut elimination' that delivered $[\Pi, \Sigma]$, an even more focused refutation—one that tells us *more narrowly* where the theoretical mistake lies, among the scientific hypotheses that were being tested.

There are four further points worth reflecting on before we conclude this discussion of the logic needed for scientific theorizing.

First, note that in the course of normalizing so as to obtain the reduct $[\Pi, \Sigma]$ the mathematical theorem φ that was proved by Π and then 'applied' as a premise of Σ might very well be *spirited away*. It might disappear altogether—that is not appear at all in $[\Pi, \Sigma]$. So, ironically, although φ might have been needed for *pragmatic* reasons, to gain the predictive insights necessary in order to test the scientific theory, that mathematical theorem φ might turn out, in the final analysis furnished by the reduct $[\Pi, \Sigma]$, to be *strictly unnecessary* for the falsification of the theory by appeal to the observational evidence. The mathematical contribution would come, ultimately, just from the mathematical *axioms*, serving as premises for the overall disproof.

Second, we need to make good on the claim that is the title of the present subsection: *Core Logic suffices for the experimental testing of scientific hypotheses* (emphasis now added). Thus far we have hedged by including the occasional parenthetical 'Classical', or by speaking of proofs 'in $\mathbb{C}$ or in $\mathbb{C}^+$'. It is time now to furnish an argument to the effect that we can in fact, *in principle*, manage entirely within the more limited confines of Core Logic $\mathbb{C}$ itself; we do *not* have to resort to its classicized extension $\mathbb{C}^+$, *even if one (or both) of the proofs Π and Σ happen to contain strictly classical steps*.

How can this be so? Well, from the *classicist's* point of view, it would matter not one jot if she were limited to the existential quantifier $\exists$ as her sole primitive quantifier, and obliged to *define* $\forall x \varphi(x)$ as the classically equivalent $\neg \exists x \neg \varphi(x)$. So suppose this is done, so that we are working with a first order language based on the logical primitives $\neg, \wedge, \vee, \rightarrow, \exists$ and $=$. For *this* language we can state, as an easy corollary of a Gödel–Glivenko–Gentzen Theorem for $\mathbb{C}$, that

$$\Delta \vdash_{\mathbf{C}} \bot \Rightarrow \Delta \vdash_{\mathbb{C}} \bot.$$

So we need look no further than $\mathbb{C}$ for regimentations of all and any empirical refutations of scientific theories.[7]

Third, the foregoing considerations reveal just how wrong Popper was to argue (in Popper [1970]) that we can justify the choice of Classical Logic $\mathbb{C}$ as the right logic on the (alleged) grounds that *only* by adopting $\mathbb{C}$ would we be able to provide the *strongest possible* tests of our scientific theories against possible observational evidence. Indeed, it is ironic that he *also* insisted that, for the natural scientist, the correct regimentation of a universal hypothesis 'All Fs are Gs' is $\neg\exists x(Fx \wedge \neg Gx)$ (with this to be read, in Popperian falsificationist fashion, as the claim that we will not find any counterexample to the generalization in question). For as we have seen, the eschewal of $\forall$ in one's language as a primitive allows for the most direct application possible of the Gödel–Glivenko–Gentzen Theorem to refute Popper's case for the choice of $\mathbb{C}$ as *the* (allegedly) single most adequate logic for scientific theory-testing.[8]

Fourth and finally, there is an important methodological reflection that has (inexplicably) been overlooked, despite the foregoing illumination of

---

[7] The Gödel–Glivenko–Gentzen Theorem for a system $\mathcal{S}$ states, for the full language $\neg, \wedge, \vee, \rightarrow, \exists, \forall$, and $=$, that

$$\Delta \vdash_{\mathbb{C}} \varphi \Rightarrow \neg\neg(\Delta^{\forall\neg\neg}) \vdash_{\mathcal{S}} \neg\neg(\varphi^{\forall\neg\neg})$$

where $\theta^{\forall\neg\neg}$ is the result of inserting an occurrence of $\neg\neg$ immediately after every universal quantifier occurrence in θ. Surprising substituends for $\mathcal{S}$ include $\mathbf{M}$, $\mathbf{I}$ and $\mathbb{C}$. The original proof of the Gödel–Glivenko–Gentzen Theorem for $\mathbf{I}$ was constructive; and the Extraction Theorem (concerning $\mathbb{C}$-proofs from $\mathbf{I}$-proofs) is also constructively provable. And these two results are all one needs in order to obtain a constructive proof of the Gödel–Glivenko–Gentzen Theorem for $\mathbb{C}$. Hence (by Extraction again, this time at the level of one's metalogic) the Gödel–Glivenko–Gentzen Theorem for $\mathbb{C}$ is provable using $\mathbb{C}$ as one's metalogic.

When the conclusion φ happens to be $\perp$, the result becomes

$$\Delta \vdash_{\mathbb{C}} \perp \Rightarrow \neg\neg(\Delta^{\forall\neg\neg}) \vdash_{\mathcal{S}} \perp.$$

When $\forall$ is then dropped as a primitive, the result becomes

$$\Delta \vdash_{\mathbb{C}} \perp \Rightarrow \neg\neg(\Delta) \vdash_{\mathcal{S}} \perp.$$

Finally, upon observing that in the system $\mathcal{S}$ any sentence θ implies its own double negation:

$$\theta \vdash_{\mathcal{S}} \neg\neg\theta$$

we obtain the corollary

$$\Delta \vdash_{\mathbb{C}} \perp \Rightarrow \Delta \vdash_{\mathcal{S}} \perp.$$

[8] These points were made in Tennant [1985].

how 'everything can be done in $\mathbb{C}$' when it comes to deriving the mathematics that is to be applied, and exploring the consequences of the evidence turning out the way it does. There has been a debate in the learned journals about the alleged need to be *classical* in one's deductive reasoning in order to prove a particular mathematical theorem that is allegedly indispensable for the derivation of the empirical and testable consequences of a particular scientific theory.[9] The theorem in question is Gleason's Theorem; and the scientific theory in question is Quantum Mechanics. What the foregoing considerations reveal is that the debate over the alleged constructive provability or non-provability of Gleason's Theorem, and over its alleged indispensability for the derivation of certain predictive consequences of Quantum Mechanics, is *entirely beside the point*. For Gleason's Theorem, even if proved *classically* (via Π) and (pragmatically) 'necessarily' involved in the derivation Σ of observational consequences of QM, can nevertheless be like the 'disappearing φ' discussed above. The ultimate disproof $[\Pi, \Sigma]$ of the scientific theory, *modulo* the observational evidence, can be rendered as a disproof *in Core Logic* $\mathbb{C}$. And Gleason's Theorem itself, regardless of its constructive or non-constructive provenance, could well prove to be a ladder that can be kicked away. These reflections, writ large across all mathematical theorems, like Gleason's, that might have only strictly classical proofs and that find pragmatic application in our scientific theorizing, provide, in effect, a *pragmatic* justification for availing oneself fully of *strictly classical* mathematical results *even if* one is a convinced constructivist as far as mathematics itself is concerned. For whatever empirical refutations one thereby achieves can always, in principle, be rendered in Core Logic $\mathbb{C}$, using only the mathematical axioms in question.

7.4 Replies to critics: an anticipation

The upshot of the foregoing considerations is that the imagined 'loss of unrestricted transitivity' that is so often adduced as an alleged defect to anyone contemplating the adoption of Core Logic is more than offset by the great epistemic gains that are potentially afforded in the two main areas of application of deductive reasoning: pure mathematics, and the empirical testing of scientific theories.

There is really no deductive loss to be suffered at all, in restricting ourselves to Core Logic. To think otherwise is to maintain that a long-held prin-

---

[9] See Hellman [1993a], Hellman [1993b], Bridges [1995], Hellman [1997], Bridges [1999], and Bridges and Richman [1999].

ciple of standard logic (namely, *absolutely unrestricted* Cut) is completely invulnerable to careful rethinking. It is not. The amount of transitivity of deduction afforded by the Cut-elimination result for Core Logic is just right: it gets the deductive job done; and it rules out, in a principled way, a lot of quite unjustifiable, irrelevantist nonsense (the First Lewis Paradox and its ilk).

The reactions thus far in the literature to the proposal that the principle of *absolutely unrestricted* Cut be suitably moderated have struck me as both uninformed as to the details proposed, and uncomprehending of the general logical and methodological issues involved. Replies to some of these critics are long overdue. They will be given in Chapter 12.

Chapter 8

Truthmakers and Consequence

Abstract

§8.1 explains and compares Tarski's notion of logical consequence (preservation of truth in all models) with that of Prawitz (transformability of warranted assertability with respect to all inferential bases). We use the latter as our point of departure for a definition of consequence in terms of the transformability of truthmakers relative to all models.

§8.2 shows that a sentence's Tarskian truth-in-M coincides with its having an M-relative truthmaker.

§8.3 shows that an M-relative truthmaker can serve as a winning strategy or game plan for player T in the 'material game' played on that sentence against the background of the model M.

§8.4 explores variations in the definition of truthmaker-based consequence resulting from various possible conditions on the domain. We enter conjectures about the soundness and completeness of Classical Core Logic with respect to the notion of consequence that results when the domain is required to be decidable.

§8.5 raises the question whether the truthmaker semantics threatens a slide to realism.

§8.6 and §8.7 work with concrete and detailed examples of core proofs whose premises are supplied with M-relative truthmakers; and show how the latter can be systematically transformed into a truthmaker for the proof's conclusion.

§8.8 takes stock by way of preparation to establish that these transformations of truthmakers can always be effected.

8.1 Defining logical consequence in terms of truthmakers

The classical conception of logical consequence, due to Tarski [1956; first published in 1936], requires only *preservation of truth*.

Definition 17 (Tarski consequence).

> ψ *is a logical consequence of* Δ
>
> > *if and only if*
>
> *for every interpretation* M*, if every member of* Δ *is true in* M*, then* ψ *is true in* M *also.*

It is worth noting that the foregoing definiens (the right-hand side of the biconditional) is often written as

> for every interpretation M, if M makes every member of Δ true, then M makes ψ true also.

This invites the reader to (mis)construe the model M itself as a truthmaker. But that would be a serious mistake, resulting from taking the gloss 'M makes φ true' much too literally. The model M is too gross an object to serve as a genuine (even M-relative!) truthmaker. Moreover, on this (mis)construal, M would be serving as *the* truthmaker for each and every sentence φ that is true in it. But the whole point of a more refined theory of truthmakers is to be able to distinguish (with respect to some same model M, whatever it may be) the *different* truthmakers that can be had by one and the same sentence φ that is true in M; and, of course, also to distinguish among the different truthmakers that can be had by different sentences. Thinking of M as the sole M-relative truthmaker would be to obliterate all such possible refinements. In the special case where M were the actual world, it would be like taking the Great Fact as the only truthmaker for each and every truth.

The foregoing model-theoretic, or Tarskian, definition of logical consequence can be stated more formally as follows:

$$\Delta \models \psi$$

$$\Leftrightarrow$$

$$\forall M([\forall \varphi \in \Delta \ M \Vdash \varphi] \Rightarrow M \Vdash \psi)$$

This definition yields no insight, however, into how the grounds of truth-in-M of the conclusion ψ might be related to, or determinable from, the grounds of truth-in-M of the premises in Δ (relative to any of the models M interpreting the extralogical vocabulary involved).[1]

There is an alternative conception of logical consequence, fashioned for Intuitionistic Logic (and for finite sets of premises), which sought to address what was taken to be this deficit of insight, and which is owed to Prawitz [1974].

Definition 18 (Prawitz consequence $\models_P$).

> ψ *is a logical consequence of* $\varphi_1, \ldots, \varphi_n$
>
> > *if and only if*
>
> *there is an effective method* f *such that for every interpretation* M, *and for all* M-*warrants* $\pi_1, \ldots, \pi_n$ *for* $\varphi_1, \ldots, \varphi_n$ *respectively,* $f(\pi_1, \ldots, \pi_n)$ *is an* M-*warrant for* ψ.

More formally:

$$\varphi_1, \ldots, \varphi_n \models_P \psi$$

$$\Leftrightarrow$$

$$\exists \textit{eff.} f \, \forall M \, \forall \pi_1 \ldots \forall \pi_n [(\mathcal{P}_M(\pi_1, \varphi_1) \wedge \ldots \wedge \mathcal{P}_M(\pi_n, \varphi_n)) \Rightarrow \mathcal{P}_M(f(\pi_1, \ldots, \pi_n), \psi)]$$

Here, $\mathcal{P}_M(\pi, \varphi)$ means that the construction π is an M-warrant for the sentence φ. On Prawitz's account, M is taken to be a so-called *basis* consisting of atomic rules of inference, which together provide some measure of interpretation of the primitive extra-logical expressions involved in them. Prawitz's conception of warrant is anti-realist in inspiration. Warrants are the canonical proofs that Dummett marshals in arguments that are intended to motivate the adoption of anti-realist semantics. Dummett argues that one should regard as licit only the *recognizable* obtaining of the conditions of truth of a sentence—such as is displayed by a canonical proof. And, given that constraint on manifestation of understanding, he argues further that the Principle of Bivalence is infirmed.[2]

So it is important to realize that Prawitz is not in the business of distinguishing metaphysical or ontological grounds for truth (in the constitutive

---

[1] The word 'determinable' here is to be taken in a constitutive, not epistemic, sense.

[2] For a critique of this 'manifestation argument', and an attempt to improve upon it, see Chapters 6 and 7 of Tennant [1997].

sense essayed here) from justificatory grounds. His warrants are epistemic objects—so they have to be finite. And for Prawitz, as for Dummett, the truth of any claim *consists in* its possessing such a warrant. As an anti-realist, Prawitz makes no distinction between the constitutive grounds that a realist might believe in, and the contrasting epistemic grounds that are the only grounds the anti-realist can (in principle, even if not in practice) recognize as obtaining. But that should not deter us from trying to appreciate an important feature of Prawitz's characterization of logical consequence, and from trying to apply it profitably in our own characterization of the role that truthmakers can be made to play in a notion of logical consequence more congenial to the realist.

On Prawitz's definition of logical consequence, one is able to interpret the normalization theorem for intuitionistic natural deductions as furnishing a proof of the *soundness* of that proof system. For the normalization procedure (call it ν) itself qualifies as the effective method f called for on the right-hand side of his definition. To see this, suppose one has an intuitionistic natural deduction Π of the conclusion ψ from (possibly logically complex) premises $\varphi_1, \ldots, \varphi_n$. We need to convince ourselves that Π will preserve warranted assertibility from its premises to its conclusion. So suppose further that one furnishes M-warrants $\pi_1, \ldots, \pi_n$ (which take the form of closed canonical proofs using atomic rules in M) for $\varphi_1, \ldots, \varphi_n$ respectively. Append each warrant to the assumption occurrences of its conclusion in Π:

$$
\begin{array}{ccc}
\pi_1 & \cdots & \pi_n \\
\varphi_1 & & \varphi_n
\end{array}
$$
$$
\underbrace{}
$$
$$
\Pi
$$
$$
\psi
$$

Then by normalizing the result one obtains an M-warrant π for the overall conclusion ψ:

$$
\begin{array}{ccc}
\pi_1 & \cdots & \pi_n \\
\varphi_1 & & \varphi_n
\end{array}
\qquad \text{normalizes to} \qquad \pi
$$
$$
\underbrace{} \qquad\qquad\qquad\quad
$$
$$
\Pi \qquad\qquad\qquad \mapsto \qquad\qquad \psi \quad , \qquad \text{where } \pi = \nu(\pi_1, \ldots, \pi_n, \Pi)
$$
$$
\psi
$$

What the foregoing reasoning shows is that

$$
\varphi_1, \ldots, \varphi_n \vdash_I \psi \quad \Rightarrow \quad \varphi_1, \ldots, \varphi_n \models_{\mathbf{P}} \psi .
$$

This is the promised soundness result for intuitionistic deduction, with respect to Prawitz's notion of logical consequence.

The converse of this result,

$$\varphi_1, \ldots, \varphi_n \models_{\mathbf{P}} \psi \;\Rightarrow\; \varphi_1, \ldots, \varphi_n \vdash_I \psi,$$

remains as Prawitz's (as yet unproved) *completeness conjecture* for Intuitionistic Logic.

A very desirable feature of Prawitz's definition of consequence is how it *Skolemizes* (as $\nu(\pi_1, \ldots, \pi_n, \Pi)$) that which is responsible for the (*M*-relative) truth of the conclusion ψ. On the classical conception, one is given to understand only that *if* the premises of the proof are true, *then* its conclusion ψ is true also. This does not yet, however, give one any idea as to how any (*M*-relative) truthmakers for the conclusion ψ might depend—via the proof in question—on (*M*-relative) truthmakers for the premises $\varphi_1, \ldots, \varphi_n$.

The desirable Skolemizing feature of Prawitz's definition can, however, be taken over by the truthmaker theorist even if in the service of a more realist rather than anti-realist conception of truth and consequence. The truthmaker theorist can define logical consequence as follows. (Recall the discussion of nomenclature in §3.4.4.)

Definition 19 (Truthmaker consequence).

> ψ *is a logical consequence of* $\varphi_1, \ldots, \varphi_n$
>
> > *if and only if*
>
> *there is a 'quasi-effective' method* f *such that for every interpretation* M, *and for all M-relative truthmakers* $\pi_1, \ldots, \pi_n$ *for* $\varphi_1, \ldots, \varphi_n$ *respectively,* $f(\pi_1, \ldots, \pi_n)$ *is an M-relative truthmaker for* ψ.

More formally:

$$\varphi_1, \ldots, \varphi_n \models_{\mathbf{T}} \psi$$

$$\Leftrightarrow$$

$$\exists q.\text{-}eff.f \,\forall M \forall \pi_1 \ldots \forall \pi_n [(\mathcal{V}_M(\pi_1, \varphi_1) \wedge \ldots \wedge \mathcal{V}_M(\pi_n, \varphi_n)) \Rightarrow \mathcal{V}_M(f(\pi_1, \ldots, \pi_n), \psi)].$$

Here, $\mathcal{V}_M(\pi, \psi)$ means that the construction π is an *M*-relative truthmaker for the sentence ψ.

We have written 'quasi-effective method f' because of problems posed by the infinite case. In the finite case, the method will certainly be effective. But in the infinite case we have to countenance definable operations on infinite sets of constructions, and these by definition cannot be effective (in the sense in which Church's Thesis states that all effective functions are recursive functions). The functions or methods that we shall employ, however, will be 'effective-looking' in the infinite case, because of the obvious, orderly way in which one 'follows the recipes' for truthmaker transformation within the infinite context. They are recipes obtained by smooth extrapolation from the finite case, preserving the main feature of well-foundedness of the relation on which the recursive definition is based. (Bear in mind: our truthmakers and falsitymakers, as abstract trees, are *finitely deep*, albeit possibly infinitely sideways branching. So the 'sub-evaluation' relation is well-founded.) In §8.6 we shall give several simple finite examples of how this truthmaker transformation is effected, by an appropriate analog of the normalization procedure for proofs.

When *M*-relative truthmakers are at hand for the premises of a formal core proof, we are tempted to think of a construction of a mixed kind—a combination of the truthmakers 'above', and the formal proof 'below'—that invites an analog of the *normalization process* of the kind familiar to a proof theorist.

The recipe for normalizing is much the same in the truthmaker setting as it is in the proof setting.[3] One treats applications of rules for evaluating-as-true as though they are applications of introduction rules, and similarly one treats applications of rules for evaluating-as-false as though they are applications of elimination rules. Only now, one has to countenance the possibility that one will be substituting *individuals* for free occurrences of variables in formulae, thereby forcing an erstwhile formal proof to involve *saturated formulae*.

It is the Skolemizing nature of the construction of grounds for the truth of the conclusion of a deductive proof, out of grounds for the truth of its premises, which underlies one's ability to focus on what is relevant. We are now in a better position to appreciate how it is that a formal deduction enables one to see how the truth of a conclusion is contained in that of the premises—not, as Frege once wrote,[4] as beams are contained in a house, but rather as a plant is contained in the seed from which it grows. And we are in a position to explicate this insight for the benefit of the Fregean *realist*, because of the fundamentally *proof-theoretic* way that we conceive of (model-relative) truthmakers and falsitymakers.

---

[3] This remark holds even when we consider *infinitary* truthmakers 'above', in combination with (perforce finitary) proofs 'below'. In the current illustration, however, we are dealing with finitary truthmakers.

[4] See Frege [1884; reprinted 1961], §88.

8.2   Truthmakers and Tarskian truth

Truthmakers and falsitymakers can contain steps involving subordinate truth- or falsitymakers, one for each individual in the domain D of the model M under consideration. For this reason, the formal definition of a truthmaker (and of a falsitymaker) should show that the notion is parametrized by D. Thus $\mathcal{V}(\Pi, \varphi, M, D)$ would be the appropriately detailed rendering of 'Π is a truthmaker for φ relative to the model M with domain D'.

Metatheorem 9. *Modulo a metatheory which contains the mathematics of $\overline{\overline{D}}$-furcating trees of finite depth, we have, for all models M with domain D,*

$$\exists \Pi \; \mathcal{V}(\Pi, \varphi, M, D) \quad \Leftrightarrow \quad \varphi \text{ is true in } M$$

where the right-hand side is in the sense of Tarski.[5]

Proof. The proof is by the obvious induction on the complexity of the sentence (or saturated formula) φ. Note that the proof is constructive, provided only that the M-relative truthmakers for universals and the M-relative falsitymakers for existentials, in the case where D is infinite, can be assumed to exist, courtesy of the background metamathematics. This observation affects the right-to-left direction in the relevant cases of the inductive step.   □

Corollary 4. *Logical consequence $\models$ in the Tarskian sense of Definition 17 coincides with logical consequence $\models_{\mathbf{T}}$ in the sense of Definition 19 in terms of quasi-effective transformability of truthmakers, but with 'quasi-effective' interpreted as imposing no restriction at all on the Skolem function involved.*

Proof. Suppose $\varphi_1, \ldots, \varphi_n \models \psi$. Let D be an arbitrary domain, and let M be a model with domain D. Let $\pi_1, \ldots, \pi_n$ be M-relative truthmakers for $\varphi_1, \ldots, \varphi_n$ respectively. By Metatheorem 9, $\varphi_1, \ldots, \varphi_n$ are true in M in the Tarskian sense. Hence ψ is true in M in the Tarskian sense. By Metatheorem 9 again, there is an M-relative truthmaker for ψ. Choose one such, by whatever method will ensure uniqueness of choice. Take that choice as the value $f(\pi_1, \ldots, \pi_n)$. We thereby 'construct' (in a possibly non-effective manner) a function f that can serve, in accordance with Definition 19, for the conclusion that $\varphi_1, \ldots, \varphi_n \models_{\mathbf{T}} \psi$.

For the converse, suppose $\varphi_1, \ldots, \varphi_n \models_{\mathbf{T}} \psi$. By Definition 19, let f be the function that will transform, for any model M, M-relative truthmakers

---

[5] We abbreviated this right-hand side as $M \Vdash \varphi$ in §8.1.

for $\varphi_1, \ldots, \varphi_n$ into an M-relative truthmaker for ψ. Let M' be any model in which $\varphi_1, \ldots, \varphi_n$ are true in M' in the Tarskian sense. By Metatheorem 9, there are M'-relative truthmakers $\pi_1, \ldots, \pi_n$, say, for $\varphi_1, \ldots, \varphi_n$. Hence $f(\pi_1, \ldots, \pi_n)$ is an M'-relative truthmaker for ψ. By Metatheorem 9 again, ψ is true in M in the Tarskian sense. Hence $\varphi_1, \ldots, \varphi_n \models \psi$. $\qquad\square$

When the domain D is infinite, the furcations within truthmakers and falsitymakers that are called for in connection with the rules $(\exists\mathcal{F})$ and $(\forall\mathcal{V})$ will involve exactly as many branches as there are individuals in the domain. Such infinitary trees are not, of course, surveyable; but they are well-defined mathematical objects that can serve as representations of truthmakers and of falsitymakers for quantified sentences speaking of the infinitely many individuals in D. They should be congenial to the metaphysical and semantic realist, who entertains no skeptical qualms about the existence of infinitary objects, or about determinate truth values for claims involving quantification over infinite, unsurveyable domains.

8.3   Truthmakers as game plans

There is another approach to defining *truth in a model*, which is reasonably well established in the literature: the *game-theoretic* approach advanced originally in Hintikka [1973]. Hintikka conceived of the game as being played *on* a sentence, by *Myself* against *Nature*. As reworked in Tennant [1978], the game is between two players, both of them (presumably) rational agents, who at any state of play are subject to a (one–one) *role assignment* R as Player T or Player F. Example: one might have $R(T) = $ John, and $R(F) = $ Mary. If John and Mary 'swap hats' as T and F, then the new role assignment is called $\overline{R}$.

A model M is assumed given, as background for their play. They proceed to play, against this background M, using sentences (or, more generally: formulae) φ. Player $R(T)$ (resp. $R(F)$) is understood to be maintaining that the sentence φ is *true* (resp. *false*) in M. More generally, if φ is a formula with free variables $x_1, \ldots, x_n$, and if f is an assignment to these variables of individuals from the domain of M, then $R(T)$ (resp. $R(F)$) is understood to be maintaining that the formula φ is *true of* (resp. *false of*) those individuals, respectively, in M. Such a 'state of play' may be characterized as $[\varphi, M, f, R]$. If play begins with a *sentence* φ (which of course has no free variables, hence no need of any assignment of individuals to variables), then the state of play is $[\varphi, M, \emptyset, R]$—using the *null assignment* $\emptyset$.

If f is an assignment of individuals to variables, then $f(x/\alpha)$ is defined to be the result of changing or extending f so that $f(x) = \alpha$.

Play proceeds by strict constitutive rules concerning who is entitled to make a move. Making a move in state of play $[\varphi, M, f, R]$ consists in making a choice—either of an immediate subformula φ, or of an individual in the domain of M. As a result of that choice, a new state of play is determined. The constitutive rules governing play are as follows.

1. In state of play $[P(t_1, \ldots, t_n), M, f, R]$, where P is a primitive predicate and $t_1, \ldots, t_n$ are singular terms respectively denoting individuals $\alpha_1, \ldots, \alpha_n$ relative to f: if $P(\alpha_1, \ldots, \alpha_n)$ holds in M, then $R(T)$ wins; otherwise, $R(F)$ wins.

2. State of play $[\neg\psi, M, f, R]$ must be succeeded by state of play $[\psi, M, f, \overline{R}]$ (so neither player gets to make any choice: they simply 'swap hats' as player T and player F).

3. In state of play $[\psi \vee \theta, M, f, R]$, player $R(T)$ chooses either $[\psi, M, f, R]$ or $[\theta, M, f, R]$ as the next state of play.

4. In state of play $[\psi \wedge \theta, M, f, R]$, player $R(F)$ chooses either $[\psi, M, f, R]$ or $[\theta, M, f, R]$ as the next state of play.

5. In state of play $[\psi \rightarrow \theta, M, f, R]$, player $R(T)$ chooses either $[\psi, M, f, \overline{R}]$ or $[\theta, M, f, R]$ as the next state of play.

6. In state of play $[\exists x \psi, M, f, R]$, player $R(T)$ chooses an individual α from the domain of M; the next state of play is $[\psi, M, f(x/\alpha), R]$.

7. In state of play $[\forall x \psi, M, f, R]$, player $R(F)$ chooses an individual α from the domain of M; the next state of play is $[\psi, M, f(x/\alpha), R]$.

Obviously and degenerately, the possessor of a winning strategy on an atomic formula A is the player who wins. We can render this as follows:

$$\mathcal{W}(A, M, f, R) = R(T) \text{ if } f \text{ satisfies } A \text{ in } M;$$
$$= R(F) \text{ otherwise.}$$

It is easy to see that

$$\mathcal{W}(\neg\psi, M, f, R) = \mathcal{W}(\psi, M, f, \overline{R}),$$

whence

$$\mathcal{W}(\neg\psi, M, f, R) = R(T) \text{ if } \mathcal{W}(\psi, M, f, \overline{R}) = \overline{R}(F);$$
$$= R(F) \text{ otherwise.}$$

We have also

$$W(\psi \vee \theta, M, f, R) = R(T) \text{ if either } W(\psi, M, f, R) = R(T)$$
$$\text{or } W(\theta, M, f, R) = R(T);$$
$$= R(F) \text{ otherwise}$$

and

$$W(\psi \wedge \theta, M, f, R) = R(F) \text{ if either } W(\psi, M, f, R) = R(F)$$
$$\text{or } W(\theta, M, f, R) = R(F);$$
$$= R(T) \text{ otherwise}$$

and

$$W(\psi \rightarrow \theta, M, f, R) = R(T) \text{ if either } W(\psi, M, f, \overline{R}) = \overline{R}(F)$$
$$\text{or } W(\theta, M, f, R) = R(T);$$
$$= R(T) \text{ otherwise.}$$

Finally, for the quantifiers we have

$$W(\exists x \psi, M, f, R) = R(T) \text{ if for some } \alpha \in M, \ W(\psi, M, f(x/\alpha), R) = R(T)$$
$$= R(F) \text{ otherwise}$$

and

$$W(\forall x \psi, M, f, R) = R(F) \text{ if for some } \alpha \in M, \ W(\psi, M, f(x/\alpha), R) = R(F)$$
$$= R(T) \text{ otherwise.}$$

It is now an easy matter to note that, whatever the role assignment ρ might be, we can replace, in the foregoing, equations of the form

$$W(\varphi, M, f, \rho) = \rho(T)$$

with

$$f \text{ satisfies } \varphi \text{ in } M,$$

and equations of the form

$$W(\varphi, M, f, \rho) = \rho(F)$$

with

$$f \text{ does not satisfy } \varphi \text{ in } M;$$

and thereby produce a Tarskian definition of satisfaction (hence of truth) in a model M.

That much has been in the folk lore for some time. We have known that 'the possessor-of-a-winning-strategy at state of play $[\varphi, M, f, R]$ is $R(T)$'

is equivalent to

'f satisfies φ in M'.

That, however, is as far as it has gone. T stands for a player in a particular role (that of asserter) at the state of play in question; and that player's having a winning strategy coincides with the formula φ's being (in M) *true of* the individuals assigned by f to its free variables. In the special case where φ is a sentence (with no free variables), the coincidence is with φ's being *true* (in M).

A player's 'having' a winning strategy is a very thin notion. It really only means that there is a winning strategy to be had by that player, even if she is unaware of it (*de re*) and unaware *that it exists* (*de dicto*). That is why we have employed hyphens in 'possessor-of-a-winning-strategy'.

Now, however, we can delete the hyphens, and get our hands on the winning strategies themselves. *They are none other than M-relative verifications of φ.* Such a verification provides a 'game plan' to the player who starts out as player T. It discloses how to make unerring choices of immediate subformulae, or of individuals to be assigned to variables, in such a way as to ensure a win by the close of play. By the same token, an M-relative *falsification* provides a 'game plan' to the player who starts out as player F.

Truthmakers, on our conception of them, turn out to be *reifications* of winning strategies for player T. They provide a 'missing link' between Tarskian truth (which consists in sentence φ having *a* verification) and Hintikkan truth (which consists in player T possessing *a* winning strategy on φ). Our truthmakers, or verifications, *are* (or at least *codify*) winning strategies.

Thus far, no formal definition has been given in the truthmaking literature, by philosophical logicians, for the quasi-technical notion of a truthmaker in analytic metaphysics.[6] It is here proposed that the pre-formal notion of truthmaker (resp., falsitymaker) is nicely explicated by our formal notion of a truthmaker (resp., falsitymaker) of a saturated formula φ, *modulo* a domain D of discourse, from the basic facts represented in a complete and coherent set M of saturated literals.

---

[6] Perhaps the best example of construction coming quite close to being truthmakers in our sense would be the infinitary proofs proposed by Carnap for arithmetic using the so-called ω-rule

$$\frac{\Psi 0 \quad \Psi s0 \quad \Psi ss0 \quad \dots \quad \Psi \underline{n} \quad \dots}{\forall n \Psi n}$$

The notion of a truthmaker for φ enables one also to isolate those possibly 'smaller parts of reality' that might suffice to determine the truth of φ. Not all of the model M need be used, nor all of its domain D be surveyed, in order to determine the truth value of φ. The exact 'basic materials' on which that truth value supervenes, so to speak, are captured by the truthmakers (or falsitymakers) available.

There could well be many different truthmakers for one and the same saturated formula under one and the same interpretation (that is, *modulo* one and the same domain D, and one and the same set M of saturated literals). These different truthmakers represent different 'constitutive routes' to the same truth value under the interpretation in question. (Similarly for falsitymakers.)

8.4   Truthmaker transformation and classical logical consequence

We were originally inspired by an interesting analogy with Prawitz's definition of intuitionistic logical consequence, when suggesting that we define a more classical notion of consequence in terms of what we called quasi-effective transformability of truthmakers for the premises of an argument into a truthmaker for its conclusion. We have seen (Corollary 4) that we obtain exactly the relation of classical logical consequence if we impose no restriction at all on the kind of method used for the transformation of truthmakers. In effect, Corollary 4 established that

$$\varphi_1, \ldots, \varphi_n \models \psi$$

$$\Leftrightarrow$$

$$\exists f \, \forall M \, \forall \pi_1 \ldots \forall \pi_n [(\mathcal{V}_M(\pi_1, \varphi_1) \wedge \ldots \wedge \mathcal{V}_M(\pi_n, \varphi_n)) \Rightarrow \mathcal{V}_M(f(\pi_1, \ldots, \pi_n), \psi)]$$

which would be the $\mathbb{N}$-relative evaluation rule of 'Universal Introduction' in our sense. It would need to be complemented, of course, by a dual rule for 'Existential Elimination':

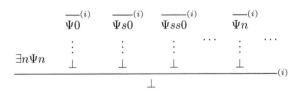

But, as we pointed out in §3.4.2, all branches in a truthmaker are of finite and *bounded* length, whereas for a proof using the ω-rule there need not be any upper bound on the lengths of its branches.

where the quantification $\exists f$ is over (set-theoretic) functions *tout court*.

The prima facie risk incurred by seeking to restrict the kind of function f that would be permissible here is that we might end up circumscribing too narrowly the resulting set of supposedly 'classically valid' arguments $\varphi_1, \ldots, \varphi_n : \psi$. The harder we make it to find such a function f, the more exigent a notion the definiens for the definiendum $\models$ becomes.

This exigency can be offset, however, by *more narrowly circumscribing* the class of models M with respect to which the f-transformations of truth-makers have to be effected. Noting that the quantification $\forall M$ is really a gloss for

$$(\forall \text{ domains } D)(\forall \text{ models } M \text{ based on domain } D),$$

we can inquire whether we might be able to constrain the kind of domain D involved, so as to offset any restriction of 'quasi-effectiveness' that might be imposed on the function f. Obvious candidates are:

(1) D is countable;

(2) D is effectively enumerable; and/or

(3) D is decidable.

Indeed, we might even consider

(4) D is the set of natural numbers.

The challenge is to find some formal delimitation of the class of permissible functions f that captures the orderliness of the transformations that are involved when one takes any *classical* natural deduction

$$\begin{array}{c} \overbrace{\varphi_1, \ldots, \varphi_n} \\ \Pi \\ \psi \end{array}$$

and accumulates on its premises respective M-relative truthmakers $\pi_1, \ldots, \pi_n$:

$$\begin{array}{c} \pi_1 \quad \cdots \quad \pi_n \\ \underbrace{\varphi_1 \qquad \varphi_n} \\ \Pi \\ \psi \end{array}$$

and thereupon seeks to 'normalize' the resulting construction so that it becomes an M-relative truthmaker for ψ:

$$
\begin{array}{cc}
\pi_1 & \pi_n \\
\varphi_1 & \cdots \quad \varphi_n \\
\underbrace{} \\
\Pi \\
\psi
\end{array}
\qquad
\begin{array}{c}
\text{normalizes to} \\
\mapsto
\end{array}
\qquad
\begin{array}{c}
\pi \\
\psi
\end{array}
\ ,
\qquad
\text{where } \pi = \nu(\pi_1, \ldots, \pi_n, \Pi)
$$

We have seen displays like these before, of course, in §8.1. But now it has to be borne in mind that the 'lower' proof Π in this context is a *classical* one; and that the 'upper' constructions $\pi_1, \ldots, \pi_n$ are (M-relative) *truthmakers*, not finitary M-warrants. We also need to remind ourselves that the actual 'grafting' of proof trees above is an expository fiction. For we are working—both in the context of deductive proof and in the context of truth- and falsity-making—with rules of elimination and falsification whose major premises have to *stand proud*. Thus our last display, in order for us to remain faithful to this constraint, should have been rendered as follows:

$$
\left[
\begin{array}{c} \Lambda_1 \\ \pi_1 \\ \varphi_1 \end{array}
\ ,\
\left[
\begin{array}{c} \Lambda_2 \\ \pi_2 \\ \varphi_2 \end{array}
\ ,\ \ldots,\
\left[
\begin{array}{cc} \Lambda_n & \overbrace{\varphi_1, \ldots, \varphi_n} \\ \varphi_n \ , & \Pi \\ \pi_n & \psi \end{array}
\right]
\cdots
\right]
\right]
\qquad = \qquad
\begin{array}{c} \Lambda\ (\subseteq \bigcup_i \Lambda_i) \\ \pi \\ \psi \end{array}
\ .
$$

For each M-relative truthmaker π_i ($1 \leq i \leq n$), Λ_i will be the set of M-literals that it exploits in order to verify φ_i. And the 'normalized' result π will be an M-relative truthmaker for ψ, exploiting some (possibly proper) subset Λ of $\bigcup_i \Lambda_i$, the 'total set of atomic M-data' used by the truthmakers $\pi_1, \ldots, \pi_n$.

Here is a suggestion, followed by a conjecture. Seek some general but interestingly restrictive characterization Ξ of the transformation functions ν that are involved in this kind of normalization process—'general' enough to encapsulate them all, but 'restrictive' enough to bring out the Skolemite flavor that was stressed above as providing illumination of the idea that the truth of the conclusion of a valid argument is somehow 'contained in' the truth of its premises. The guess is that Ξ will characterize some class of functions reasonably low in the arithmetical hierarchy.

Now recall Prawitz's soundness-via-normalization result, and its converse —his completeness conjecture—involving the relations $\models_I$ and $\models_{\mathbf{P}}$. We submit that there are analogs to be had for each of these, concerning the system of Classical Core Logic.

To see this, recall that we have just talked above of a normalization process involving *classical* natural deductions, and truthmakers for their undischarged assumptions.

First, on the assumption that Ξ has been chosen so as to capture correctly the kind of method implicit in this normalization process, we shall have an analog, for the classical case, of Prawitz's soundness theorem. And this will be the case regardless of any restriction that might be imposed on the domains D—in fact, the stronger any such restriction might be, the easier it will be for functions f with property Ξ to effect what is required of them.

We propose imposing the restriction that the domain D be *decidable*. Of the four options listed above, this one appears to be the most natural. So the normalization claim that would constitute the soundness theorem for classical core proofs will be stated as the following conjecture:

Conjecture 3.

$$\varphi_1, \ldots, \varphi_n \vdash_{C+} \psi$$

$$\Rightarrow$$

$$\exists f (\Xi f \wedge \forall \text{ decidable domains } D \ \forall \text{ models } M \text{ based on } D$$
$$\forall \pi_1 \ldots \forall \pi_n [(\mathcal{V}_M(\pi_1, \varphi_1) \wedge \ldots \wedge \mathcal{V}_M(\pi_n, \varphi_n)) \Rightarrow \mathcal{V}_M(f(\pi_1, \ldots, \pi_n), \psi)]) \, .$$

That normalizability will hold, in this requisite sense, is made more plausible by Metatheorem 11 below.

Second, we venture to assert the converse—now in the form of a completeness conjecture. It is to be hoped that the proposed (and as yet unspecified) restriction Ξ on our transformation methods f will exactly offset the restriction that the domain D be decidable.[7]

Conjecture 4 (Completeness Conjecture for Classical Core Logic).

$$\exists f (\Xi f \wedge \forall \text{ decidable domains } D \ \forall \text{ models } M \text{ based on } D$$
$$\forall \pi_1 \ldots \forall \pi_n [(\mathcal{V}_M(\pi_1, \varphi_1) \wedge \ldots \wedge \mathcal{V}_M(\pi_n, \varphi_n)) \Rightarrow \mathcal{V}_M(f(\pi_1, \ldots, \pi_n), \psi)])$$

$$\Rightarrow$$

$$\varphi_1, \ldots, \varphi_n \vdash_{C+} \psi \, .$$

---

[7] We know already by the downward Löwenheim–Skolem Theorem that restricting domains D to be both countable and decidable will not result in any unwanted curtailment of the classical logical consequence relation.

We therefore commend to the reader the research problem of formulating a suitable property Ξ and establishing (or making highly plausible) the ensuing completeness conjecture in the form just given. One has to confront the problem, of course—which confronts Prawitz's original completeness conjecture for the intuitionistic case as well—that the use of any *informal* terms such as 'effective' and 'decidable' renders the conjecture incapable of truly *formal* proof. (In the same way, Church's Thesis to the effect that every effective function on the naturals is recursive is incapable of truly formal proof.) But it is to be hoped that, despite this, some sort of 'progress in persuasion' might be made, concerning both Prawitz's completeness conjecture for the intuitionistic case, and the current completeness conjecture for the classical core case that is modeled, by analogy, upon it.

8.5   Resisting a threatened slide to realism

Our treatment of truthmakers might provide the realist with some new resources in his debate with the anti-realist. It is time now to reveal how this is so, and also to indicate how this new line of realist argument might be resisted.

The $\mathcal{V}$- and $\mathcal{F}$-rules are a straightforward transcription, into a recursive recipe for building constructions of the appropriate kinds, of the familiar inductive clauses in the Tarskian definition of truth or satisfaction. (This observation lies behind the obvious proof of Metatheorem 9.) Since the latter clauses, as already stressed, are acceptable to the anti-realist, it might be thought that the $\mathcal{V}$- and $\mathcal{F}$-rules would be acceptable also.

On the assumption (for the time being) that they are indeed acceptable, let us explore some further consequences.

Metatheorem 10 (Principle of Bivalence).

For all φ, either $\exists \Pi\ \mathcal{V}(\Pi, \varphi, M, D)$ or $\exists \Sigma\ \mathcal{F}(\Sigma, \varphi, M, D)$.

Proof. The proof is laid out with anti-realist commentary. It (the proof) is reasonably obvious (by induction on the complexity of φ), but it is important to realize that it is strictly classical.

The basis step of the inductive proof requires the completeness of M, in the sense explained above. This is the first point at which a classical conception obtrudes.

Classical metareasoning obtrudes further, at two more places.

First, in the inductive step dealing with saturated formulae of the form $\forall x \psi x$ we need, in the metalogic, the assurance that either every D-instance $\psi\alpha$ has an M-relative truthmaker, or some D-instance $\psi\alpha$ does not have any M-relative truthmaker.

The first disjunct will allow one to construct an M-relative truthmaker for $\forall x\psi$ from the M-relative truthmakers for all its instances. By appeal to the Inductive Hypothesis the second disjunct will imply

some D-instance $\psi\alpha$ has an M-relative falsitymaker,

and this will allow one to construct an M-relative falsitymaker for $\forall x\psi$. Thus we obtain either an M-relative truthmaker or an M-relative falsitymaker for $\forall x\psi$.

Second, in the inductive step dealing with saturated formulae of the form $\exists x\psi x$ we need, in the metalogic, the assurance that

either every D-instance $\psi\alpha$ has an M-relative falsitymaker, or some D-instance $\psi\alpha$ does not have any M-relative falsitymaker.

The first disjunct will allow one to construct an M-relative falsitymaker for $\exists x\psi$ from the M-relative falsitymakers for all its instances. By appeal to the Inductive Hypothesis the second disjunct will imply

some D-instance $\psi\alpha$ has an M-relative truthmaker,

and this will allow one to construct an M-relative truthmaker for $\exists x\psi$. Thus we obtain either an M-relative falsitymaker or an M-relative truthmaker for $\exists x\psi$.

In handling these two quantifier clauses in our co-inductive definition of truthmakers and falsitymakers, the assurance that is needed has the logical form

$$\forall x\Psi x \; \vee \; \exists x\neg\Psi x.$$

If this is taken to be available as a theorem of the metalogic, then we are dealing with a *strictly classical* metalogic.[8] For this principle is *not* acceptable to the intuitionist. So the intuitionist is unable to deliver Metatheorem 10—which is as it should be. □

Metatheorem 11. *Any M-relative truthmaker for $\neg\neg\varphi$ contains an M-relative truthmaker for φ. Hence in the object language we have $\neg\neg\varphi \models \varphi$.*

---

[8] This point applies equally well to the case where the classically needed but intuitionistically disputed principle has the logical form $\forall x\neg\Psi x \; \vee \; \exists x\Psi x$.

Proof. Suppose Π is an *M*-relative truthmaker for $\neg\neg\varphi$. According to the closure clause of our co-inductive definition of truthmakers and falsitymakers, Π must have been constructed by appeal to the rule $(\neg\mathcal{V})$, and accordingly have the form

$$\dfrac{\rule{1.2cm}{0.4pt}}{\neg\varphi}{\scriptstyle(i)}$$
$$\Sigma$$
$$\dfrac{\bot}{\neg\neg\varphi}{\scriptstyle(i)}$$

where Σ is an *M*-relative falsitymaker for $\neg\varphi$. According to the closure clause of our co-inductive definition of truthmakers and falsitymakers, Σ must have been constructed by appeal to the rule $(\neg\mathcal{F})$, and accordingly have the form

$$\Theta$$
$$\dfrac{\neg\varphi \quad \varphi}{\bot}$$

where Θ is an *M*-relative truthmaker for φ. $\square$

The anti-realist's immediate objection to this result will involve challenging the framing of the co-inductive definition of truthmakers and falsitymakers. For the anti-realist envisages situations where it is known that $\neg\varphi$ itself leads to absurdity, without that in itself guaranteeing that there is warrant for the assertion of φ. So the anti-realist's real complaint will be against the closure clause of the inductive definition. For this closure clause, particularly as it applies to falsitymakers for negations, makes the construction rule $(\neg\mathcal{F})$, with its built-in guarantee of a truthmaker for φ, the sole route to the rejection of a negation. While not having space here to develop this criticism on behalf of the anti-realist, we nevertheless wish at least to *identify* the exact issue that has to be joined in any further debate with the realist over this all-too-easy-looking validation of the strictly classical rule of Double Negation Elimination.

We turn now to the implementation, over the course of several instructive examples, of the theoretical ideas bruited above. We shall be examining various core proofs, along with truthmakers for their premises (relative to a small, hence manageable, chosen model M). We shall see, step by step, how to transform those materials into a truthmaker for the proof's conclusion. In the course of doing so, we shall encounter the need to frame convenient (and contextually rather obvious) procedures (called distribution conversions, and reductions) whereby we simplify what is in hand and work towards the desired terminus. Our examples will throw up all the procedures needed for an eventual proof that these transformations can always be carried out to a successful completion—whatever core proof one has, whatever model M one chooses, and whatever M-relative truthmakers one is given for the premises of the proof in question. This result is, as it were, the proof theorist's analog of a soundness theorem for Core Logic.

8.6 Any proof in Core Logic turns truthmakers for its premises into a truthmaker for its conclusion

8.6.1 Example of preservation of truthmakers

Consider the model M whose domain is $\{\alpha, \beta, \gamma\}$ and whose *atomic diagram* $\mathcal{B}$ is the complete set

$$\left\{ \overline{L\alpha\alpha}\,,\, \overline{L\alpha\beta}\,,\, \overline{L\alpha\gamma}\,,\, \overline{L\beta\alpha}\,,\, \overline{L\beta\beta}\,,\, \overline{L\beta\gamma}\,,\, \frac{L\gamma\alpha}{\bot}\,,\, \frac{L\gamma\beta}{\bot}\,,\, \frac{L\gamma\gamma}{\bot}\,,\, \frac{\alpha=\beta}{\bot}\,,\, \frac{\alpha=\gamma}{\bot}\,,\, \frac{\beta=\gamma}{\bot} \right\}$$

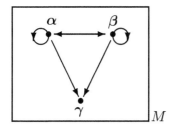

The sentence $\forall y \exists x Lxy$ has eight verifications in M:

$$\frac{\overline{\qquad}^M}{L\left\{\begin{matrix}\alpha\\\beta\end{matrix}\right\}\alpha} \quad \frac{\overline{\qquad}^M}{L\left\{\begin{matrix}\alpha\\\beta\end{matrix}\right\}\beta} \quad \frac{\overline{\qquad}^M}{L\left\{\begin{matrix}\alpha\\\beta\end{matrix}\right\}\gamma}$$
$$\frac{\exists x Lx\alpha \qquad \exists x Lx\beta \qquad \exists x Lx\gamma}{\forall y \exists x Lxy}{}^M$$

generated by the choices of witnesses that can be made from within the curly parentheses.

By contrast with $\forall y \exists x Lxy$, the sentence $\exists x \forall y Lxy$ has only two verifications in M:

$$\dfrac{\dfrac{\dfrac{}{L\alpha\alpha}{\scriptstyle M} \qquad \dfrac{}{L\alpha\beta}{\scriptstyle M} \qquad \dfrac{}{L\alpha\gamma}{\scriptstyle M}}{\forall y L\alpha y}}{\exists x \forall y Lxy} \qquad\qquad \dfrac{\dfrac{\dfrac{}{L\beta\alpha}{\scriptstyle M} \qquad \dfrac{}{L\beta\beta}{\scriptstyle M} \qquad \dfrac{}{L\beta\gamma}{\scriptstyle M}}{\forall y L\beta y}}{\exists x \forall y Lxy}$$

Consider now the following (formal, core) proof Σ of $\forall y \exists x Lxy$ from $\exists x \forall y Lxy$. Remember that a and b are *parameters*, not designations of any particular individuals—let alone of individuals in the domain of M.

$$\Sigma : \quad \dfrac{\dfrac{\exists x \forall y Lxy \qquad \dfrac{{\scriptstyle (2)}\dfrac{}{\forall y Lay} \qquad \dfrac{\dfrac{}{Lab}{\scriptstyle (1)}}{\exists x Lxb}}{\exists x Lxb}{\scriptstyle (1)}}{\exists x Lxb}{\scriptstyle (2)}}{\forall y \exists x Lxy}$$

What happens if one supplies for the premise $\exists x \forall y Lxy$ of the proof Σ an M-relative verification (π, say)? Answer: The proof Σ encodes a method of transforming π into an M-relative verification of $\forall y \exists x Lxy$, the conclusion of Σ. Or, to put the matter slightly differently: there is a method of transforming both π and Σ together, so that the transformed result is an M-relative verification $[\pi, \Sigma]$, say, of $\forall y \exists x Lxy$. Formally:

$$\left.\begin{array}{c} \pi \\ \exists x \forall y Lxy \\ \\ \exists x \forall y Lxy \\ \Sigma \\ \forall y \exists x Lxy \end{array}\right\} \quad \rightsquigarrow \quad \begin{array}{c} [\pi, \Sigma] \\ \forall y \exists x Lxy \end{array} .$$

Another formal way of putting the matter is to say that the transform, or *reduct*

$$\left[\begin{array}{c} \pi \\ \exists x \forall y Lxy \end{array} , \begin{array}{c} \exists x \forall y Lxy \\ \Sigma \\ \forall y \exists x Lxy \end{array}\right]$$

(whose final form is to be calculated, according to certain transformation rules) is an M-relative verification of $\forall y \exists x Lxy$.

It remains, then, so to define the binary operation indicated by the square brackets that this last claim about the reduct is true. Note that the first operand, π, in $[\pi, \Sigma]$ is an M-relative *verification* of $\exists x \forall y Lxy$ (a *truth-maker*); while the second operand, Σ, is a *core proof* of $\forall y \exists x Lxy$ from the single premise $\exists x \forall y Lxy$. This makes the situation formally analogous to the one we faced when establishing the admissibility of Cut for (Classical) Core Logic. The only new feature is that both π and $[\pi, \Sigma]$ are M-*relative verifications*, rather than core proofs. But note that every step in a verification is an application of a rule in Core Logic, with the exceptions only of steps of $(\forall \mathcal{V})$ and $(\exists \mathcal{F})$. We shall see in due course how to take these deviations from the previous pattern of 'Cut-elimination' in our stride.

Definition 20. *We define* $\{t(x)\}_{x \in X}$ *to be* $\{y | \exists x (x \in X \wedge y = t(x))\}$.

8.6.2 A distribution conversion to obtain M-relative verifications

Where D is the domain of M, π is an M-relative verification of φ, and the final step of Σ (the second operand) is an application of $(\forall I)$, with parameter a and with immediate subproof Θ:

$$
\left[\begin{array}{cc} & \overbrace{\varphi, \Gamma} \\ \pi & \Theta \\ \varphi & \psi \\ & \forall x \psi_x^a \end{array} \right] =_{df} (\forall\mathcal{V})\frac{\left\{ \left[\begin{array}{cc} & \overbrace{\varphi, \Gamma} \\ \pi & \Theta_\delta^a \\ \varphi & \psi_\delta^a \end{array} \right]_{\delta \in D} \right\}}{\forall x \psi x} \, {}_M
$$

Note that the final step in the reduct on the right is an application of the M-relative *verification* rule $(\forall\mathcal{V})$.

8.6.3 A distribution conversion to obtain M-relative falsifications

Where $\varphi \neq \exists x \psi$, we have the following important distribution conversion, dual to the one for universals in §8.6.2. Remember that π is an M-relative truthmaker for φ, and that the domain of M is D.

$$
\left[\begin{array}{cc} & \overset{(i)}{\overbrace{\psi_a^x, \varphi, \Gamma}} \\ \pi & \Sigma \\ \varphi & \frac{\exists x \psi \quad \bot}{\bot}{}^{(i)} \end{array} \right] =_{df} (\exists\mathcal{F})\frac{\left\{ \left[\begin{array}{cc} & \overset{(i)}{\overbrace{\psi_\delta^x, \varphi, \Gamma}} \\ \pi & \Sigma \\ \varphi & \bot \end{array} \right]_{\delta \in D}^{(i)} \right\}}{\exists x \psi \quad \bot}{}_M
$$

Note that the final step in the reduct on the right is an application of the M-relative *falsification* rule $\exists\mathcal{F}$.

8.6.4 The reduction procedure for existentials

Where the final step of π (the first operand) is an application of ($\exists$-$\mathcal{V}$) with witness α, and the final step of Σ (the second operand) is an application of ($\exists$E), with parameter a and with immediate subproof Θ:

$$
\left[\begin{array}{ccc}
\pi' & & \overbrace{\psi_a^x}^{\text{\hspace{0.5em}}}\,,\;\Gamma \\[2pt]
\dfrac{\psi_\alpha^x}{\exists x\psi} & , & \dfrac{\Theta}{\exists x\psi \quad \theta} \\[-2pt]
& & \rule{2.5cm}{0.4pt}^{(1)} \\
& & \theta
\end{array}\right]
=_{df}
\left[\begin{array}{cc}
\pi' & \overbrace{\psi_\alpha^x}^{\text{\hspace{0.5em}}}\,,\;\Gamma \\[2pt]
\dfrac{\psi_\alpha^x}{} & , & \dfrac{\Theta_\alpha^a}{\theta}
\end{array}\right]
$$

We are making the simplifying assumption that $\exists x\psi$ does not occur again as a major premise for $\exists$E, within the subproof Θ. But if it does, then we take care of such further MPE occurrences by stipulating

$$
\left[\begin{array}{ccc}
\pi' & & \overset{(1)}{\overline{\psi_a^x}}\,,\;\Gamma \\[2pt]
\dfrac{\psi_\alpha^x}{\exists x\psi} & , & \dfrac{\Theta}{\exists x\psi \quad \theta} \\[-2pt]
& & \rule{2.5cm}{0.4pt}^{(1)} \\
& & \theta
\end{array}\right]
=_{df}
\left[\begin{array}{c}
\pi' \\ \dfrac{\psi_\alpha^x}{} \end{array}\,,\;
\left[\begin{array}{cc}
\pi' & \overbrace{\psi_\alpha^x}^{\text{\hspace{0.5em}}}\,,\;\Gamma \\[2pt]
\dfrac{\psi_\alpha^x}{\exists x\psi} & , & \dfrac{\Theta_\alpha^a}{\theta}
\end{array}\right]\right]
$$

8.6.5 Back to our example

Let π be the first of our two M-relative verifications above of $\exists x \forall y Lxy$. Then

$$
\left[\begin{array}{cc}
\pi & \begin{array}{c}\exists x\forall y Lxy \\ \Sigma \\ \forall y\exists x Lxy\end{array} \\
\exists x\forall y Lxy &
\end{array}\right]
$$

$$
=
\left[\begin{array}{cc}
\dfrac{\dfrac{}{L\alpha\alpha}M \quad \dfrac{}{L\alpha\beta}M \quad \dfrac{}{L\alpha\gamma}M}{\dfrac{\forall y L\alpha y}{\exists x\forall y Lxy}} &\; , \;
\dfrac{\exists x\forall y Lxy \quad \dfrac{(2)\dfrac{}{\forall y Lay} \quad \dfrac{\overline{Lab}^{(1)}}{\exists x Lxb}}{\dfrac{\exists x Lxb}{}}^{(1)}}{\dfrac{\exists x Lxb}{\forall y\exists x Lxy}}^{(2)}
\end{array}\right]
$$

$$= \left\{ \left[\frac{\dfrac{\overline{L\alpha\alpha}^M \quad \overline{L\alpha\beta}^M \quad \overline{L\alpha\gamma}^M}{\forall y L\alpha y}}{\exists x \forall y L x y} \, , \quad \underset{\exists x L x y}{(2)\dfrac{\forall y L\alpha y \quad (1)\dfrac{\overline{L\alpha\delta}^{(1)}}{\exists x L x \delta}}{\exists x L x \delta}{}^{(2)}} \right] \right\}_{\delta\in\{\alpha,\beta,\gamma\}}^M $$
$$\dfrac{}{\forall y \exists x L x y}$$

$$= \left\{ \left[\frac{\overline{L\alpha\alpha}^M \quad \overline{L\alpha\beta}^M \quad \overline{L\alpha\gamma}^M}{\forall y L\alpha y} \, , \quad \dfrac{\forall y L\alpha y \;\; \dfrac{\overline{L\alpha\delta}^{(1)}}{\exists x L x \delta}}{\exists x L x \delta}{}^{(1)} \right] \right\}_{\delta\in\{\alpha,\beta,\gamma\}}^M $$
$$\dfrac{}{\forall y \exists x L x y}$$

(suppressing subscripts M)

$$= \frac{\left[\dfrac{\overline{L\alpha\alpha}\;\overline{L\alpha\beta}\;\overline{L\alpha\gamma}}{\forall y L\alpha y}, \forall y L\alpha y \dfrac{\overline{L\alpha\alpha}^{(1)}}{\exists x L x\alpha}{}^{(1)}\right]\left[\dfrac{\overline{L\alpha\alpha}\;\overline{L\alpha\beta}\;\overline{L\alpha\gamma}}{\forall y L\alpha y}, \forall y L\alpha y \dfrac{\overline{L\alpha\beta}^{(1)}}{\exists x L x\beta}{}^{(1)}\right]\left[\dfrac{\overline{L\alpha\alpha}\;\overline{L\alpha\beta}\;\overline{L\alpha\gamma}}{\forall y L\alpha y}, \forall y L\alpha y \dfrac{\overline{L\alpha\gamma}^{(1)}}{\exists x L x\gamma}{}^{(1)}\right]}{\forall y \exists x L x y}{}_M$$

(restoring subscripts M)

$$= \frac{\left[\dfrac{\overline{L\alpha\alpha}^M}{L\alpha\alpha}, \dfrac{L\alpha\alpha}{\exists x L x\alpha}\right]\left[\dfrac{\overline{L\alpha\beta}^M}{L\alpha\beta}, \dfrac{L\alpha\beta}{\exists x L x\beta}\right]\left[\dfrac{\overline{L\alpha\gamma}^M}{L\alpha\gamma}, \dfrac{L\alpha\gamma}{\exists x L x\gamma}\right]_M}{\forall y \exists x L x y}$$

$$= \frac{\dfrac{\overline{L\alpha\alpha}^M}{\exists x L x\alpha} \quad \dfrac{\overline{L\alpha\beta}^M}{\exists x L x\beta} \quad \dfrac{\overline{L\alpha\gamma}^M}{\exists x L x\gamma}{}_M}{\forall y \exists x L x y}$$

We have found, then, that

$$\left[\frac{\dfrac{\overline{L\alpha\alpha}^M \quad \overline{L\alpha\beta}^M \quad \overline{L\alpha\gamma}^M}{\forall y L\alpha y}}{\exists x \forall y L x y} \, , \quad \frac{(2)\dfrac{\forall y L\alpha y \quad (1)\dfrac{\overline{Lab}^{(1)}}{\exists x L x b}}{\exists x L x b}{}^{(2)}}{\forall y \exists x L x y} \right]$$

is the verification

$$
\cfrac{\cfrac{}{L\alpha\alpha}M \quad \cfrac{}{L\alpha\beta}M \quad \cfrac{}{L\alpha\gamma}M}{\cfrac{\exists xLx\alpha \quad \exists xLx\beta \quad \exists xLx\gamma}{\forall y\exists xLxy}M}
$$

Likewise, if we were to use the other M-relative verification of $\exists x\forall yLxy$, we would find that

$$
\left[\;
\cfrac{\cfrac{}{L\beta\alpha}M \quad \cfrac{}{L\beta\beta}M \quad \cfrac{}{L\beta\gamma}M}{\cfrac{\forall yL\beta y}{\exists x\forall yLxy}}
\;,\;
\cfrac{\exists x\forall yLxy \quad \cfrac{(2)\cfrac{}{\forall yLay} \quad \cfrac{\cfrac{}{Lab}(1)}{\exists xLxb}(1)}{\exists xLxb}(2)}{\cfrac{\exists xLxb}{\forall y\exists xLxy}}
\;\right]
$$

is the verification

$$
\cfrac{\cfrac{}{L\beta\alpha}M \quad \cfrac{}{L\beta\beta}M \quad \cfrac{}{L\beta\gamma}M}{\cfrac{\exists xLx\alpha \quad \exists xLx\beta \quad \exists xLx\gamma}{\forall y\exists xLxy}M}
$$

Recall that there were *eight* distinct M-relative verifications of $\forall y\exists xLxy$. But only *two* of them feature as reducts of the form $[\pi, \Sigma]$. This is because there were only *two* possibilities for π as an M-relative verification of $\exists x\forall yLxy$. We see that the proof Σ 'channels' the truth-making of its premise into a particular truthmaker for its conclusion. The six other 'ways of seeing that $\forall y\exists xLxy$ is true in M' remain 'hidden from view'.

This is no surprise. For the two M-relative verifications of $\exists x\forall yLxy$ exploited, respectively, only the following restricted sets M_1, M_2 of atomic information about M:

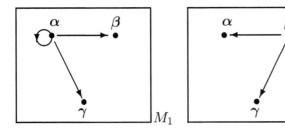

So the proof Σ, in deducing $\forall y \exists x Lxy$ from $\exists x \forall y Lxy$, has to 'make do', in each of these two cases, with just the atomic information respectively revealed, in enabling one to construct an M-relative verification of $\forall y \exists x Lxy$.

8.7 More examples of how to produce truthmakers for conclusions of proofs, from truthmakers for their premises

8.7.1 Example 1

Let the model M have the domain $\{\alpha, \beta, \gamma\}$, and let M assign to each of the primitive monadic predicates F and G the whole domain as its extension. We have the following M-relative truthmakers, which we shall call π_F and π_G:

$$\pi_F \quad = \quad \cfrac{\cfrac{}{F\alpha}M \quad \cfrac{}{F\beta}M \quad \cfrac{}{F\gamma}M}{\forall x Fx}M$$

$$\pi_G \quad = \quad \cfrac{\cfrac{}{G\alpha}M \quad \cfrac{}{G\beta}M \quad \cfrac{}{G\gamma}M}{\forall x Gx}M$$

Consider also the following proof Σ:

$$\cfrac{\forall x Fx \quad \cfrac{\forall x Gx \quad \cfrac{\cfrac{}{Fa}(1) \quad \cfrac{}{Ga}(2)}{\cfrac{Fa \wedge Ga}{Fa \wedge Ga}(2)}}{Fa \wedge Ga}(1)}{\cfrac{Fa \wedge Ga}{\forall x(Fx \wedge Gx)}}$$

Our task is to compute the truthmaker

$$\left[\begin{array}{c} \pi_F \\ \forall x Fx \end{array} \, , \, \left[\begin{array}{c} \pi_G \\ \forall x Gx \end{array} \, , \, \begin{array}{c} \forall x Fx \, , \, \forall x Gx \\ \Sigma \\ \forall x(Fx \wedge Gx) \end{array} \right] \right]$$

To alert the reader: the beast we are trying to corner is the *M*-relative truthmaker

$$
\cfrac{\cfrac{\overline{}^{M}\;\; \overline{}^{M}}{Fa \quad Ga}}{\cfrac{Fa \wedge Ga \quad\quad Fb \wedge Gb \quad\quad Fc \wedge Gc}{\forall x(Fx \wedge Gx)}}
$$

$$
\frac{\dfrac{\overline{F\alpha}^{M}\;\;\overline{G\alpha}^{M}}{F\alpha \wedge G\alpha} \qquad \dfrac{\overline{F\beta}^{M}\;\;\overline{G\beta}^{M}}{F\beta \wedge G\beta} \qquad \dfrac{\overline{F\gamma}^{M}\;\;\overline{G\gamma}^{M}}{F\gamma \wedge G\gamma}}{\forall x(Fx \wedge Gx)}{}^{M}
$$

Corner it we shall, but in stages. We start by working out the innermost reduct, namely

$$
\left[\begin{array}{c} \pi_G \\ \forall x Gx \end{array} \;,\; \begin{array}{c} \forall x Fx,\ \forall x Gx \\ \Sigma \\ \forall x(Fx \wedge Gx) \end{array} \right]
$$

More explicitly, this is

$$
\left[\dfrac{\overline{G\alpha}^{M}\;\;\overline{G\beta}^{M}\;\;\overline{G\gamma}^{M}}{\forall x Gx}{}^{M} \;,\; \dfrac{\forall x Fx \quad \dfrac{\forall x Gx \quad \dfrac{\overline{Fa}^{(1)}\;\;\overline{Ga}^{(2)}}{Fa \wedge Ga}{}^{(2)}}{Fa \wedge Ga}{}^{(1)}}{\dfrac{Fa \wedge Ga}{\forall x(Fx \wedge Gx)}} \right]
$$

Recall the following general transformation, which we formulated above and which applies here:

$$
\left[\pi \;,\; \dfrac{\overbrace{\varphi,\ \Gamma}\quad \Theta}{\dfrac{\psi}{\forall x \psi_x^a}} \atop \varphi \right] = \left\{ \left[\pi \;,\; \dfrac{\overbrace{\varphi,\ \Gamma}\quad \Theta_\delta^a}{\psi_\delta^a} \atop \varphi \right] \right\}_{\delta \in D} \Big/ \forall x \psi x \quad {}_M
$$

Upon applying this transformation, we obtain

$$
\left\{ \left[\dfrac{\overline{G\alpha}^{M}\;\;\overline{G\beta}^{M}\;\;\overline{G\gamma}^{M}}{\forall x Gx}{}^{M} \;,\; \dfrac{\forall x Fx \quad \dfrac{\forall x Gx \quad \dfrac{\overline{F\delta}^{(1)}\;\;\overline{G\delta}^{(2)}}{F\delta \wedge G\delta}{}^{(2)}}{F\delta \wedge G\delta}{}^{(1)}}{F\delta \wedge G\delta} \right] \right\}_{\delta \in \{\alpha,\beta,\gamma\}} \Big/ \forall x(Fx \wedge Gx) \quad {}_M
$$

Distributing π_G over the major premise $\forall x F x$, this becomes

$$
\cfrac{
\left\{
\cfrac{
\begin{array}{c}
\cfrac{\dfrac{}{G\alpha}M \quad \dfrac{}{G\beta}M \quad \dfrac{}{G\gamma}M}{\forall x G x}M
\end{array}
, \quad
\cfrac{\forall x G x \quad \cfrac{\dfrac{}{F\delta}(1) \quad \dfrac{}{G\delta}(2)}{F\delta \wedge G\delta}}{\cfrac{F\delta \wedge G\delta}{}(2)}(1)
}{F\delta \wedge G\delta}
\right\}\ \delta \in \{\alpha,\beta,\gamma\}\ M
}{\forall x (F x \wedge G x)}
$$

We are now in a position to apply the $\mathcal{V}$-analog of a $\forall$I-reduction, since $\forall x G x$ is the conclusion of the verification that is the first argument of the embedded reduct, and is also the major premise of the terminal step (an application of $\forall$E) in the proof that is the second argument of the reduct. Doing so we obtain

$$
\cfrac{
\left\{
\cfrac{\forall x F x \quad \cfrac{\dfrac{}{F\delta}(1) \quad \dfrac{}{G\delta}M}{F\delta \wedge G\delta}}{F\delta \wedge G\delta}(1)
\right\}\ \delta \in \{\alpha,\beta,\gamma\}\ M
}{\forall x (F x \wedge G x)}
$$

—that is,

$$
\cfrac{
\cfrac{\forall x F x \quad \cfrac{\dfrac{}{F\alpha}(1) \quad \dfrac{}{G\alpha}M}{F\alpha \wedge G\alpha}}{F\alpha \wedge G\alpha}(1)
\qquad
\cfrac{\forall x F x \quad \cfrac{\dfrac{}{F\beta}(1) \quad \dfrac{}{G\beta}M}{F\beta \wedge G\beta}}{F\beta \wedge G\beta}(1)
\qquad
\cfrac{\forall x F x \quad \cfrac{\dfrac{}{F\gamma}(1) \quad \dfrac{}{G\gamma}M}{F\gamma \wedge G\gamma}}{F\gamma \wedge G\gamma}(1)\ M
}{\forall x (F x \wedge G x)}
$$

This, then, is the construct that we have computed as the reduct

$$
\left[\ \pi_G \atop \forall x G x\ ,\ {\forall x F x\,,\ \forall x G x \atop {\Sigma \atop \forall x (F x \wedge G x)}}\ \right].
$$

Note that this reduct is neither a truthmaker nor a core proof. To be sure, it appeals to the atomic (or *basic*) M-facts $G\alpha$, $G\beta$, and $G\gamma$; but it also appeals to a logically complex premise, namely $\forall x F x$. It might best be described as an M-relative truthmaker premised on the (as yet unverified) $\forall x F x$. But of course the latter premise *does* have an *M*-relative verification π_F, as we have seen.

It remains now to finish our computation of the sought truthmaker for $\forall x(Fx \wedge Gx)$:

$$
\left[
\begin{array}{c}
\pi_F \\
\overline{\forall x Fx}
\end{array}
,
\begin{array}{c}
\dfrac{\forall x Fx \quad \dfrac{\dfrac{}{F\alpha}^{(1)} \quad \dfrac{}{G\alpha}M}{\dfrac{F\alpha \wedge G\alpha}{}}}{F\alpha \wedge G\alpha}^{(1)}
\quad
\dfrac{\forall x Fx \quad \dfrac{\dfrac{}{F\beta}^{(1)} \quad \dfrac{}{G\beta}M}{F\beta \wedge G\beta}}{F\beta \wedge G\beta}^{(1)}
\quad
\dfrac{\forall x Fx \quad \dfrac{\dfrac{}{F\gamma}^{(1)} \quad \dfrac{}{G\gamma}M}{F\gamma \wedge G\gamma}}{\dfrac{F\gamma \wedge G\gamma}{}}^{(1)}M \\
\hline
\forall x(Fx \wedge Gx)
\end{array}
\right]
$$

—or, going back to our convenient δ-notation,

$$
\left[
\begin{array}{c}
\pi_F \\
\overline{\forall x Fx}
\end{array}
,
\begin{array}{c}
\left\{
\dfrac{\forall x Fx \quad \dfrac{\dfrac{}{F\delta}^{(1)} \quad \dfrac{}{G\delta}M}{F\delta \wedge G\delta}}{F\delta \wedge G\delta}^{(1)}
\right\}_{\delta \in \{\alpha, \beta, \gamma\}} \\
\hline
\forall x(Fx \wedge Gx)
\end{array} M
\right]
$$

The left argument π_F distributes over the terminal step of $\forall \mathcal{V}$ in the second argument of this reduct, whereupon we have

$$
\left\{
\left[
\begin{array}{c}
\pi_F \\
\overline{\forall x Fx}
\end{array}
,
\begin{array}{c}
\dfrac{\forall x Fx \quad \dfrac{\dfrac{}{F\delta}^{(1)} \quad \dfrac{}{G\delta}M}{F\delta \wedge G\delta}}{F\delta \wedge G\delta}^{(1)} \\
\hline
\forall x(Fx \wedge Gx)
\end{array}
\right]
\right\}_{\delta \in \{\alpha, \beta, \gamma\}} M
,
$$

that is

$$
\left\{
\left[
\begin{array}{c}
\dfrac{\dfrac{}{F\alpha}M \quad \dfrac{}{F\beta}M \quad \dfrac{}{F\gamma}M}{\forall x Fx}M \\
\end{array}
,
\begin{array}{c}
\dfrac{\forall x Fx \quad \dfrac{\dfrac{}{F\delta}^{(1)} \quad \dfrac{}{G\delta}M}{F\delta \wedge G\delta}}{F\delta \wedge G\delta}^{(1)} \\
\hline
\forall x(Fx \wedge Gx)
\end{array}
\right]
\right\}_{\delta \in \{\alpha, \beta, \gamma\}} M
$$

The obvious $\forall \mathcal{V}$-reductions (one for each of the three values that δ can take) produce

$$\cfrac{\left\{\cfrac{\overline{F\delta}^{\,M} \quad \overline{G\delta}^{\,M}}{\overline{F\delta \wedge G\delta}}\right\}_{\delta \in \{\alpha,\,\beta,\,\gamma\}}}{\forall x(Fx \wedge Gx)}_M$$

that is

$$\cfrac{\cfrac{\overline{F\alpha}^{\,M} \quad \overline{G\alpha}^{\,M}}{\overline{F\alpha \wedge G\alpha}} \quad \cfrac{\overline{F\beta}^{\,M} \quad \overline{G\beta}^{\,M}}{\overline{F\beta \wedge G\beta}} \quad \cfrac{\overline{F\gamma}^{\,M} \quad \overline{G\gamma}^{\,M}}{\overline{F\gamma \wedge G\gamma}}}{\forall x(Fx \wedge Gx)}_M$$

The beast has been cornered.

8.7.2 Example 2

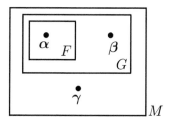

The model M makes the sentence $\forall x(Fx \rightarrow Gx)$ true. Here is one of its M-relative truthmakers:

$$\pi : \quad \cfrac{\cfrac{\overline{G\alpha}^{\,M}}{F\alpha \rightarrow G\alpha} \quad \cfrac{\overline{G\beta}^{\,M}}{F\beta \rightarrow G\beta} \quad \cfrac{\cfrac{\overline{F\gamma}^{\,(1)}}{\bot}^{\,M}}{F\gamma \rightarrow G\gamma}^{(1)}}{\forall x(Fx \rightarrow Gx)}_M$$

Note that the basic M-facts used in π are:

$$\overline{G\alpha}^{\,M} \;\; ; \quad \overline{G\beta}^{\,M} \;\; ; \quad \cfrac{F\gamma}{\bot}^{\,M} \;\; ; \quad \alpha, \beta, \gamma \text{ are all the individuals}$$

Consider now the following proof of $\forall x(\neg Gx \rightarrow \neg Fx)$ from $\forall x(Fx \rightarrow Gx)$:

$$\Sigma: \quad \cfrac{\cfrac{\forall x(Fx \rightarrow Gx)}{\cfrac{(2)\,\overline{\qquad\qquad}}{Fa \rightarrow Ga} \quad \overline{Fa}^{(3)} \quad \cfrac{\cfrac{}{\neg Ga}^{(4)} \quad \cfrac{}{Ga}^{(1)}}{\cfrac{\bot}{\qquad\qquad}^{(1)}}}{\cfrac{\bot}{\cfrac{\bot}{\cfrac{\neg Fa}{\cfrac{\neg Ga \rightarrow \neg Fa}{\forall x(\neg Gx \rightarrow \neg Fx)}}^{(4)}}^{(3)}}^{(2)}}}{}}$$

We seek now the reduct $[\pi, \Sigma]$, which will be an M-relative truthmaker for $\forall x(\neg Gx \rightarrow \neg Fx)$. We shall calculate it in stages, as we did with the previous example.

Consider the immediate subproof $\Sigma_1(a)$ of Σ, parametric in a:

$$\cfrac{\Sigma_1(a)}{\neg Ga \rightarrow \neg Fa} \quad : \quad \cfrac{\forall x(Fx \rightarrow Gx) \quad \cfrac{\cfrac{(2)\,\overline{\qquad\qquad}}{Fa \rightarrow Ga} \quad \overline{Fa}^{(3)} \quad \cfrac{\cfrac{}{\neg Ga}^{(4)} \quad \cfrac{}{Ga}^{(1)}}{\cfrac{\bot}{\qquad\qquad}^{(1)}}}{\bot}^{(2)}}{\cfrac{\bot}{\cfrac{\neg Fa}{\neg Ga \rightarrow \neg Fa}^{(4)}}^{(3)}}}{}$$

We first observe that, since Σ proves a universal,

$$[\pi, \Sigma] \quad = \quad \cfrac{\left\{ \left[\pi, \; \cfrac{\Sigma_1(\delta)}{\neg G\delta \rightarrow \neg F\delta} \right] \right\}_{\delta \in \{\alpha, \beta, \gamma\}}}{\forall x(\neg Gx \rightarrow \neg Fx)}_M$$

Let us first simplify the embedded reduct

$$\left[\pi, \; \cfrac{\Sigma_1(\delta)}{\neg G\delta \rightarrow \neg F\delta} \right].$$

Distributing π across the bottom two steps in $\Sigma_1(\delta)$, this becomes

$$
\left[\pi\;,\quad
\dfrac{\dfrac{\forall x(Fx \to Gx) \qquad \dfrac{\dfrac{(2)\overline{\quad\quad}}{F\delta \to G\delta}\;\;\dfrac{(3)\overline{\quad}}{F\delta}}{\bot}\;\;\dfrac{\dfrac{(4)\overline{\neg G\delta}\;\;\overline{G\delta}\!\!^{(1)}}{\bot}}{}\!\!^{(1)}}{\bot}\!\!^{(2)}}{\bot}\right]
$$

$$
\dfrac{\dfrac{\bot}{\neg F\delta}\!\!^{(3)}}{\neg G\delta \to \neg F\delta}\!\!^{(4)}
$$

—that is,

$$
\left[\dfrac{\dfrac{\dfrac{\overline{Ga}\!\!^{M}}{Fa \to Ga}\;\;\dfrac{\overline{G\beta}\!\!^{M}}{F\beta \to G\beta}\;\;\dfrac{\dfrac{\overline{F\gamma}\!\!^{(1)}}{\bot}\!\!^{(1)}}{F\gamma \to G\gamma}\!\!^{M}}{\forall x(Fx \to Gx)}\;,\;\; \dfrac{\forall x(Fx \to Gx)\quad\dfrac{\dfrac{(2)\overline{\quad}}{F\delta \to G\delta}\;\dfrac{(3)\overline{\quad}}{F\delta}}{\bot}\;\dfrac{\dfrac{(4)\overline{\neg G\delta}\;\overline{G\delta}\!\!^{(1)}}{\bot}}{}\!\!^{(1)}}{\bot}\!\!^{(2)}\right]
$$

$$
\dfrac{\dfrac{\bot}{\neg F\delta}\!\!^{(3)}}{\neg G\delta \to \neg F\delta}\!\!^{(4)}
$$

When we take α for δ, and apply the called-for reduction procedure for $\forall$, this becomes

$$
\left[\dfrac{\dfrac{\overline{Ga}\!\!^{M}}{Fa \to Ga}\;,\;\;\dfrac{\dfrac{Fa \to Ga\;\;\dfrac{(3)\overline{\quad}}{Fa}}{\bot}\;\dfrac{\dfrac{(4)\overline{\neg Ga}\;\overline{Ga}\!\!^{(1)}}{\bot}}{}\!\!^{(1)}}{}\right]
$$

$$
\dfrac{\dfrac{\bot}{\neg Fa}\!\!^{(3)}}{\neg Ga \to \neg Fa}\!\!^{(4)}
$$

which, upon applying the called-for reduction procedure for $\to$, simplifies to

$$
\dfrac{\dfrac{(4)\overline{\neg Ga}\;\;\overline{Ga}\!\!^{M}}{\bot}}{\neg Ga \to \neg Fa}\!\!^{(4)}
$$

Similarly for the case where β replaces δ. But for the case where γ replaces δ, matters are slightly different. By the same sequence of reductions (for $\forall$ and then for →) we arrive at the simplification

$$
\cfrac{\cfrac{\cfrac{\overline{}^{(3)}}{F\gamma}\,M}{\cfrac{\bot}{\neg F\gamma}^{(3)}}^{\textstyle}}{\neg G\gamma \to \neg F\gamma}{}^{(4)}
$$

The sought value for $[\pi, \Sigma]$, accordingly, is

$$
\cfrac{\cfrac{\cfrac{\overline{}^{(4)}}{\neg G\alpha} \quad \overline{}\,M}{\cfrac{\bot}{\neg G\alpha \to \neg F\alpha}^{(4)}} \quad \cfrac{\cfrac{\overline{}^{(4)}}{\neg G\beta} \quad \overline{}\,M}{\cfrac{\bot}{\neg G\beta \to \neg F\beta}^{(4)}} \quad \cfrac{\cfrac{\cfrac{\overline{}^{(3)}}{F\gamma}\,M}{\cfrac{\bot}{\neg F\gamma}^{(3)}}}{\neg G\gamma \to \neg F\gamma}{}^{(4)}\,M}{\forall x(\neg Gx \to \neg Fx)}
$$

Note that it uses the same basic facts about M that were used by π.

The core proof Σ of $\forall x(\neg Gx \to \neg Fx)$ from $\forall x(Fx \to Gx)$ has turned the M-relative truthmaker π for its premise $\forall x(Fx \to Gx)$ into the M-relative truthmaker just exhibited above for its conclusion $\forall x(\neg Gx \to \neg Fx)$. We did not have to 'consult the world' (i.e. marshall basic M-facts) afresh in order to verify this conclusion. Rather, we were able to work a priori with the given M-relative truthmaker π and the core proof Σ, re-arranging the basic M-facts responsible for the truth of the premise of Σ so as to obtain a truthmaker for its conclusion. The truth (in M) of the conclusion of this proof is indeed contained in the truth (in M) of its premise. More generally, and put another way: an M-relative truthmaker for the conclusion of any proof can be constructed from any M-relative truthmakers for its premises, no matter what model M may be.

Consider now the following alternative M-relative truthmaker π':

$$
\pi': \qquad \cfrac{\cfrac{\overline{}\,M}{G\alpha}}{F\alpha \to G\alpha} \quad \cfrac{\cfrac{\cfrac{\overline{}^{(1)}}{F\beta}\,M}{\bot}}{F\beta \to G\beta}{}^{(1)} \quad \cfrac{\cfrac{\cfrac{\overline{}^{(1)}}{F\gamma}\,M}{\bot}}{F\gamma \to G\gamma}{}^{(1)}\,M \over \forall x(Fx \to Gx)
$$

Note that the basic M-facts used in π' are:

$$
\cfrac{\overline{}}{G\alpha}\,M \ ; \quad \cfrac{F\beta}{\bot}\,M \ ; \quad \cfrac{F\gamma}{\bot}\,M \ ; \quad \alpha, \beta, \gamma \text{ are all the individuals}
$$

The reader can easily check that $[\pi', \Sigma]$ is

$$
\frac{\dfrac{}{\neg G\alpha \qquad G\alpha}\ M}{\dfrac{\bot}{\neg G\alpha \to \neg F\alpha}(4)}(4)
\qquad
\frac{\dfrac{\dfrac{\overline{\qquad}(3)}{F\beta}\ M}{\dfrac{\bot}{\neg F\beta}(3)}}{\neg G\beta \to \neg F\beta}(4)
\qquad
\frac{\dfrac{\dfrac{\overline{\qquad}(3)}{F\gamma}\ M}{\dfrac{\bot}{\neg F\gamma}(3)}}{\neg G\gamma \to \neg F\gamma}(4)\ M
$$

$$
\forall x(\neg Gx \to \neg Fx)
$$

which uses the same basic M-facts as π'.

This is a general feature of an M-relative truthmaker for the conclusion of a core proof that is computed from M-relative truthmakers for that proof's premises. It uses only basic M-facts that were used by those truthmakers for the premises.

Note the careful statement of 'inclusion only':

> Any M-relative truthmaker for the conclusion of a core proof that is computed from M-relative truthmakers for that proof's premises uses *only* basic M-facts that were used by those truthmakers for the premises.

We refrain from going so far as to say

> Any M-relative truthmaker for the conclusion of a core proof that is computed from M-relative truthmakers for that proof's premises uses *all and only the* basic M-facts that were used by those truthmakers for the premises.

The reader will easily verify that this last, stronger statement, is infirmed by examples such as the following. For these counterexamples we need mention only instances of the sequent established by the proof Σ. This is enough to make the point at hand, regardless of the detailed structure of Σ, and regardless also of the choice of a truthmaker π for Σ's premise:

$$
\exists x(Fx \wedge Gx) : \exists x Fx
$$

$$
\forall x(Fx \wedge Gx) : \forall x Fx
$$

These two examples are single-premise sequents. But the reader will easily come up with multi-premise sequents $\varphi_1, \ldots, \varphi_n : \psi$ $(n \geq 2)$ that can be used to make the same point:

given respective M-relative truthmakers $\pi_1, \ldots, \pi_n$ and a proof Σ of that sequent, the truthmaker $[\pi_1, [\pi_2, \ldots [\pi_n, \Sigma] \ldots]]$ need not make use of *all* the basic M-facts used by $\pi_1, \ldots, \pi_n$.

This multi-premise sequent is one such:

$$\forall x Fx, \exists x Gx : \exists x (Fx \wedge Gx).$$

Neither of the two premises suffices for the conclusion; but it is also the case that their joint truth requires strictly more than is required for the conclusion. In the model M':

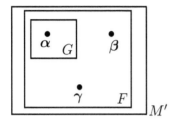

the obvious truthmakers for $\forall x Fx$ and $\exists x Gx$ use, between them, the following collection of basic M-facts:

$$\overline{F\alpha}^M \;\; ; \;\; \overline{F\beta}^M \;\; ; \;\; \overline{F\gamma}^M \;\; ; \;\; \overline{G\alpha}^M \;\; ; \;\; \alpha, \beta, \gamma \text{ are all the individuals.}$$

The truthmaker for the conclusion, however, uses only the proper subcollection

$$\overline{F\alpha}^M \;\; ; \;\; \overline{G\alpha}^M \; .$$

8.7.3   Example 3

We continue with the model M that was used in Example 2:

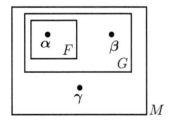

This time we shall work with the same M-relative truthmaker π for $\forall x(Fx \rightarrow Gx)$:

$$\pi: \quad \cfrac{\cfrac{\overline{}\,M}{Ga}}{Fa \rightarrow Ga} \quad \cfrac{\overline{}\,M}{F\beta \rightarrow G\beta} \quad \cfrac{\cfrac{\overline{}\!\!^{(1)}}{\cfrac{F\gamma}{\bot}\,M}}{F\gamma \rightarrow G\gamma}\!\!^{(1)}}{\forall x(Fx \rightarrow Gx)}\,M$$

but with a different proof Σ, this time of the new conclusion $\neg \exists x(Fx \wedge \neg Gx)$ from the premise $\forall x(Fx \rightarrow Gx)$:

$$\Sigma: \quad \cfrac{\cfrac{}{\exists x(Fx \wedge \neg Gx)}\!\!^{(5)} \quad \cfrac{\cfrac{}{\forall x(Fx \rightarrow Gx)}\!\!^{(4)} \quad \cfrac{\cfrac{\cfrac{}{Fa \rightarrow Ga}\!\!^{(3)} \quad \cfrac{}{Fa}\!\!^{(2)}}{\bot}\quad \cfrac{\cfrac{}{\neg Ga}\!\!^{(2)} \quad \cfrac{}{Ga}\!\!^{(1)}}{\bot}\!\!^{(1)}}{\bot}\!\!^{(1)}}{\bot}\!\!^{(3)}}{\cfrac{Fa \wedge \neg Ga}{\bot}\!\!^{(4)}}}{\cfrac{\bot}{\neg \exists x(Fx \wedge \neg Gx)}\!\!^{(5)}}$$

Once again we seek the reduct $[\pi, \Sigma]$, which will be an M-relative truthmaker for $\neg \exists x(Fx \wedge \neg Gx)$. Let Σ_1 be the immediate subproof of Σ. It is to the overall deductive effect

$$\underbrace{\forall x(Fx \rightarrow Gx),\ \exists x(Fx \wedge \neg Gx)}_{\Sigma_1}$$
$$\bot$$

and its final step is an application of $\exists E$, with major premise $\exists x(Fx \wedge \neg Gx)$.

We commence our calculation of the reduct $[\pi, \Sigma]$ by performing a distribution conversion over the final step of $\neg I$ in Σ. The result is

$$\cfrac{\left[\pi,\ \cfrac{\underbrace{\forall x(Fx \rightarrow Gx),\ \overline{\exists x(Fx \wedge \neg Gx)}\!\!^{(5)}}_{\Sigma_1}}{\bot}\right]}{\cfrac{\bot}{\neg \exists x(Fx \wedge \neg Gx)}\!\!^{(5)}}$$

We now reveal another important layer of structure within Σ_1. The foregoing construct becomes

$$
\left[
\begin{array}{c}
\begin{array}{cc}
\begin{array}{c}
\pi \\
\forall x(Fx \to Gx)
\end{array} \quad , \quad (5)\dfrac{}{\exists x(Fx \land \neg Gx)}
\end{array}
\quad
\begin{array}{c}
(4)\overline{\quad\quad\quad\quad}\quad\overline{\quad\quad\quad}(1) \\
\underbrace{Fa \land \neg Ga \quad , \quad Fa \to Ga} \\
\Sigma_2(a) \\
\dfrac{\forall x(Fx \to Gx) \quad \bot}{\dfrac{\bot}{\bot}(1)}(4)
\end{array} \\
\dfrac{\bot}{\bot}
\end{array}
\right]
$$

$$(5)\dfrac{\bot}{\neg\exists x(Fx \land \neg Gx)}$$

where

$$
\Sigma_2(a) \;=\;
\dfrac{
Fa \to Ga \quad \dfrac{}{Fa}(2)
}{
\dfrac{Fa \land \neg Ga}{\bot}
}(2)
\quad
\dfrac{
(2)\dfrac{}{\neg Ga} \quad \dfrac{}{Ga}(1)
}{
\dfrac{\bot}{\bot}(1)
}(1)
$$

Note the following features of the embedded reduct (in square brackets): the conclusion of π is not the MPE of the final step of Σ_1, but is a more deeply embedded premise within Σ_1; and the final step of Σ_1 is an application of $\exists$E with conclusion $\bot$. *Under such circumstances, in general,* the reduct in question must end with an application of the M-relative falsification rule $\exists \mathcal{F}$, as explained in §8.6.3. So our embedded reduct will become

$$
\dfrac{\exists x(Fx \land \neg Gx) \quad
\left\{
\left[
\begin{array}{c}
\pi \quad , \quad
\begin{array}{c}
(4)\underbrace{\overline{F\delta \land \neg G\delta} \quad , \quad \overline{F\delta \to G\delta}}^{(1)} \\
\Sigma_2(\delta) \\
\dfrac{\forall x(Fx \to Gx) \quad \bot}{\bot}(1)
\end{array}
\end{array}
\right]
\right\}_{\delta \in \{\alpha,\beta,\gamma\}}
}{\bot}(4)
$$

We can now go to work on the embedded reduct in the construct immediately above, maintaining generality with the placeholder δ for the three individuals α, β, γ. Substitutions of each of these for δ will come later. We are seeking now to simplify

$$
\left[\pi \ , \quad \cfrac{\Vdash x(Fx \to Gx) \qquad \cfrac{\cfrac{\underbrace{\overset{(4)}{\rule{2cm}{0.4pt}} \qquad \overset{(1)}{\rule{2cm}{0.4pt}}}_{\Sigma_2(\delta)}}{F\delta \wedge \neg G\delta \ , \ F\delta \to G\delta}}{\bot}}{\bot}{\scriptstyle(1)} \right]
$$

—that is,

$$
\left[\cfrac{\cfrac{\overline{G\alpha}^{M}}{F\alpha \to G\alpha} \quad \cfrac{\overline{G\beta}^{M}}{F\beta \to G\beta} \quad \cfrac{\cfrac{\overset{(1)}{\overline{F\gamma}}^{M}}{\bot}^{(1)}}{F\gamma \to G\gamma}^{M}}{\forall x(Fx \to Gx)} \ , \quad \cfrac{\forall x(Fx \to Gx) \qquad \cfrac{\underbrace{\overset{(4)}{\rule{2cm}{0.4pt}} \qquad \overset{(1)}{\rule{2cm}{0.4pt}}}_{\Sigma_2(\delta)}}{F\delta \wedge \neg G\delta \ , \ F\delta \to G\delta}}{\bot}{\scriptstyle(1)} \right]
$$

For each of the three possible values of δ, there is a $\forall$-reduction to be performed. Taking α for δ, the result of the $\forall$-reduction is

$$
\left[\cfrac{\overline{G\alpha}^{M}}{F\alpha \to G\alpha} \ , \ \Sigma_2(\alpha) \right]
$$

—that is,

$$
\left[\cfrac{\overline{G\alpha}^{M}}{F\alpha \to G\alpha} \ , \quad \cfrac{F\alpha \wedge \neg G\alpha \qquad \cfrac{\cfrac{F\alpha \to G\alpha \quad \overset{(2)}{F\alpha}}{\overset{(2)}{\overline{\neg G\alpha}} \quad \overset{(1)}{\overline{G\alpha}}}^{(2)} \quad \bot}{\bot}^{(1)}}{\bot}{\scriptstyle(1)} \right]
$$

Distributing the truthmaker on the left over the terminal step of $\wedge$E in $\Sigma_2(\alpha)$, and performing the requisite $\to$-reduction, this reduct becomes

$$
\cfrac{F\alpha \wedge \neg G\alpha \qquad \cfrac{\overset{(2)}{\overline{\neg G\alpha}} \quad \overline{G\alpha}^{M}}{\bot}}{\bot}{\scriptstyle(2)}
$$

Likewise, the reduct involving β becomes

$$
\cfrac{F\beta \wedge \neg G\beta \qquad \cfrac{(2)\cfrac{\quad}{\neg G\beta} \quad \cfrac{\quad}{G\beta}\,M}{\bot}\,(2)}{\bot}
$$

Finally, we need to calculate the reduct involving γ:

$$
\left[\cfrac{\cfrac{(1)\cfrac{\quad}{F\gamma}\,M}{\bot}\,(1)}{F\gamma \rightarrow G\gamma} \quad , \quad \cfrac{F\gamma \wedge \neg G\gamma \qquad \cfrac{\cfrac{F\gamma \rightarrow G\gamma \quad F\gamma}{\bot}\,(2) \qquad \cfrac{(2)\cfrac{\quad}{\neg G\gamma} \quad (1)\cfrac{\quad}{G\gamma}}{\bot}\,(1)}{\bot}\,(1)}{\bot} \right]
$$

As before, we distribute the truthmaker on the left over the terminal step of $\wedge E$ in $\Sigma_2(\alpha)$, and perform the requisite $\rightarrow$-reduction. The reduct thereby becomes

$$
\cfrac{F\gamma \wedge \neg G\gamma \qquad \cfrac{(1)\cfrac{\quad}{F\gamma}\,M}{\bot}\,(2)}{\bot}
$$

If the reader takes a fresh look at the schema above that involved large curly parentheses subscripted by '$\delta \in \{\alpha, \beta, \gamma\}$', she will appreciate that all that remains to be done is to supply the three reducts just calculated for α, β, and γ in order to obtain the sought truthmaker for $\neg\exists x(Fx \wedge \neg Gx)$ along the lines detailed above:

$$
\cfrac{\cfrac{\exists x(Fx \wedge \neg Gx) \qquad \cfrac{(4)\cfrac{F\alpha \wedge \neg G\alpha \qquad \cfrac{(2)\cfrac{\quad}{\neg G\alpha} \quad \cfrac{\quad}{G\alpha}\,M}{\bot}\,(2)}{\bot}}{}}{} \quad \cfrac{(4)\cfrac{F\beta \wedge \neg G\beta \qquad \cfrac{(2)\cfrac{\quad}{\neg G\beta} \quad \cfrac{\quad}{G\beta}\,M}{\bot}\,(2)}{\bot}}{} \quad \cfrac{(4)\cfrac{F\gamma \wedge \neg G\gamma \qquad \cfrac{(1)\cfrac{\quad}{F\gamma}\,M}{\bot}\,(2)}{\bot}}{}\,(4)}{\neg\exists x(Fx \wedge \neg Gx)}\,(5)
$$

We have cornered another beast.

8.7.4 Example 4

We continue with the model M that was used in Examples 2 and 3:

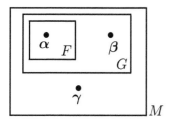

We shall use once more the M-relative truthmaker π for $\forall x(Fx \to Gx)$ that was used in those examples:

$$\pi: \qquad \cfrac{\cfrac{\overline{}\,M}{Ga}}{Fa \to Ga} \qquad \cfrac{\cfrac{\overline{}\,M}{G\beta}}{F\beta \to G\beta} \qquad \cfrac{\cfrac{\cfrac{\overline{}\,\text{(1)}}{F\gamma}\,M}{\bot}}{F\gamma \to G\gamma}\text{(1)}\,M$$
$$\overline{\qquad\qquad\qquad\qquad \forall x(Fx \to Gx) \qquad\qquad\qquad\qquad}$$

as well as the following obvious truthmaker π' for $\exists x Fx$:

$$\pi': \qquad \cfrac{\cfrac{\overline{}\,M}{Fa}}{\exists x Fx}$$

For the present example, the proof Σ will be of the conclusion $\exists x Gx$ from the two premises $\forall x(Fx \to Gx)$ and $\exists x Fx$:

$$\cfrac{\exists x Fx \qquad \cfrac{\cfrac{\forall x(Fx \to Gx) \qquad \cfrac{\overset{(2)}{\overline{}} \qquad \overset{(3)}{\overline{Fa}} \quad \overset{(1)}{\overline{Ga}}}{\cfrac{Ga}{}\text{(1)}}}{\cfrac{Ga}{\exists x Gx}\text{(2)}}}{\qquad}\text{(3)}}{\exists x Gx}$$

The task now is to calculate the truthmaker $[\pi, [\pi', \Sigma]]$. We do this in the obvious way, by working inside out. The reduct $[\pi', \Sigma]$ is

$$\left[\;\cfrac{\overline{}\,M}{\cfrac{Fa}{\exists x Fx}}\;,\; \cfrac{\exists x Fx \qquad \cfrac{\cfrac{\forall x(Fx \to Gx) \qquad \cfrac{\overset{(2)}{\overline{}} \quad \overset{(3)}{\overline{Fa}} \quad \overset{(1)}{\overline{Ga}}}{\cfrac{Ga}{}\text{(1)}}}{\cfrac{Ga}{\exists x Gx}\text{(2)}}}{\exists x Gx}\text{(3)}\;\right]$$

Applying ∃-reduction we obtain

$$
\left[
\begin{array}{c}
\dfrac{\quad}{F\alpha}\,M \;,\quad
\dfrac{\displaystyle \forall x(Fx \to Gx) \qquad \dfrac{(2)\dfrac{\quad}{F\alpha \to G\alpha}\ F\alpha \quad \dfrac{\quad}{G\alpha}{}^{(1)}}{G\alpha}{}^{(1)}}{\dfrac{G\alpha}{\exists xGx}}{}^{(2)}
\end{array}
\right]
$$

A distribution conversion turns this into

$$
\dfrac{\displaystyle \forall x(Fx \to Gx) \qquad \left[\dfrac{\dfrac{\quad}{F\alpha}\,M \;,\quad \dfrac{(2)\dfrac{\quad}{F\alpha \to G\alpha}\ F\alpha \quad \dfrac{\quad}{G\alpha}{}^{(1)}}{G\alpha}{}^{(1)}}{}\right]{}^{(2)}}{\dfrac{G\alpha}{\exists xGx}}
$$

which simplifies to

$$
\dfrac{\displaystyle \forall x(Fx \to Gx) \qquad \dfrac{(2)\dfrac{\quad}{F\alpha \to G\alpha} \quad \dfrac{\quad}{F\alpha}\,M \quad \dfrac{\quad}{G\alpha}{}^{(1)}}{G\alpha}{}^{(1)}}{\dfrac{G\alpha}{\exists xGx}}{}^{(2)}
$$

Having thus completed our calculation of $[\pi', \Sigma]$, we turn now to complete the calculation of $[\pi, [\pi', \Sigma]]$. This final task begins, then, with

$$
\left[
\dfrac{\dfrac{\dfrac{\quad}{G\alpha}\,M}{F\alpha \to G\alpha} \quad \dfrac{\dfrac{\quad}{G\beta}\,M}{F\beta \to G\beta} \quad \dfrac{\dfrac{\dfrac{\quad}{F\gamma}{}^{(1)}}{\bot}\,M}{F\gamma \to G\gamma}{}^{(1)}}{\forall x(Fx \to Gx)}\,M \;,\quad \dfrac{\forall x(Fx \to Gx) \qquad \dfrac{(2)\dfrac{\quad}{F\alpha \to G\alpha}\ F\alpha \quad \dfrac{\quad}{G\alpha}\,M\ {}^{(1)}}{G\alpha}{}^{(1)}}{\dfrac{G\alpha}{\exists xGx}}{}^{(2)}
\right]
$$

which of course calls for a distribution conversion so as to become

$$
\dfrac{\left[\dfrac{\dfrac{\dfrac{\quad}{G\alpha}\,M}{F\alpha \to G\alpha} \quad \dfrac{\dfrac{\quad}{G\beta}\,M}{F\beta \to G\beta} \quad \dfrac{\dfrac{\dfrac{\quad}{F\gamma}{}^{(1)}}{\bot}\,M}{F\gamma \to G\gamma}{}^{(1)}}{\forall x(Fx \to Gx)}\,M \;,\quad \dfrac{\forall x(Fx \to Gx) \qquad \dfrac{(2)\dfrac{\quad}{F\alpha \to G\alpha}\ F\alpha \quad \dfrac{\quad}{G\alpha}\,M\ {}^{(1)}}{G\alpha}{}^{(1)}}{G\alpha}{}^{(2)}\right]}{\dfrac{G\alpha}{\exists xGx}}
$$

Now we can apply $\forall$-reduction inside the square brackets to obtain

$$
\cfrac{\left[\ \cfrac{\overline{\ \ \ }\ M}{\cfrac{G\alpha}{F\alpha \to G\alpha}}\ ,\ \cfrac{F\alpha \to G\alpha \quad F\alpha \quad \overline{\ \ \ }^{-(1)}\!\!\!G\alpha}{G\alpha}^{-(1)}\ \right]}{\cfrac{G\alpha}{\exists x G x}}
$$

which immediately simplifies, by $\to$-reduction, to

$$
\cfrac{\overline{\ \ \ }\ M}{\cfrac{G\alpha}{\exists x F x}}
$$

Note how this reduct uses only the basic M-fact

$$
\overline{\ \ \ }\ M \ ;
$$
$$
G\alpha
$$

the process remains blind to the basic M-fact

$$
\overline{\ \ \ }\ M
$$
$$
G\beta
$$

even though *that* information was embedded within the truthmaker π. It was bypassed, however, by a reduction process proceeding algorithmically, which happened to accord dominant importance to α over β. This was because the other truthmaker used in the calculation, namely π', used only the basic M-fact involving α. Had we used, in place of π', the alternative M-relative truthmaker

$$
\pi'' : \quad \cfrac{\overline{\ \ \ }\ M}{\cfrac{F\beta}{\exists x F x}}\ ,
$$

for the premise $\exists x F x$ of Σ, then the M-relative truthmaker calculated for the conclusion $\exists x G x$ of Σ would have been

$$
\cfrac{\overline{\ \ \ }\ M}{\cfrac{G\beta}{\exists x F x}}\ ,
$$

and it would have been the F-ness of α (in M) that got ignored.

This provides a nice explication of the idea that 'the' individual that witnesses the truth of the conclusion $\exists x G x$ *is* 'the' individual that witnesses the truth of the premise $\exists x F x$—even though there may be more than one candidate for the witnessing of both.

8.7.5   Example 5

Suppose that the core proof Σ is a *dis*proof, and has just two premises, φ_1 and φ_2:

$$
\overbrace{\varphi_1, \varphi_2}
$$
$$
\Sigma
$$
$$
\bot
$$

Then any *M*-relative truthmaker (verification) for φ_1 can be transformed (using Σ) into an *M*-relative *falsification* of φ_2. In order to illustrate this, we shall consider the obvious core proofs of the following sequents:

$$
\forall x Fx, \exists x \neg Fx : \bot \qquad \exists x Fx, \forall x \neg Fx : \bot
$$

along with truthmakers (relative, of course, to different models) for each of their premises, and go through the simple calculations involved in extracting a falsification (relative to the model in question) of the other premise. In each of these illustrations, one should assume that the model M does actually furnish the basic facts invoked in the truthmaker that is supplied.

$\exists x \neg Fx, \forall x Fx : \bot$, **along with a truthmaker for** $\exists x \neg Fx$.

The reduct

$$
\left[
\begin{array}{cc}
\dfrac{\dfrac{\rule{2em}{0.4pt}}{F\alpha}{}^{(1)}_{M}}{\dfrac{\dfrac{\bot}{\neg F\alpha}{}^{(1)}}{\exists x \neg Fx}} &
\dfrac{\exists x \neg Fx \quad \dfrac{\forall x Fx \quad \dfrac{\overset{(2)}{\underline{\neg Fa}} \quad \overset{(1)}{\underline{Fa}}}{\bot}}{\dfrac{\bot}{\bot}{}^{(1)}}{}^{(2)}}{\bot}
\end{array}
\right]
$$

becomes, by $\exists$-reduction,

$$
\left[
\begin{array}{cc}
\dfrac{\dfrac{\rule{2em}{0.4pt}}{F\alpha}{}^{(1)}_{M}}{\dfrac{\bot}{\neg F\alpha}{}^{(1)}} & , \quad
\dfrac{\forall x Fx \quad \dfrac{\overline{\neg F\alpha} \quad \overset{(1)}{\overline{F\alpha}}}{\bot}}{\bot}{}^{(1)}
\end{array}
\right] ;
$$

whereupon a distribution conversion yields

$$
\cfrac{
\left[
\cfrac{\cfrac{\overline{}{\scriptstyle(1)}}{F\alpha}\,{\scriptstyle M}}{\cfrac{\bot}{\neg F\alpha}{\scriptstyle(1)}}
\quad,\quad
\cfrac{\neg F\alpha \quad \overline{F\alpha}^{\scriptstyle(1)}}{\bot}
\right]
}{
\cfrac{\forall x F x \qquad\qquad\qquad\qquad \bot}{\bot}{\scriptstyle(1)}
}\;;
$$

so that the obvious ¬-reduction finally gives us the falsitymaker

$$
\cfrac{\forall x F x \qquad \cfrac{\cfrac{\overline{}{\scriptstyle(1)}}{F\alpha}\,{\scriptstyle M}}{\bot}}{\bot}{\scriptstyle(1)}
$$

$\exists x \neg F x,\ \forall x F x : \bot$, **along with a truthmaker for** $\forall x F x$.

The reduct

$$
\left[
\cfrac{\cfrac{\overline{}\,{\scriptstyle M}\quad\overline{}\,{\scriptstyle M}}{F\alpha \qquad F\beta}}{\forall x F x}\,{\scriptstyle M}
\quad,\quad
\cfrac{\exists x \neg F x \qquad \cfrac{\forall x F x \quad \cfrac{\overline{\neg F a}^{\scriptstyle(2)} \quad \overline{F a}^{\scriptstyle(1)}}{\bot}{\scriptstyle(1)}}{\bot}}{\bot}{\scriptstyle(2)}
\right]
$$

calls for a distribution conversion with the mandatory terminal step of $\exists \mathcal{F}$ described in §8.6.3:

$$
\cfrac{
\left\{
\left[
\cfrac{\cfrac{\overline{}\,{\scriptstyle M}\quad\overline{}\,{\scriptstyle M}}{F\alpha \qquad F\beta}}{\forall x F x}\,{\scriptstyle M}
\quad,\quad
\cfrac{\forall x F x \quad \cfrac{\overline{\neg F\delta}^{\scriptstyle(2)} \quad \overline{F\delta}^{\scriptstyle(1)}}{\bot}{\scriptstyle(1)}}{\bot}
\right]
\right\}_{\delta \in \{\alpha,\beta\}}
}{
\exists x \neg F x \qquad\qquad\qquad\qquad\qquad\qquad \bot
}{\scriptstyle(2)}\;;
$$

whereupon, by the obvious steps of ∀-reduction called for in the two instances of the square-bracketed reduct, we obtain the falsitymaker

$$
\cfrac{\exists x \neg F x \qquad \cfrac{\overline{\neg F\alpha}^{\scriptstyle(2)} \quad \overline{F\alpha}\,{\scriptstyle M}}{\bot} \qquad \cfrac{\overline{\neg F\beta}^{\scriptstyle(2)} \quad \overline{F\beta}\,{\scriptstyle M}}{\bot}}{\bot}{\scriptstyle(2)}
$$

$\exists x F x, \forall x \neg F x : \bot$, **along with a truthmaker for** $\exists x F x$.

The reduct

$$
\left[
\begin{array}{ccc}
\dfrac{\quad}{Fa}M & , & \dfrac{\quad}{\exists x F x}
\end{array}
\quad
\begin{array}{c}
\dfrac{\dfrac{(1)\overline{\quad}}{\neg F a} \quad \dfrac{\overline{\quad}}{F a}(2)}{\dfrac{\bot}{\bot}(2)} \\
\forall x \neg F x \quad \dfrac{\quad}{\bot}(1)
\end{array}
\right]
$$

calls for an immediate $\exists$-reduction, which yields

$$
\left[
\dfrac{\quad}{Fa}M \quad , \quad
\begin{array}{c}
\forall x \neg F x \quad \dfrac{\dfrac{(1)\overline{\quad}}{\neg F\alpha} \quad \dfrac{\quad}{F\alpha}}{\bot}\\
\dfrac{\quad}{\bot}(1)
\end{array}
\right] ;
$$

whereupon the obvious distribution conversion produces the falsitymaker

$$
\begin{array}{c}
\forall x \neg F x \quad \dfrac{\dfrac{(1)\overline{\quad}}{\neg F\alpha} \quad \dfrac{\overline{\quad}}{F\alpha}M}{\bot}\\
\dfrac{\quad}{\bot}(1)
\end{array}
$$

$\exists x F x, \forall x \neg F x : \bot$, **along with a truthmaker for** $\forall x \neg F x$.

The reduct

$$
\left[
\begin{array}{c}
\dfrac{\dfrac{\dfrac{(1)\overline{\quad}}{F\alpha}M}{\bot}(1)}{\neg F\alpha} \quad \dfrac{\dfrac{\dfrac{(2)\overline{\quad}}{F\beta}M}{\bot}(2)}{\neg F\beta}M \\
\forall x \neg F x
\end{array}
\quad , \quad
\begin{array}{c}
\dfrac{\dfrac{(1)\overline{\quad}}{\neg F a} \quad \dfrac{\overline{\quad}}{F a}(2)}{\dfrac{\bot}{\bot}(2)} \\
\exists x F x \quad \dfrac{\quad}{\bot}(1)
\end{array}
\right] ,
$$

like the second example in this subsection, calls for a distribution conversion with the mandatory terminal step of $\exists\mathcal{F}$ described in §8.6.3:

$$\cfrac{\exists xFx \quad \left\{\left[\cfrac{\cfrac{\overline{}^{(1)}}{F\alpha}_M \quad \cfrac{\overline{}^{(2)}}{F\beta}_M}{\cfrac{\cfrac{\perp}{\neg F\alpha}^{(1)} \quad \cfrac{\perp}{\neg F\beta}^{(1)}}{\forall xFx}_M} \quad , \quad \cfrac{\forall xFx \quad \cfrac{(2)\cfrac{}{\neg F\delta} \quad \cfrac{}{F\delta}^{(1)}}{\perp}}{\perp}{}^{(1)}\right]\right\}{}_{\delta\in\{\alpha,\beta\}}}{\perp}{}^{(2)} \quad ;$$

whereupon, by the obvious steps of $\forall$-reduction called for in the two instances of the square-bracketed reduct, we obtain

$$\cfrac{\exists xFx \quad \left[\cfrac{\cfrac{\overline{}^{(1)}}{F\alpha}_M}{\cfrac{\perp}{\neg F\alpha}^{(1)}} \quad , \quad \cfrac{\neg F\alpha \quad \overline{F\alpha}^{(2)}}{\perp}\right]\left[\cfrac{\cfrac{\overline{}^{(1)}}{F\beta}_M}{\cfrac{\perp}{\neg F\beta}^{(1)}} \quad , \quad \cfrac{\neg F\beta \quad \overline{F\beta}^{(2)}}{\perp}\right]}{\perp}{}^{(2)} \quad ,$$

which, upon the two obvious steps of $\neg$-reduction respectively called for in the two square-bracketed reducts, yields the falsitymaker

$$\cfrac{\exists xFx \quad \cfrac{\overline{}^{(2)}}{F\alpha}_M \quad \cfrac{\perp}{} \quad \cfrac{\overline{}^{(2)}}{F\beta}_M \quad \cfrac{\perp}{}}{\perp}{}^{(2)}$$

8.7.6   Example 6

In §8.7.5 we made use of core proofs of the sequents

$$\forall xFx, \exists x\neg Fx : \perp \qquad \exists xFx, \forall x\neg Fx : \perp .$$

Each of these proofs can be extended by one step of $\neg I$ to prove, respectively, the sequents

$$\exists x\neg Fx : \neg\forall xFx \qquad \forall x\neg Fx : \neg\exists xFx .$$

Here we look at the remaining core-provable quantifier-duality, namely

$$\neg\exists xFx : \forall x\neg Fx ,$$

and we see how its proof can turn a truthmaker for the premise $\neg\exists x Fx$ into a truthmaker for the conclusion $\forall x\neg Fx$. We shall keep matters simple, as we have done before, by considering only the two-element model M whose domain is $\{\alpha, \beta\}$, and which makes the premise $\neg\exists x Fx$ true.

The reduct to simplify is

$$
\left[
\begin{array}{c}
\dfrac{\overline{}^{(1)}\quad \overline{}^{(1)}}{} \\
(2)\dfrac{}{\exists x Fx}\quad \dfrac{F\alpha}{\bot}{}_M\quad \dfrac{F\beta}{\bot}{}_M \qquad \neg\exists x Fx\quad \dfrac{\overline{}^{(1)}}{\exists x Fx}\dfrac{\overline{Fa}^{(1)}}{} \\
\dfrac{\bot}{\neg\exists x Fx}(2) \qquad\qquad \dfrac{\bot}{\neg F a}(1) \\
\hspace{6cm}\forall x\neg Fx
\end{array}
\right] \quad,
$$

This calls for a distribution conversion with the mandatory terminal step of $\forall\mathcal{V}$ described in §8.6.2:

$$
\left\{
\left[
\begin{array}{c}
(2)\dfrac{}{\exists x Fx}\quad \dfrac{\overline{F\alpha}^{(1)}}{\bot}{}_M\quad \dfrac{\overline{F\beta}^{(1)}}{\bot}{}_M \qquad \neg\exists x Fx\quad \dfrac{\overline{F\delta}^{(1)}}{\exists x Fx} \\
\dfrac{\bot}{\neg\exists x Fx}(2) \qquad\qquad \dfrac{\bot}{\neg F\delta}(1) \\
\hspace{5cm}\forall x\neg Fx
\end{array}
\right]
\right\}_{\delta\in\{\alpha,\beta\}}
$$

One more distribution conversion yields

$$
\left\{
\left[
\begin{array}{c}
(2)\dfrac{}{\exists x Fx}\quad \dfrac{\overline{F\alpha}^{(1)}}{\bot}{}_M\quad \dfrac{\overline{F\beta}^{(1)}}{\bot}{}_M \qquad \neg\exists x Fx\quad \dfrac{\overline{F\delta}^{(2)}}{\exists x Fx} \\
\dfrac{\bot}{\neg\exists x Fx}(2) \qquad\qquad \dfrac{\bot}{\neg F\delta}(2) \\
\dfrac{\bot}{\neg F\delta}(2) \\
\forall x\neg Fx
\end{array}
\right]
\right\}_{\delta\in\{\alpha,\beta\}} \quad,
$$

which, upon applying ¬-reduction, becomes

$$
\left\{
\left[
\begin{array}{c}
\dfrac{\overline{}^{(2)}}{\dfrac{F\delta}{\exists x Fx}}
\;,\;
\dfrac{\exists x Fx \quad \dfrac{\overline{}^{(1)}}{\dfrac{F\alpha\,_M}{\bot}} \quad \dfrac{\overline{}^{(1)}}{\dfrac{F\beta\,_M}{\bot}}}{\dfrac{\bot}{\dfrac{\bot}{\neg F\delta}\,^{(2)}}}\,^{(1)}
\end{array}
\right]
\right\}_{\delta \in \{\alpha,\beta\}}
\,,
$$

$$
\dfrac{}{\forall x \neg Fx}
$$

which, upon making the obvious ∃-reductions called for in the α- and β-instances, becomes the truthmaker

$$
\dfrac{\dfrac{\overline{}^{(1)}}{\dfrac{F\alpha\,_M}{\dfrac{\bot}{\neg F\alpha}\,^{(1)}}} \quad \dfrac{\overline{}^{(2)}}{\dfrac{F\beta\,_M}{\dfrac{\bot}{\neg F\beta}\,^{(2)}}}\,_M}{\forall x \neg Fx}
$$

8.8 Taking stock

Thus far, in our discussion of how truthmakers for the premises of a proof can be transformed into a truthmaker for its conclusion, we have stated various operations for carrying out these transformations, and have confirmed our hunch that they work as intended, by applying them to various simple examples. We are now in a position to advance to a more general and abstract result, to the effect that these transformations can *always* be effected. The technique of proof of this agreeable soundness result comes, ultimately, from proof theory; but the payoff for formal semantics is considerable. We turn to this task in Chapter 9.

Chapter 9

Transmission of Truthmakers

Abstract

Section-by-section summary of this chapter would be unnecessarily labored. We begin by introducing the formal genus 'conditional M-relative construct', of which M-relative truthmakers and falsitymakers, and core proofs, are species. Fortunately they can stand in symbiotic relations, even though they cannot hybridize. The remaining sections are best understood as follows. They aim to generalize the earlier method we used in order to prove Cut-Elimination, so that the inputs Π for the binary operation $[\Pi, \Sigma]$ can be truthmakers (whereas Σ remains a core proof); and so that the reduct itself, when it is finally determined by recursive application of all the transformations called for, is a truthmaker for the conclusion of Σ. This result can be understood as revealing that formal semantics can be carried out in a kind of infinitary proof theory. Core proof transmits truth courtesy of normalization.

9.1 Conditional evaluations

In §8.7.1 we saw an example of a 'hybrid' construct—part truthmaker, part deductive proof—that we said might best be described as an M-relative truthmaker premised on the as yet unverified sentence $\forall x F x$. The model M in question had the domain $\{\alpha, \beta, \gamma\}$, and assigned to each of the primitive monadic predicates F and G the whole domain as its extension. The hybrid construct in question was

$$
\begin{array}{ccc}
\cfrac{\cfrac{\quad}{F\alpha}(1) \quad \cfrac{\quad}{G\alpha}M}{\forall xFx \quad \cfrac{F\alpha \land G\alpha}{F\alpha \land G\alpha}(1)} &
\cfrac{\cfrac{\quad}{F\beta}(1) \quad \cfrac{\quad}{G\beta}M}{\forall xFx \quad \cfrac{F\beta \land G\beta}{F\beta \land G\beta}(1)} &
\cfrac{\cfrac{\quad}{F\gamma}(1) \quad \cfrac{\quad}{G\gamma}M}{\forall xFx \quad \cfrac{F\gamma \land G\gamma}{F\gamma \land G\gamma}(1)}
\end{array}
$$
$$
\cfrac{}{\forall x(Fx \land Gx)}M
$$

This construct, we want to say, is an *M*-relative verification of its conclusion $\forall x(Fx \land Gx)$ conditional upon the set $\{\forall xFx\}$ of as-yet-unverified premises. More generally, it is straightforward to imagine how to produce hybrid constructs that are *M*-relative verifications *or falsifications* of a saturated formula ψ conditional upon a set Δ of saturated formulae.[1]

In §4.6 we defined the purely deductive notion $\mathcal{P}(\Pi, \varphi, \Delta)$—'$\Pi$ is a proof of the conclusion φ from the set Δ of undischarged assumptions'. And in §3.3 we defined the two intercalating evaluative notions $\mathcal{V}(V, \varphi, \Lambda)$ and $\mathcal{F}(F, \varphi, \Lambda)$—respectively, '$V$ is a verification of φ relative to the set Λ of literals', and 'F is a falsification of φ relative to the set Λ of literals'.

All we need to do now is extend the ideas involved in these two kinds of definition. We shall thoroughly hybridize evaluations (truthmakers and falsitymakers) with ordinary (core) deductive proofs. The suggestion to be entertained is that we consider, and take seriously as theoretical objects to work with, what we shall call *conditional M-relative constructs*.

Definition 21. Conditional *M*-relative constructs *are defined inductively in the obvious way by appeal to the rules of Core Logic, supplemented by the two M-relative rules $\forall\mathcal{V}$ and $\exists\mathcal{F}$.*

Recall that all the other $\mathcal{V}$- and $\mathcal{F}$-rules are already rules of Core Logic.

Observation 11. *M-relative truthmakers, M-relative falsitymakers, and core proofs are all conditional M-relative constructs.*

We shall proceed to establish the appropriate analogue of Metatheorem 4, for conditional *M*-relative constructs. To this end we seek an appropriate partition of Π–Σ construct-pairs for the 'non-grounding' cases (as we did in Chapter 6), so that we can provide a definition of the recursive function $[\Pi, \Sigma]^M$ (where the superscript *M* reminds one of the *M*-relativity involved). All our earlier definitions of such notions as *target sequent, subsequent*, etc. carry over in the obvious way to this more general setting.

---

[1] Recall that sentences are special cases of saturated formulae.

9.2 Reducts for *M*-relative conditional constructs: grounding cases

The reader will see from the last line in the table that follows that we have to be able to cater, in our definition of $[\Pi, \Sigma]^M$, for cases where the sentence that Π verifies (in M) is *not* a premise of Σ.

Cases	Sub-cases	in which	$[\Pi, \Sigma]^M$	$p[\Pi, \Sigma]^M$	$c[\Pi, \Sigma]^M$
$c\Pi \in p\Sigma$	Π trivial	$p\Pi = \{c\Pi\}$	$=_{df} \overline{A}^M$	$= \{A\} = \{A\} \cup \emptyset$	$= A$
	and	$= p\Sigma = \{c\Sigma\}$		$= \{A\} \cup (\{A\} \setminus \{A\})$	$= c\Sigma$
	Σ trivial	$= \{A\}$		$= p\Pi \cup (p\Sigma \setminus \{c\Pi\})$	
	Π trivial	$p\Pi = \{c\Pi\}$	$=_{df} \Sigma[A/\overline{A}^M]$	$= p\Sigma$	$= c\Sigma$
	and			$= \{c\Pi\} \cup (p\Sigma \setminus \{c\Pi\}$	
	Σ non-trivial			$= p\Pi \cup (p\Sigma \setminus \{c\Pi\})$	
	Π non-trivial	$\{c\Pi\} = p\Sigma$	$=_{df} \Pi$	$= p\Pi$	$= c\Pi$
	and	$= \{c\Sigma\};$ and		$= p\Pi \cup \emptyset$	$= c\Sigma$
	Σ trivial	$c\Pi = c\Sigma$		$= p\Pi \cup (p\Sigma \setminus \{c\Pi\})$	
$c\Pi \notin p\Sigma$			$=_{df} \Sigma$	$= p\Sigma$	$= c\Sigma$
				$= p\Sigma \setminus \{c\Pi\}$	
				$\subseteq p\Pi \cup (p\Sigma \setminus \{c\Pi\})$	

As the table shows, we have the following transformations in grounding cases.

1. Π and Σ are trivial, so $\Pi = \Sigma = A$. Here $[\Pi, \Sigma]^M =_{df} \overline{A}^M$

2. Π is trivial, Σ is non-trivial, and $c\Pi \in p\Sigma$. Here $[\Pi, \Sigma]^M =_{df} \Sigma$

3. Π is non-trivial, Σ is trivial and $c\Pi \in p\Sigma$. Here $[\Pi, \Sigma]^M =_{df} \Pi$

4. $c\Pi \notin p\Sigma$. Here $[\Pi, \Sigma]^M =_{df} \Sigma$

Lemma 7. *In each grounding case,*
(i) $p[\Pi, \Sigma]^M \subseteq p\Pi \cup (p\Sigma \setminus \{c\Pi\})$ and
(ii) $c[\Pi, \Sigma]^M = c\Sigma$.

Proof. Clear by inspection of the table above. $\qquad\square$

9.3 Non-grounding Π–Σ pairs

All non-grounding Π–Σ pairs satisfy the following conditions.

(i) Π is non-trivial;

(ii) Σ is non-trivial; and

(iii) Π connects with Σ, i.e. $c\Pi \in p\Sigma$

Recall that α is the dominant operator of A.

In a non-grounding case the arguments Π, Σ for the operation $[\Pi, \Sigma]$ can be represented graphically by the annotated natural-deduction-style schemata

$$
\begin{array}{ccc}
& \Delta & \\
\Pi \; \textit{non-trivial} \;\longrightarrow\; & \Pi \, \rho & \qquad\qquad \Sigma \; \textit{non-trivial} \;\longrightarrow\; \; \begin{array}{c} A\,,\,\Gamma \\ \Sigma \, \sigma \\ \theta \end{array}\\
\rho \neq \neg\text{-}E & A &
\end{array}
$$

Note: the last step of Π is called ρ; the last step of Σ is called σ

9.4 Transformations for reducts in non-grounding cases

Remember, σ is the final step of the construct Σ, and ρ is the final step of the construct Π. We partition the 'non-grounding' possibilities for the ordered pair $\langle \Pi, \Sigma \rangle$ as follows. We start with a major dichotomy—whether the cut sentence A stands as the major premise of the final step σ of the proof Σ. If, on one hand, A does *not* so stand, then we look at a fourfold partition of possibilities, depending on whether σ is an application of an I-rule, or an E-rule, or $\forall\mathcal{V}$ or $\exists\mathcal{F}$. If, on the other hand, A *does* so stand, then we look at a different partition of possibilities, determined this time by the final step ρ of the construct Π, which can only be an application of an I-rule or of $\forall\mathcal{V}$.

The partition of possibilities, the names of the appropriate kinds of transformations they call for, and detailed statements of the individual transformations (using our compact notation), are summarized in §9.4.1.

Our transformations for conditional constructs are very much like the earlier ones for core proofs, with the exception that they are designed to make reducts truthmaker-like rather than proof-like. Recall that we used two special distribution conversions dealing, respectively, with proofs (here, more generally, constructs) Σ ending either with Universal Introductions (steps of $\forall$I), or with Existential Eliminations (steps of $\exists$E) for $\bot$ as conclusion. Such steps had to, respectively, become steps of universal *verification* $\forall\mathcal{V}$, or existential *falsification* $\exists\mathcal{F}$, when the first argument Π in $[\Pi, \Sigma]$ was a truthmaker. We intend here to maintain this pressure, whatever the exact nature of the constructs Π and Σ (truthmaker, proof, or some hybrid thereof), to replace steps of $\forall$I with ones of $\forall\mathcal{V}$, and steps of $\exists$E with ones of $\exists\mathcal{F}$. Pay careful attention, accordingly, to our definitions of $[\Pi, \forall_I\Sigma_1]$ and of $[\Pi, \exists_E\Sigma_1\bot]$ below.

Note that a is the parameter for the application of the quantifier rule in question—$\forall$I or $\exists$E. The interpretations to be given to the parenthetically enclosed '$\{\forall_\mathcal{V}\}$' and '$\{\exists\mathcal{F}\}$' are clear in the circumstances: apply the rule in the parentheses if, but only if, none of the reducts involving immediate subconstructs (reducts of the form $[\Pi, \Sigma_1{}^a_\delta]$) already establishes a subsequent of the target sequent; otherwise, choose an appropriate $[\Pi, \Sigma_1{}^a_\delta]$ and be done. Either way, the resulting construct establishes a subsequent of the target sequent.

9.4.1   List of transformations

First we consider those cases where the final step in Σ does *not* have the cut sentence as its major premise for elimination or falsification.

Distribution conversions

$[\Pi, \neg_I\Sigma_1]$	$\{\neg_I\}[\Pi, \Sigma_1]$
$[\Pi, \wedge_I\Sigma_1\Sigma_2]$	$\{\wedge_I\}[\Pi, \Sigma_1][\Pi, \Sigma_2]$
$[\Pi, \vee_I\Sigma_1]$	$\{\vee_I\}[\Pi, \Sigma_1]$
$[\Pi, \rightarrow_I\Sigma_1]$	$\{\rightarrow_I\}[\Pi, \Sigma_1]$
$[\Pi, \exists_I\Sigma_1]$	$\{\exists_I\}[\Pi, \Sigma_1]$
$[\Pi, \forall_I\Sigma_1]$	$\{\forall_\mathcal{V}\}\{[\Pi, \Sigma_1{}^a_\delta]\}_{\delta\in D}$
$[\Pi, \forall\mathcal{V}\{\Sigma_\delta(\delta)\}_{\delta\in D}]$	$\{\forall_\mathcal{V}\}\{[\Pi, \Sigma_\delta(\delta)]\}_{\delta\in D}$
$[\Pi, \neg_E\Sigma_1]$	$\{\neg_E\}[\Pi, \Sigma_1]$
$[\Pi, \wedge_E\Sigma_1]$	$\{\wedge_E\}[\Pi, \Sigma_1]$
$[\Pi, \vee_E\Sigma_1\Sigma_2]$	$\{\vee_E\}[\Pi, \Sigma_1][\Pi, \Sigma_2]$
$[\Pi, \rightarrow_E\Sigma_1\Sigma_2]$	$\{\rightarrow_E\}[\Pi, \Sigma_1][\Pi, \Sigma_2]$
$[\Pi, \exists_E\Sigma_1\theta] \ (\theta \neq \bot)$	$\{\exists_E\}[\Pi, \Sigma_1]$
$[\Pi, \exists_E\Sigma_1\bot]$	$\{\exists_\mathcal{F}\}\{[\Pi, \Sigma_1{}^a_\delta]\}_{\delta\in D}$
$[\Pi, \exists_\mathcal{F}\{\Sigma_\delta(\delta)\}_{\delta\in D}]$	$\{\exists_\mathcal{F}\}\{[\Pi, \Sigma_\delta(\delta)]\}_{\delta\in D}$
$[\Pi, \forall_E\Sigma_1]$	$\{\forall_E\}[\Pi, \Sigma_1]$

Next are the cases where the final step in Σ *does* have the cut sentence as its major premise for elimination or falsification. $d(t)$ is the denotation of t.

Permutative conversions

Final step of Π is $\exists\mathcal{F}$ or βE, $\beta \neq \neg$, β not dominant in the cut sentence

$[\exists_\mathcal{F}\{\Pi_\delta(\delta)\}_{\delta\in D}, \Sigma]$ $\{\exists_\mathcal{F}\}\{[\Pi_\delta(\delta), \Sigma]\}_{\delta\in D}$

$[\wedge_E\Pi_1, \Sigma]$ $\{\wedge_E\}[\Pi_1, \Sigma]$

$[\vee_E\Pi_1\Pi_2, \Sigma]$ $\{\vee_E\}[\Pi_1, \Sigma][\Pi_2, \Sigma]$

$[\rightarrow_E\Pi_1\Pi_2, \Sigma]$ $\{\rightarrow_E\}\Pi_1[\Pi_2, \Sigma]$

$[\exists_E\Pi_1, \Sigma]$ $\{\exists_E\}[\Pi_1, \Sigma]$

$[\forall_E\Pi_1, \Sigma]$ $\{\forall_E\}[\Pi_1, \Sigma]$

I-reductions

Final step of Π is αI, where α is dominant in the cut sentence. (NB: $\alpha \neq \boldsymbol{\alpha}$.)

$[\neg_I\Pi_1, \neg_E\Sigma_1]$ $[[\Pi, \Sigma_1], \Pi_1]$

$[\wedge_I\Pi_1\Pi_2, \wedge_E\Sigma_1]$ $[\Pi_1, [\Pi_2, [\Pi, \Sigma_1]]]$

$[\vee_I\Pi_1, \vee_E\Sigma_1\Sigma_2]$ $[\Pi_1, [\Pi, \Sigma_i]], (i = 1, 2)$

$[\rightarrow_{I(a)}\Pi_1, \rightarrow_E\Sigma_1\Sigma_2]$ $[[\Pi, \Sigma_1], \Pi_1]$

$[\rightarrow_{I(b)}\Pi_1, \rightarrow_E\Sigma_1\Sigma_2]$ $[[\Pi, \Sigma_1], [\Pi_1, [\Pi, \Sigma_2]]]$

$[\exists_I\Pi_1(t), \exists_E\Sigma_1(a)]$ $[\Pi_1, [\Pi, \Sigma_{1t}^a]]$

$[\exists_I\Pi_1(t), \exists_\mathcal{F}\{\Sigma_\delta(\delta)\}_{\delta\in D}]$ $[\Pi_1(d(t)), [\Pi, \Sigma_{d(t)}(d(t))]]$

$[\forall_I\Pi_1(a), \forall_E\Sigma_1(t_1,...,t_n)]$ $[\Pi_{1t_1}^a, \ldots, [\Pi_{1t_n}^a, [\Pi, \Sigma_1]] \ldots]$

$[\forall_I\Pi_1(a), \forall_\mathcal{F}\Sigma_1(\boldsymbol{\alpha})]$ $[\Pi_1(\boldsymbol{\alpha}), [\Pi, \Sigma_1(\boldsymbol{\alpha})]]$ (for $\boldsymbol{\alpha} \in D$)

$\mathcal{V}$-Reductions

Final step of Π is $\alpha\mathcal{V}$, where α is dominant in the cut sentence. (NB: $\alpha \neq \boldsymbol{\alpha}$.)

$[\exists_\mathcal{V}\Pi_1(\boldsymbol{\alpha}), \exists_E\Sigma_1(a)]$ $[\Pi_1, [\Pi, \Sigma_{1\alpha}^a]]$ (for $\boldsymbol{\alpha} \in D$)

$[\exists_\mathcal{V}\Pi_1(\boldsymbol{\alpha}), \exists_\mathcal{F}\{\Sigma_\delta(\delta)\}_{\delta\in D}]$ $[\Pi_1(\boldsymbol{\alpha}), [\Pi, \Sigma_\alpha(\boldsymbol{\alpha})]]$ (for $\boldsymbol{\alpha} \in D$)

$[\forall_\mathcal{V}\{\Pi_\delta(\delta)\}_{\delta\in D}, \forall_E\Sigma_1(t_1,...,t_n)]$ $[\Pi_{\delta d(t_1)}^\delta, \ldots, [\Pi_{\delta d(t_n)}^\delta, [\Pi, \Sigma_1]] \ldots]$

$[\forall_\mathcal{V}\{\Pi_\delta(\delta)\}_{\delta\in D}, \forall_\mathcal{F}\Sigma_1(\boldsymbol{\alpha})]$ $[\Pi_\alpha(\boldsymbol{\alpha}), [\Pi, \Sigma_1(\boldsymbol{\alpha})]]$ (for $\boldsymbol{\alpha} \in D$)

Note that the list of the forms of $[\Pi, \Sigma]$ that need to be considered for conversions or reductions does not have to be combinatorially exhaustive by including (among reductions, say) each and every one of the forms

$$[@_I \ldots, @_E \ldots]$$
$$[@_I \ldots, @_\mathcal{F} \ldots]$$
$$[@_\mathcal{V} \ldots, @_E \ldots]$$
$$[@_\mathcal{V} \ldots, @_\mathcal{F} \ldots]$$

This is because

$@_I$ subsumes $@_V$ for $@ = \neg, \wedge, \vee, \rightarrow, \exists$;

$@_E$ subsumes $@_F$ for $@ = \neg, \wedge, \vee, \rightarrow, \forall$.

9.5   Results

Observation 12. *In the foregoing inspectably exhaustive table of transformations for non-grounding cases of pairs Π–Σ, the definiendum reduct $[\Pi, \Sigma]$ —whose non-grounding cases are listed in the first column—is defined in terms of other reducts in the respective ways listed in the second column. Each of the latter reducts is of one of the forms $[\Pi', \Sigma']$, $[\Pi, \Sigma']$, or $[\Pi', \Sigma]$, where Π' is an immediate subconstruct of Π, and Σ' is an immediate subconstruct of Σ. Thus the pairs involved in reducts within the definiens are of lower 'combined height' complexity than the definiendum reduct—and infinitary furcations are irrelevant to this complexity measure.*

Lemma 8. *In each non-grounding case for conditional evaluation constructs, the property*

$p[\Pi, \Sigma] \subseteq p\Pi \cup (p\Sigma \setminus \{c\Pi\})$ *and* $c[\Pi, \Sigma] = c\Sigma$
(whence $s[\Pi, \Sigma] \sqsubseteq s(\Pi * \Sigma)$)

is preserved from the reducts in the definiens to the reduct being defined.

Proof. By straightforward inspection of cases.   □

Metatheorem 12. *There is a method $[\ , \ ]_M$ such that:*

1. *if M is finite, then $[\ , \ ]_M$ transforms any two conditional M-relative constructs*

$$
\begin{array}{cc}
\Delta & A, \Gamma \\
\Pi & \Sigma \\
A & \theta
\end{array}
\qquad \text{(where } A \notin \Gamma \text{ and } \Gamma \text{ may be empty)}
$$

into an M-relative conditional construct proof $[\Pi, \Sigma]_M$ of θ or of $\perp$ from (some subset of) $\Delta \cup \Gamma$; and

2. *if M is infinite, then the method $[\ , \ ]_M$ is to the same effect, and is at least (in an obvious sense) 'quasi-effective'.*

Proof. By induction on the complexity of Π–Σ pairs, bearing in mind Observation 12, and applying Lemma 8. □

Observation 13. *Metatheorem 4 is the special case of Metatheorem 12 that one obtains by simply dropping the rules $\forall\mathcal{V}$ and $\exists\mathcal{F}$ from consideration.*

Corollary 5. *For all models M, for all M-relative truthmakers*

$$
\begin{array}{ccc}
\Lambda_1 & & \Lambda_n \\
\pi_1 & ,\ldots, & \pi_n \\
\varphi_1 & & \varphi_n
\end{array} ,
$$

and for all $\mathbb{C}$-*proofs*

$$
\underbrace{\varphi_1,\ldots,\varphi_n}_{} \\
\Pi \\
\psi
$$
,

the reduct

$$
\left[\, \pi_1 \,,\, \left[\, \pi_2 \,,\ldots,\, \left[\, \pi_n \,,\, \Pi \,\right]\ldots\right]\right],
$$

i.e.

$$
\left[\begin{array}{c} \Lambda_1 \\ \pi_1 \\ \varphi_1 \end{array} \,,\, \begin{array}{c} \Lambda_2 \\ \pi_2 \\ \varphi_2 \end{array} \,,\ldots,\, \left[\begin{array}{cc} \Lambda_n & \overbrace{\varphi_1,\ldots,\varphi_n} \\ \pi_n \,, & \Pi \\ \varphi_n & \psi \end{array} \,\right]\cdots\right]
$$

is an M-relative truthmaker for ψ; and its premises are in $\bigcup_i \Lambda_i$.

Proof. Immediate by Metatheorem 12, Observation 11, and the consideration that for any model M the set of M-literals is coherent. □

9.6 Some reflections on methodology

The two special marks of model-relative evaluations (truthmakers and falsity-makers) are (i) they can involve individuals from the domain of the model, via saturated formulae at their nodes; and (ii) they can contain applications of the quantifier rules $\forall\mathcal{V}$ and $\exists\mathcal{F}$, which involve premise instances for each and every member of the domain. This in turn means that, if the domain of the model is infinite, such evaluations will, at those rule applications, branch 'infinitely sideways'. Evaluations track 'the real'—as constituted by, or consisting of, the individuals that exist in the domain of the model, and the atomic properties and relations that they enjoy (or lack) therein.

The two contrasting special marks of (core) proofs, *qua* proofs, are (respectively) that they are purely syntactic, and that they are finite (hence, in principle, can be inspected for formal correctness). Proofs track 'the ideal' or 'the imaginary' lines of representation and thought that have validity over all possible ways—actual or only imagined—that 'the real' might happen to be.

By hybridizing evaluations with core proofs, we have arrived at constructs that partake of both the real and the ideal. Our choice of terms here is intended to impress upon the reader an obvious analogy with the mathematics of real and complex numbers. It is well known that a result about the real numbers can be obtained by making an excursus into the theory of the complex numbers. One thereby makes feasible the proof of a theorem about the reals that has only infeasible proofs within the strict confines of real number theory. The suggestion here is that our conditional constructs have played a role analogous to that played by the complex numbers. We have established Corollary 5, which tells us 'what goes on at the level of the real' (truthmakers for the premises mapping to a truthmaker for the conclusion); and we have established this result by making an excursus into the more complex theory of conditional constructs.

All the meta-reasoning culminating in the proof of Corollary 5 can be feasibly formalized by a *core proof* in the theory of iterated inductive definitions that furnishes definitions of the notions of truthmaker, falsitymaker, core proof and conditional construct. The definitions are given by basis and inductive clauses codifiable as rules of introduction for the definienda, with the closure clauses being codifiable as the corresponding elimination rules for the same.

9.7 Formal semantics reworked in terms of truthmakers

Recall Definition 19 of the consequence relation $\models_{\mathbf{T}}$ defined in terms of truthmakers:

$$\varphi_1, \ldots, \varphi_n \models_{\mathbf{T}} \psi$$

$$\Leftrightarrow_{df}$$

$$\exists \text{q.-eff.} f \, \forall M \, \forall \pi_1 \ldots \forall \pi_n$$

$$[((\mathcal{V}_M(\pi_1, \varphi_1) \wedge \ldots \wedge \mathcal{V}_M(\pi_n, \varphi_n)) \Rightarrow \mathcal{V}_M(f(\pi_1, \ldots, \pi_n), \psi)]$$

One way to realize the quasi-effective f that is needed in order to establish logical consequence is to construct a thing—call it Π—'out of' $\varphi_1, \ldots, \varphi_n, \psi$, in terms of which one can specify an operation f_Π on both M-relative truth-makers for the premises $\varphi_1, \ldots, \varphi_n$ and on Π itself, so that $f_\Pi(\pi_1^M, \ldots, \pi_n^M, \Pi)$ is the sought truthmaker π^M for ψ:

$$
\left.
\begin{array}{c}
\dfrac{\pi_1^M}{\varphi_1} \quad , \ldots, \quad \dfrac{\pi_n^M}{\varphi_n} \\[2mm]
\underbrace{\varphi_1 \quad , \ldots, \quad \varphi_n}_{\Pi} \\[2mm]
\psi
\end{array}
\right\}
\quad
\begin{array}{cc}
f_\Pi & \pi^M \\
\rightsquigarrow & \psi
\end{array}
$$

Conjecture 5. Π *can always be taken to be a core proof, with f_Π as the normalization operation of Corollary 5.*

Chapter 10

The Relevance Properties of Core Logic

Abstract

We begin by contrasting our approach to relevance with that of Anderson and Belnap. In §10.1 we point out a reflexive irony, which is that they argue for the rejection of Disjunctive Syllogism by means of an argument that appears to employ it. We then prepare the ground for a best possible variable-sharing result, which will be established for Classical Core Logic $\mathbb{C}^+$, and which will therefore hold for any of its subsystems, including Core Logic $\mathbb{C}$, the Logic of Evaluation $\mathbb{E}$, Truth-Table Logic $\mathbb{T}$, *and* the Anderson–Belnap system **R**.

§10.2 is devoted to positive and negative occurrences of primitive expressions and subformulae within formulae, and various syntactic notions that can be defined in terms of such occurrences. The focus is usually on some same atom (say) enjoying occurrences of the same or of opposite parity within two sentences. The concept-building culminates in the definition of a very exigent relevance condition $\mathcal{R}(\Delta, \varphi)$ on the premise set Δ and the conclusion φ of any proof. It should be emphasized that this *includes* first-order proofs, since the variable-sharing tradition is so often concerned only with the propositional case.

§10.3 explains why $\mathcal{R}$ is a best possible explication of the sought notion of relevance.

§10.4 reports the main result of Tennant [2015d] on relevance: for every proof of φ from Δ in Classical Core Logic $\mathbb{C}^+$, we have $\mathcal{R}(\Delta, \varphi)$.

§10.5 argues that our main result is optimal, and challenges relevantists in the Anderson–Belnap tradition to identify *any strengthening* of the relation $\mathcal{R}(\Delta, \varphi)$ that can be shown to hold for some subsystem of **R** but that can be shown to *fail* for $\mathbb{C}^+$.

Core Logic $\mathbb{C}$ can be thought of as the result of relevantizing Intuitionistic Logic **I** by rethinking the forms of Introduction and Elimination rules and banning the rule of *Ex Falso Quodlibet*. Classical Core Logic $\mathbb{C}^+$ can be thought of as the result of relevantizing Classical Logic **C** in the same way. In each case (intuitionistic or classical) the relevantized system furnishes, as we have seen, proofs for all theorems, all inconsistencies, and all consequences of consistent sets of premises of the 'parent' system that it relevantizes.

The Core logician relevantizes, however, at the 'level of the turnstile', and not by dramatically altering the logical behavior of the object language conditional $\rightarrow$. The Core logician seeks to *preserve* the usual meanings of the logical operators. The aim is only to cleanse both Classical and Intuitionistic logic of certain irrelevancies. These are the irrelevancies that arise, essentially, from applications—in their systems of natural deduction—of the Absurdity Rule and of certain forms of introduction and elimination rules that permit 'vacuous discharge' of assumptions. In the corresponding sequent calculi, the culprit rules are those of Thinning and of (unrestricted) Cut, as well as those Right and Left rules that have hitherto countenanced the corresponding vacuousness of certain discharges.

Usually, relevance logicians seek to demonstrate the success of their systems (in *relevantizing*) by proving 'variable-sharing' results. (See Tennant [2015d] for a survey.) So the question arises: *what sort of variable-sharing results might one be able to prove for the Core systems* $\mathbb{C}$ *and* $\mathbb{C}^+$? With Core Logic so effectively shadowing Intuitionistic Logic, and with Classical Core Logic doing the same with Classical Logic—as shown by the aforementioned (meta)theorems—one might expect the prospects to be rather dim for establishing a strong variable-sharing result in the propositional case, and a strong extralogical vocabulary-sharing result in the first-order case. This chapter reports a theorem that decisively confutes such an expectation. The main result of §10.4 (for the propositional case) can be extended to both the classical and the first-order cases. (For the full details, see Tennant [2015d].)

In automated deduction the aim is to write programs to solve deductive problems of the form 'Is there a proof of φ from Δ?'. It is very useful to have 'relevance filters' to weed out as many 'no-hopers' as possible before even embarking on a serious search for a proof. The main results of this chapter furnish a strongest possible relevance filter. They should provide other relevance logicians with a stimulus to investigate the following question:

> Can extant variable-sharing results that have already been established by those logicians for the various systems of relevance logic that they favor or find of interest be strengthened *beyond* the variable-sharing result that is stated in §10.4 for the system of Core Logic (and its extension to Classical Core Logic)?

An extensive argument against this possibility is provided in §10.3.1. Note that under the circumstances it is rather ironic that our variable-sharing result for Classical Core Logic (containing, as it does, Disjunctive Syllogism, the *bête noire* of relevantists in the Anderson–Belnap tradition) holds also for the Anderson–Belnap system **R** of relevance logic, and of course for any of the latter's subsystems. This is because every sequent provable in **R** is provable in $\mathbb{C}^+$. The result holds also for the Logic of Evaluation $\mathbb{E}$, since all its deducibilities are Core deducibilities.

10.1   A reflexive irony in the motivation for R

We have not yet entered into any polemics over the difference between our approach to relevantizing, and that of Anderson and Belnap, as epitomized in their system **R**. In this section we raise a problem of reflexive stability for their approach.

This much is common to their approach and mine: *we have to avoid the First Lewis Paradox*. That is, we cannot allow the sequent $A, \neg A : B$ to be provable. So both approaches confront the problem of how best to ward off Lewis's own clever little derivation of B from the premises $A, \neg A$. To remind the reader (or to introduce the reader who is not well-versed in the minutiae of relevant logics), Lewis's informal derivation ran as follows.

> Suppose A. Then by ($\lor$I) we have $A \lor B$. Now suppose $\neg A$. By *Disjunctive Syllogism*, we have B. Hence, by *Unrestricted Transitivity of Deduction*, we have B following from $A, \neg A$.

There is another small thing we—Anderson and Belnap, and I—hold in common: the rule ($\lor$I) is irreproachable. So we are all left contemplating this predicament:

> We have two principles in play, namely *Disjunctive Syllogism* and *Unrestricted Transitivity of Deduction*. Together they lead to disaster. So, one of them has to go.

The question is: *which* one? Here is how Anderson and Belnap, in effect, reasoned:

> Either we can give up *Unrestricted Transitivity of Deduction* or we can give up *Disjunctive Syllogism*. But we cannot give up *Unrestricted Transitivity of Deduction*. So, we can (hence, in these circumstances, must) give up *Disjunctive Syllogism*.

Not to put too fine a point on it: this move has a gratifyingly familiar ring. One discerns in it the logical form

> Either A or B; but not-A; so, B.

So Anderson and Belnap have employed Disjunctive Syllogism in order to argue that we ought to give up ... *Disjunctive Syllogism!* Consider, by contrast, the way the Core logician argues out of the same predicament, but for a different reformist conclusion:

> Either we can give up *Unrestricted Transitivity of Deduction* or we can give up *Disjunctive Syllogism*. But we cannot give up *Disjunctive Syllogism*. So, we can (hence, in these circumstances, must) give up *Unrestricted Transitivity of Deduction*.

Here, the legitimacy of the use of Disjunctive Syllogism within the argument is affirmed by one of its premises. So there is no reflexive instability, the way there is with the Anderson–Belnap argument.

Now, one can disprove a proposition φ by assuming φ for the sake of argument, and then deriving from it the conclusion $\neg\varphi$. But I am not at all sure that one can argue for the rejection of a logical principle by using that very principle in one's argument for the conclusion that one should reject it. I shall leave the reader to mull over this, and the subtle difference that might be involved between propositions and rules. In the meantime, we need to revert to our exposition of the exceptionally nice form of relevance that can be established as obtaining between premises and conclusions of (classical) core proofs.

10.2 Definitions of concepts, and easy results

This section characterizes various syntactic properties and relations that will be used in order to formulate precisely the variable-sharing property enjoyed by (Classical) Core Logic (in the propositional case, where the 'variables' in question are the atomic sentences of the language). The treatment can be extended to first-order logic, with the quantifiers $\forall$ and $\exists$.[1] The language

---

[1] In the first-order case it is no longer appropriate to speak of 'variable' sharing, especially since there is the risk of confusion with the individual variables that can enjoy both free and bound occurrences within first-order formulae. Rather, we can speak in the first-order case of *predicate*-sharing, since all atomic formulae of a first-order language are formed from primitive predicates. But that is beside the point; here we focus on the propositional case, and the usual terminology of (propositional) variable-sharing.

in the propositional case is based on a countable set of *atoms A, B, C, ...*
Sentences can be built up from them in the usual way by means of the
connectives ¬, ∧, ∨, and →. These must all be taken as primitive, of course,
since Core Logic is contained within Intuitionistic Logic, and in the latter
system none of these connectives can be defined in terms of the others.

Definition 22. *An* atom *is a propositional variable.*

Remark 1. ⊥ *is not an atom.* ⊥ *never occurs as a subformula; it is merely
a punctuation device in proofs, used in order to register absurdity.*

10.2.1   On positive and negative occurrences of subsentences

We are about to define the notion of a positive (resp., negative) occurrence
of a subsentence within a sentence. This is a very useful notion, exploited,
for example, in Lyndon's strengthening of Craig's Interpolation Theorem
(see Craig [1957] and Lyndon [1959]). For the reader unfamiliar with signed
occurrences of subsentences, a good motivation for the signing is provided
by a discipline that emerged only later, in the 1960s—automated deduction.
Keeping track of positive and negative occurrences of subsentences within
premises and conclusion of a deductive problem helps to determine which
rules to try to apply in a search for a proof. To take a simple example: if
one is searching for a proof of $P \to Q$, by →-Introduction (so-called Conditional Proof), from premises Δ, then one will set oneself the *subproblem* of
finding a proof of the consequent Q of the conditional from its antecedent P
(along with Δ). Because sought conclusions are taken as 'positive', and the
assumptions or premises that one is using are taken as 'negative', this means
that the consequent Q is a positive subsentence occurrence within $P \to Q$,
while its antecedent P is a negative one. (Indeed, the terminology of 'consequent part' and 'antecedent part' is sometimes adopted instead of the talk
of positive and negative subsentence occurrences, respectively.) The considerations just entered iterate. In the sentence $(R \to S) \to Q$, considered
as a conclusion to be deduced, the subsentence R enjoys a *positive* occurrence: for, in order for R to be used along with the assumption $(R \to S)$ for
Conditional Proof of Q, R will have to be deduced as the *conclusion* of an
appropriate subproof.

Definition 23. *We use the expression* $\chi \preceq^+ \theta$ *to mean that the subsentence
occurrence* χ *is a positive one within the sentence* θ. *Likewise,* $\chi \prec^- \theta$ *means
that the subsentence occurrence* χ *is a negative one within the sentence* θ.
The co-inductive definition of these relations is as follows.

1. *Every sentence is a positive subsentence occurrence in itself:*
$$\varphi \preceq^+ \varphi$$

2. *All positive [resp., negative] subsentence occurrences in φ are negative [resp., positive] subsentence occurrences in $\neg\varphi$:*
$$\chi \preceq^+ \varphi \Rightarrow \chi \prec^- \neg\varphi$$
$$\chi \prec^- \varphi \Rightarrow \chi \preceq^+ \neg\varphi$$

3. *All positive [resp., negative] subsentence occurrences in φ and ψ are positive [resp., negative] subsentence occurrences in $(\varphi \wedge \psi)$:*
$$\chi \preceq^+ \varphi \Rightarrow \chi \preceq^+ (\varphi \wedge \psi) \qquad \chi \prec^- \varphi \Rightarrow \chi \prec^- (\varphi \wedge \psi)$$
$$\chi \preceq^+ \psi \Rightarrow \chi \preceq^+ (\varphi \wedge \psi) \qquad \chi \prec^- \psi \Rightarrow \chi \prec^- (\varphi \wedge \psi)$$

4. *All positive [resp., negative] subsentence occurrences in φ and ψ are positive [resp., negative] subsentence occurrences in $(\varphi \vee \psi)$:*
$$\chi \preceq^+ \varphi \Rightarrow \chi \preceq^+ (\varphi \vee \psi) \qquad \chi \prec^- \varphi \Rightarrow \chi \prec^- (\varphi \vee \psi)$$
$$\chi \preceq^+ \psi \Rightarrow \chi \preceq^+ (\varphi \vee \psi) \qquad \chi \prec^- \psi \Rightarrow \chi \prec^- (\varphi \vee \psi)$$

5. *All positive [resp., negative] subsentence occurrences in ψ are positive [resp., negative] subsentence occurrences in $(\varphi \rightarrow \psi)$:*
$$\chi \preceq^+ \psi \Rightarrow \chi \preceq^+ (\varphi \rightarrow \psi)$$
$$\chi \prec^- \psi \Rightarrow \chi \prec^- (\varphi \rightarrow \psi)$$

6. *All positive [resp., negative] subsentence occurrences in φ are negative [resp., positive] subsentence occurrences in $(\varphi \rightarrow \psi)$:*
$$\chi \preceq^+ \varphi \Rightarrow \chi \prec^- (\varphi \rightarrow \psi)$$
$$\chi \prec^- \varphi \Rightarrow \chi \preceq^+ (\varphi \rightarrow \psi)$$

7. *If $\chi \preceq^+ \theta$ then this can be shown by clauses (1)–(6)*

8. *If $\chi \prec^- \theta$ then this can be shown by clauses (1)–(6)*

When displaying a formula, we shall use the overhead annotation $\oplus$ for positive atom-occurrences within it, and $\ominus$ for negative ones.

Examples (with overhead parity labeling):

$$\overset{\ominus}{\neg D} \qquad \overset{\ominus}{A} \rightarrow \overset{\oplus}{C} \qquad (\overset{\ominus}{A} \vee \overset{\ominus}{C}) \rightarrow \overset{\oplus}{B}$$

$$\overset{\oplus}{\neg\neg A} \qquad \overset{\oplus}{D} \vee \overset{\oplus}{A} \qquad (\overset{\oplus}{D} \rightarrow \overset{\ominus}{A}) \rightarrow \overset{\oplus}{B}$$

$$\overset{\ominus}{\neg\neg\neg C} \qquad \neg(\overset{\oplus}{B} \rightarrow \overset{\ominus}{C}) \qquad (\overset{\ominus}{A} \vee \neg\neg\overset{\ominus}{C}) \rightarrow \overset{\ominus}{\neg B}$$

$$\overset{\oplus}{A} \wedge \overset{\oplus}{C} \qquad \neg((\overset{\oplus}{B} \wedge \overset{\oplus}{D}) \rightarrow \overset{\ominus}{A}) \qquad (\overset{\oplus}{D} \rightarrow \overset{\oplus}{\neg A}) \rightarrow \overset{\oplus}{B}$$

10.2.2 Further syntactic properties and relations defined in terms of signed occurrences

Definition 24. $\varphi \underset{A}{\approx} \psi \equiv_{df}$ *the atom A has occurrences of the same* parity *in φ and in ψ; that is, either $(A \preceq^{+} \varphi$ and $A \preceq^{+} \psi)$ or $(A \prec^{-} \varphi$ and $A \prec^{-} \psi)$.*

Definition 25. $\varphi \approx \psi \equiv_{df}$ *for some atom A we have $\varphi \underset{A}{\approx} \psi$.*

Examples (with overhead parity labeling for sufficient witnessing):

$$\neg \overset{\ominus}{D} \approx \neg\neg\neg \overset{\ominus}{D} \qquad \overset{\ominus}{A} \to C \approx C \vee \neg \overset{\ominus}{A} \qquad\qquad (\overset{\ominus}{A}\vee C) \to D \approx (D \to \overset{\ominus}{A}) \to B$$

$$\neg\neg \overset{\oplus}{A} \approx \overset{\oplus}{A} \wedge C \qquad \neg(\overset{\oplus}{B} \to A) \approx \neg((\overset{\oplus}{B} \wedge D) \to A) \qquad (A\vee C) \to \overset{\oplus}{D} \approx (\overset{\oplus}{D} \to A) \to B$$

$$A \to \overset{\oplus}{C} \approx \overset{\oplus}{C} \vee \neg A \qquad \neg(B \to \overset{\ominus}{A}) \approx \neg((B\wedge D) \to \overset{\ominus}{A})$$

Definition 26. $\varphi \approx \Delta \equiv_{df}$ *for some ψ in Δ, we have $\varphi \approx \psi$; that is, some atom has the* same *parity (positive or negative, at some occurrence) in φ as it has at some occurrence in* some *member of Δ.*

Examples:

$$\neg\neg \overset{\oplus}{A} \approx \{\overset{\oplus}{A} \wedge C,\ \neg D\} \qquad\qquad \neg \overset{\ominus}{A} \approx \{\overset{\ominus}{A} \to C,\ D \wedge A\}$$

$$\neg \overset{\ominus}{A} \approx \{(D \to \overset{\ominus}{A}) \to B,\ A \vee D\} \qquad \neg \overset{\ominus}{A} \approx \{(\overset{\ominus}{A} \vee C) \to B,\ A \vee D\}$$

$$\neg(\overset{\oplus}{B} \to C)\ \approx\ \{\neg((\overset{\oplus}{B} \wedge D) \to A),\ \neg\neg C\}$$

Remark 2. *For atoms A, we have $A \preceq^{+} \varphi$ if and only if $A \approx \varphi$.*

Definition 27. $\varphi \bowtie \psi \underset{A}{} \equiv_{df}$ *the atom A has occurrences of* opposite *parities in φ and in ψ; that is, either $(A \preceq^{+} \varphi$ and $A \prec^{-} \psi)$ or $(A \prec^{-} \varphi$ and $A \preceq^{+} \psi)$.*

Definition 28. $\varphi \bowtie \psi \equiv_{df}$ *for some atom A we have $\varphi \underset{A}{\bowtie} \psi$.*

Examples (with overhead parity labeling for sufficient witnessing):

$$\overset{\oplus}{A} \wedge \neg A \bowtie A \wedge \neg \overset{\ominus}{A} \qquad \neg \overset{\ominus}{A} \bowtie \overset{\oplus}{A} \wedge C \qquad \neg \overset{\ominus}{A} \bowtie \neg(\overset{\oplus}{A} \wedge B) \to C$$

$$\neg(\overset{\oplus}{B} \to C) \bowtie \overset{\ominus}{B} \to A \qquad \neg \overset{\ominus}{A} \bowtie B \to (\overset{\oplus}{A} \vee C) \qquad \neg\neg \overset{\oplus}{A} \bowtie (D \to \overset{\ominus}{A}) \to B$$

$$A \to \overset{\oplus}{B} \bowtie \overset{\ominus}{B} \to C$$

Definition 29. $\pm\varphi \equiv_{df} \varphi\bowtie\varphi$.

Examples (with overhead parity labeling for sufficient witnessing):

$$\pm(\overset{\ominus}{A}\to\overset{\oplus}{A}) \qquad \pm\neg(\overset{\oplus}{A}\to\overset{\ominus}{A}) \qquad \pm(\overset{\oplus}{A}\wedge\neg\overset{\ominus}{A})$$

$$\pm\neg(\overset{\ominus}{A}\wedge\neg\overset{\oplus}{A}) \qquad \pm((A\to\overset{\oplus}{B})\wedge(\overset{\ominus}{B}\to C)) \qquad \pm(\neg\neg\overset{\ominus}{A}\to(\overset{\oplus}{A}\wedge C))$$

Remark 3. *For atoms A, we have $A\prec^{-}\varphi$ if and only if $A\bowtie\varphi$.*

Definition 30. *A sequence $\varphi_1,\ldots,\varphi_n$ ($n>1$) of pairwise distinct sentences is a $\bowtie$-path connecting φ_1 to φ_n in Δ $\equiv_{df}$ for $1\le i\le n$, φ_i is in Δ, and for $1\le i<n$, $\varphi_i\bowtie\varphi_{i+1}$.*

Example: $\overset{\oplus}{A}$, $\overset{\ominus}{A}\to\overset{\oplus}{B}$, $\overset{\ominus}{B}\to\overset{\oplus}{C}$, $\overset{\ominus}{C}\to\overset{\oplus}{D}$, $\overset{\ominus}{D}\to E$ is a $\bowtie$-path connecting A to $D\to E$ in any set containing all these sentences. For we have the $\bowtie$-path

$$A \underset{A}{\bowtie} (A\to B) \underset{B}{\bowtie} (B\to C) \underset{C}{\bowtie} (C\to D) \underset{D}{\bowtie} (D\to E).$$

Definition 31. *φ and ψ are $\bowtie$-connected in Δ (in symbols: $\varphi\underline{\bowtie}\psi$) $\equiv_{df}$ if $\varphi\ne\psi$, then there is a $\bowtie$-path connecting φ to ψ in Δ.*

Remark 4. *It is an immediate consequence of Definition 31 that if $\varphi\bowtie\psi$ (in any set of sentences), then $\varphi\underline{\bowtie}\psi$.*

Remark 5. *It is an immediate consequence of Definition 31 that $\varphi\underline{\bowtie}\varphi$ (in any set of sentences containing φ), even if it is not the case that $\pm\varphi$.*

Definition 32. *A set Δ of formulae is $\bowtie$-connected $\equiv_{df}$ for all φ, ψ in Δ, if $\varphi\ne\psi$, then $\varphi\underline{\bowtie}\psi$.*

Remark 6. *It is an immediate consequence of Definition 32 that the singleton $\{\varphi\}$ of any sentence φ is $\bowtie$-connected even if it is not the case that $\pm\varphi$.*

Definition 33. *A $\bowtie$-component of Δ is a non-empty, inclusion-maximal $\bowtie$-connected subset of Δ (where the $\bowtie$-connections are established via members of Δ).*

Remark 7. *If φ is in Δ but it is not the case that $\varphi\bowtie\psi$ for any other member ψ of Δ, then $\{\varphi\}$ is a $\bowtie$-component of Δ, even if it is not the case that $\pm\varphi$.*

Observation 14. *Any two $\bowtie$-components of Δ are disjoint.*

Observation 15. *Every sentence in* Δ *is in exactly one $\bowtie$-component of* Δ.

Lemma 9. *The $\bowtie$-components of* Δ *are uniquely determined, and they partition* Δ.

Proof. Immediate from Remark 7 and Observations 14 and 15. □

Example of a set partitioned into its $\bowtie$-components (of which there are three, separated by the two vertical lines):

$$(\overset{\oplus}{A} \wedge H) \bowtie (\overset{\ominus}{A} \to \overset{\oplus}{B}) \bowtie (\overset{\ominus}{B} \to E) \mid \overset{\oplus}{C} \bowtie (\overset{\ominus}{C} \to \overset{\oplus}{D}) \bowtie \neg \overset{\ominus}{D} \mid F \wedge G$$

Note that it is not in general the case that each non-singleton component consists of but a single, linear, $\bowtie$-path. The last example, if altered so that the occurrence of H is replaced by $\neg E$ (say), becomes

$$(\overset{\oplus}{A} \wedge \neg \overset{\ominus}{E}) \bowtie (\overset{\ominus}{A} \to \overset{\oplus}{B}) \bowtie (\overset{\ominus}{B} \to \overset{\oplus}{E}) \mid \overset{\oplus}{C} \bowtie (\overset{\ominus}{C} \to \overset{\oplus}{D}) \bowtie \neg \overset{\ominus}{D} \mid F \wedge G$$

in which the first component consists of a $\bowtie$-path forming a trilateral loop. Another structural possibility for a component is that it be 'spoked' from a 'hub', as with

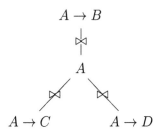

The $\bowtie$-connectedness of components allows for a great variety of '$\bowtie$-structuration' of components. It is essentially the same variety of components that one encounters in finite non-directed graphs. Treat the sentences in Δ as nodes, and let '$\varphi \bowtie \psi$' mean that there is an (undirected) edge in the graph joining the two (distinct) nodes φ and ψ. Any component with n nodes must have at least $(n-1)$ $\bowtie$-edges (in order to ensure $\bowtie$-connectedness), and at most $n(n-1)$ of them (since edges join only distinct nodes). Those are the only constraints on the $\bowtie$-structuration of a component, over and above its $\bowtie$-connectedness.[2]

---

[2] We ignore here the further possibility that for any node φ, we might have $\varphi \bowtie \varphi$.

Definition 34. $\varphi \blacktriangleleft \Delta \equiv_{df}$ *for every $\bowtie$-component Γ of Δ, $\varphi \approx \Gamma$.*

It is worth remarking here that Definition 34 does not by itself provide our sought explication of relevance among premises Δ and conclusion φ of a proof formalizing a good argument. Our explication of that target notion of relevance is given below by Definition 36, which draws on the definiendum $\blacktriangleleft$ of Definition 34 in but one of its three case-clauses.

Remark 8. *We cannot have $\bot \blacktriangleleft \Delta$, since $\bot$ is not a sentence.*

Observation 16. $\varphi \blacktriangleleft \{\varphi\}$.

Definition 35. *Suppose $\Delta \neq \emptyset$. Then*

$$\natural\Delta \equiv_{df} \begin{cases} \text{if } \Delta \text{ is a singleton, say } \{\delta\}, \text{ then } \pm \delta; \\ \text{and} \\ \text{if } \Delta \text{ is not a singleton, then } \Delta \text{ is } \bowtie\text{-connected} \end{cases}$$

10.2.3 The explication of relevance of premises to conclusion of a relevantly valid sequent

Definition 36. *We shall say that a set Δ of premises is* relevantly connected both within itself and to a conclusion φ (in symbols: $\mathcal{R}(\Delta, \varphi)$) *just in case exactly one of the following three conditions is satisfied:*

1. *Δ is non-empty, φ is $\bot$, and $\natural\Delta$.*

2. *Δ is non-empty, φ is not $\bot$, and $\varphi \blacktriangleleft \Delta$.*

3. *Δ is empty, φ is not $\bot$, and $\pm\varphi$.*

A helpful graphic representation of these three cases is Figure 10.1 on the next page.

The case that is crossed out would correspond to the derivability of the 'totally empty' sequent $\emptyset : \emptyset$ (also representable as $\emptyset : \bot$)—which is of course impossible, by inspection of the rules of Core Logic. It would be similarly impossible in *any* proof system aspiring to capture only valid arguments. That crossing-out will always be there, regardless of what deductive system one is considering.

We are therefore left with only the remaining three cases to consider as genuine possibilities for any sequent $\Delta : \varphi$ that has a core proof. These three cases partition the space of syntactic possibilities for $\Delta : \varphi$.

$\mathcal{R}(\Delta, \varphi)$	$\varphi = \perp$	$\varphi \neq \perp$
$\Delta = \emptyset$		$\pm \varphi$ Case 3
$\Delta \neq \emptyset$	$\natural \Delta$ Case 1	$\varphi \blacktriangleleft \Delta$ Case 2

Figure 10.1: The three cases of relevance

Cases 1 and 3 cover the two logical extremes.

Metatheorem 13 (see §10.4) tells us that in Case 1, in which Δ is inconsistent, Δ itself is the only $\bowtie$-component of Δ. Moreover, if Δ is a singleton $\{\delta\}$, then $\pm\delta$, i.e. δ contains some atom both positively and negatively.

Metatheorem 13 tells us that in Case 3, in which φ is a logical theorem, we have $\pm\varphi$, i.e. φ contains some atom both positively and negatively.

Case 2 covers the 'middle range', so to speak, in which Δ is consistent and φ is not a logical theorem. It is in this case that Metatheorem 13 reveals the most interesting structure involving both Δ and φ. We know by Lemma 9 quite generally that any set Δ of sentences admits of a unique partition into its $\bowtie$-components $\Delta_1, \ldots, \Delta_n$ $(n \geq 1)$. But when there is a core proof of φ from Δ, each $\bowtie$-component Δ_i bears a special relation to φ, to wit: some atom occurs with the same parity in φ as it does in some member of Δ_i.

It is this partitioning of the set Δ of premises, in Case 2, that makes our 'variable-sharing property' $\mathcal{R}(\Delta, \varphi)$ so much more exigent than any of the extant variable-sharing properties that have been formulated within the Anderson–Belnap tradition.

The three 'live' cases in the diagram above are worth spelling out in greater logical detail, unpacking the compiled concepts $\pm$, $\natural$, and $\blacktriangleleft$. Their general descriptions are as follows, each illustrated with a suggestive diagram and with instructive examples.

Case 1: $\Delta \neq \emptyset$; $\varphi = \perp$
If Δ is a singleton $\{\delta\}$, then some atom occurring in δ has both a positive

and a negative occurrence therein. If Δ has more than one member, then any two of them are connected by a $\bowtie$-path, that is a sequence of members of Δ in which, for any two immediate neighbors, there is some atom that has a positive occurrence in one of them and a negative occurrence in the other.

In Figure 10.2 below, the rectangle represents the set Δ of premises, which in turn are represented by the dots in the rectangle. $\bowtie$-connections are represented by the lines annotated with '$\bowtie$'. The proof from the premises is represented by the triangle below, with its conclusion (in this case, $\bot$) at its bottom vertex. Note that Δ is $\bowtie$-connected; hence is its own sole $\bowtie$-component.

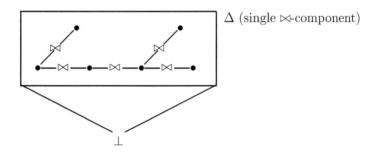

Figure 10.2: A suggestive diagram for Case 1

Examples of Case 1.

Singleton examples are $\overset{\oplus}{A} \wedge \overset{\ominus}{\neg A} : \bot$; $\neg(\overset{\oplus}{A}{\to}\overset{\ominus}{A}) : \bot$

Non-singleton examples are $\neg\overset{\ominus}{A}, \overset{\oplus}{A} : \bot$; $\overset{\oplus}{A} \wedge \overset{\oplus}{B}, \overset{\ominus}{A}{\to}C, \overset{\ominus}{B}{\to}\neg C : \bot$

Case 2: $\Delta \neq \emptyset$; $\varphi \neq \bot$

For the unique partition $\Delta_1|\ldots|\Delta_n$ of Δ into its $\bowtie$-components:
for $1 \leq i \leq n$ there is some member θ of Δ_i such that $\theta \approx \varphi$, that is there is some atom A and such that A has an occurrence in θ of the same parity as one of its occurrences in φ.

Figure 10.3 on the next page illustrates the sort of situation invoved in Case 2. The proof from the premises ends with conclusion φ ($\neq \bot$), shown at the bottom vertex. The partition of the premise set into its $\bowtie$-components is indicated by vertical lines within the rectangle. Note that *every* $\bowtie$-component contains at least one premise bearing the $\approx$-relation to the conclusion φ.

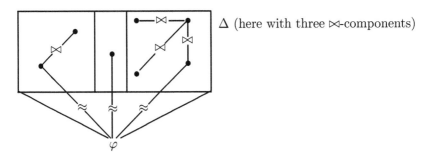

Figure 10.3: A suggestive diagram for Case 2

Examples of Case 2.
With just one component in the partition of premises:

$$A, A \rightarrow B : B \qquad A, \neg A \lor B : B \qquad A \lor C, A \rightarrow B, C \rightarrow D : B \lor D$$

With two components in the partition of premises:

$$A, A \rightarrow B \mid C, C \rightarrow D \; : \; B \land D$$

Case 3: $\Delta = \emptyset; \varphi \neq \bot$
Some atom occurring in φ has both a positive and a negative occurrence
therein.

Figure 10.4 shows what is involved in Case 3. The proof from the
premises ends with conclusion φ ($\neq \bot$), shown at the bottom vertex. The
premise set is empty. (All assumptions have been discharged.) So φ is a
logical theorem.

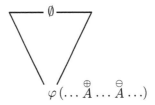

Figure 10.4: A suggestive diagram for Case 3

Examples of Case 3: $: A \rightarrow A \; ; \; : \neg(A \land \neg A) \; ; \; : (\neg A \land \neg B) \rightarrow \neg(A \lor B)$

10.3 Reasons why $\mathcal{R}(\Delta, \varphi)$ is a good explication of relevance

Observation 17. *Suppose $\mathcal{R}(\Delta, \varphi)$. Suppose that a uniform substitution turns Δ and φ into Δ' and φ' respectively. Then $\mathcal{R}(\Delta', \varphi')$. That is, relevance is preserved under uniform substitution.*

We cannot require that all atoms that occur in the premises occur in the conclusion (with or without any parity conditions). An immediate counterexample to this overly strong requirement is $A \to B, B \to C : A \to C$.

We cannot require either that all atoms in the conclusion occur in (some) premise (again, with or without any parity conditions). For this would rule out $A : A \lor B$.

Nor can we require that the premises, if there are more than one, share an atom (not even merely pairwise). For this would rule out $A, B : A \land B$.

10.3.1 Would-be strengthenings of $\mathcal{R}(\Delta, \varphi)$ to which there are counterexamples

One might still wonder, however, whether there are any other logical strengthenings of our defined relevance condition $\mathcal{R}(\Delta, \varphi)$ that might prove to be satisfied by the premise sets Δ and conclusions φ of proofs in any logical system $\mathcal{S}$ deserving the title of a 'relevance logic'. Any such logical strengthenings would have to be effected on one or more of the three cases identified above. For we know that, given any $\mathcal{S}$-proof of φ from Δ, Cases 1, 2, and 3 are mutually exclusive and jointly exhaustive. They partition the logical space of possibilities for Δ and φ.

Case 1: $\Delta \neq \emptyset; \varphi = \bot$
Consider the would-be strengthening

> If Δ is a singleton $\{\delta\}$, then *every* atom occurring in δ has both a positive and a negative occurrence therein.

This succumbs to the counterexample of the intuitively relevantly valid sequent

$$(A \land \neg A) \land B : \bot$$

in which the atom B enjoys only one occurrence.

Now consider the would-be strengthening

> If Δ has more than one member, then for all distinct δ_1, δ_2 in Δ we have $\delta_1 \bowtie \delta_2$, that is there is some atom that has a positive occurrence in one of them and a negative occurrence in the other.

This succumbs to

$$A, A \to C, B, B \to \neg C : \bot$$

by taking A for δ_1 and taking B for δ_2.

Case 2: $\Delta \neq \emptyset;\ \varphi \neq \bot$

Recall that $\mathcal{R}(\Delta, \varphi)$ provides that in this case

> for the unique partition $\Delta_1 | \ldots | \Delta_n$ of Δ into its $\bowtie$-components: for $1 \leq i \leq n$ there is some member θ of Δ_i and there is some atom A such that A has an occurrence in θ of the same parity as one of its occurrences in φ.

The logical form of this provision, which one might seek to strengthen somehow, is

$$\forall i [1 \leq i \leq n]\ \exists \theta \in \Delta_i\ \exists A\ \theta \underset{A}{\approx} \varphi .$$

The only possible ways of strengthening this claim would be (i) to uniformize—that is, to change the quantifier-block $\forall i \exists \theta$ to $\exists \theta \forall i$—or (ii) to strengthen one of the existentials to a universal.

The quantifier switch in suggestion (i) is impossible, because components are pairwise disjoint, whence θ belongs to a unique component.

As for suggestion (ii), strengthening $\exists \theta$ to $\forall \theta$ has already been ruled out by the counterexample $A, A \to B : B$. One might then consider strengthening $\exists A$—either to (a) '$\forall A$ occurring in θ' or to (b) '$\forall A$ occurring in φ'.

The proposed strengthening by (a) would yield

$$\forall i [1 \leq i \leq n]\ \exists \theta \in \Delta_i\ \forall A \text{ occurring in } \theta \text{ we have } \theta \underset{A}{\approx} \varphi .$$

A counterexample to this proposed strengthening is $A \wedge B : A$. The only candidate for θ is $A \wedge B$, whose atom B does not occur at all in the conclusion, let alone positively.

The proposed strengthening by (b) would yield

$$\forall i[1 \leq i \leq n] \; \exists \theta \in \Delta_i \; \forall A \text{ occurring in } \varphi \text{ we have } \theta \underset{A}{\approx} \varphi$$

A counterexample to this proposed strengthening is $A : A \vee B$. For the atom B in the conclusion does not occur in any premise (let alone positively).

Case 3: $\Delta = \emptyset$; $\varphi \neq \perp$

Recall that $\mathcal{R}(\Delta, \varphi)$ provides that in this case

some atom occurring in φ has both a positive and a negative occurrence therein.

The only conceivable strengthening of this would be

every atom occurring in φ has both a positive and a negative occurrence therein.

But this falls to the counterexample $\emptyset : (A \rightarrow A) \vee B$, in which the atom B enjoys only one occurrence.

We conclude, then, that the relevance condition $\mathcal{R}(\Delta, \varphi)$ for provable sequents $\Delta : \varphi$ cannot readily be improved upon. It would appear to be a 'Goldilocks' condition—not too weak, and not too strong; indeed, just right.

10.4   Main result

As the inductive proof of Metatheorem 13 reveals (see Tennant [2015d]), the application of any rule of inference of Core Logic to appropriate subproofs each of whose premise set and conclusion are related by $\mathcal{R}$ ensures that the set of premises (i.e. undischarged assumptions) of the proof resulting from that rule application also stands in the relation $\mathcal{R}$ to its conclusion. In other words, $\mathcal{R}$ *transmits under licit application of any rule of Core Logic*. Moreover, since it is obvious that $\mathcal{R}(\{\varphi\}, \varphi)$, we shall have, by induction on core proofs, that $\mathcal{R}(\Delta, \varphi)$ for any core proof Π of φ from Δ.

Note that in any core proof Π, any *sub*proof Π' will have been formed prior to the formation of Π itself; and if Δ' is the set of premises of Π', and φ' is its conclusion, then $\mathcal{R}(\Delta', \varphi')$.

The reader might wonder whether a core proof of φ from Δ can itself somehow be divided up into subproofs whose sets of premises correspond respectively to the components of Δ.[3] The answer is negative, but instructively so. Consider, for example, the core-provable sequent

$$A \wedge B, A \to C, B \to D : C \wedge D \,.$$

The three premises here form a *single* $\bowtie$-component. But in the obvious proof

$$
\cfrac{
A \wedge B \qquad
\cfrac{
\cfrac{A \to C \quad \overline{\quad}(1) \quad \overline{\quad}(2)}{C} \qquad
\cfrac{B \to D \quad \overline{\quad}(1) \quad \overline{\quad}(3)}{D}
}{C \wedge D}
\; (1)
}{C \wedge D}
$$

the subproof for the final step of $\wedge$-Elimination is

$$
\cfrac{
\cfrac{A \to C \quad A \quad \overline{\quad}(2)}{C}\;(2) \qquad
\cfrac{B \to D \quad B \quad \overline{\quad}(3)}{D}\;(3)
}{C \wedge D}
$$

whose premise set is partitioned into two $\bowtie$-components:

$$A \to C, A \mid B \to D, B \,.$$

What this example shows us is that by applying an elimination rule, two components of the premise set of a subordinate proof can be 'merged into one', via the major premise for the elimination, whose immediate constituents, *qua* premises of the subordinate proof, are subsumed into the major premise when they are discharged. This theme recurs repeatedly in the various cases in the inductive step for the proof of Metatheorem 13 that concern applications of elimination rules.

Metatheorem 13 (Relevance).
For every classical core proof whose set of undischarged assumptions is Δ and whose conclusion is φ, we have $\mathcal{R}(\Delta, \varphi)$.

Proof. See Tennant [2015d].   □

---

[3] This question was raised by an anonymous referee.

10.5   Final reflections

We have formulated an exigent relevance property, $\mathcal{R}(\Delta, \varphi)$, that is exhibited by the premise set Δ and conclusion φ of any proof in the first-order system $\mathbb{C}^+$ of Classical Core logic. Recall that the relevance logic $\mathbf{R}$ of Anderson and Belnap is included in $\mathbb{C}^+$. *We have therefore established* $\mathcal{R}(\Delta, \varphi)$ *for proofs in* $\mathbf{R}$ *and, a fortiori, for proofs in any subsystems of* $\mathbf{R}$. Moreover, the relevance property $\mathcal{R}(\Delta, \varphi)$ is more exigent than any such property formulated thus far, and shown to hold, for $\mathbf{R}$ or any of its subsystems.

This reflects well on the proof-theoretic methods that we have chosen to deploy in our investigation of just how tightly the relevance-relation can be drawn between premises and conclusions of relevant proofs. For $\mathbb{C}^+$ *properly* includes $\mathbf{R}$. Indeed, $\mathbb{C}^+$ *even* contains Disjunctive Syllogism $(A \vee B, \neg A : B)$, which, as already remarked, is the *bête noire* of $\mathbf{R}$. Thus, prima facie, it should have been much more difficult to accomplish a proof of strong relevance for $\mathbb{C}^+$ rather than for (some subsystem of) $\mathbf{R}$.

We saw in §10.3.1 that none of the obvious ways of strengthening the relation $\mathcal{R}(\Delta, \varphi)$ survives the weight of simple counterexamples from among the intuitively relevant, valid sequents. The counterexamples in question are the following:

$$A \to B, B \to C : A \to C \quad A : A \vee B \qquad\qquad A, B : A \wedge B$$
$$(A \wedge \neg A) \wedge B : \bot \qquad A, A \to C, B, B \to \neg C : \bot \quad A, A \to B : B$$
$$A \wedge B : A \qquad\qquad \emptyset : (A \to A) \vee B$$

All these sequents are provable in $\mathbf{R}$. So the prospects seem very dim for any strengthening of $\mathcal{R}(\Delta, \varphi)$ that might hold for $\mathbf{R}$. Relevantists in the Anderson–Belnap school might, however, take issue with this claim. So we lay down the following challenge for future research:

> Identify *any strengthening* of $\mathcal{R}(\Delta, \varphi)$ that can be shown to hold for some subsystem of $\mathbf{R}$ but that can be shown to *fail* for $\mathbb{C}^+$.

If there proves to be no such strengthening, then the advocate of $\mathbf{R}$ (or any of its subsystems) is left with the untoward conclusion that considerations of vocabulary-sharing alone cannot justify their choice of system over the rival system $\mathbb{C}^+$ (in the classical case) or $\mathbb{C}$ (in the intuitionistic case).

But against any argument in favor of an Anderson–Belnap type of relevance logic over $\mathbb{C}^+$ or $\mathbb{C}$ we raise the parting objection that the Core systems $\mathbb{C}$ and $\mathbb{C}^+$, but not the Anderson–Belnap systems, allow the faithful regimentation of deductive reasoning in intuitionistic and classical mathematics respectively. The Core systems are truer to Frege's founding purpose in formalizing a system of deductive logic.

Chapter 11

Core Logic and the Paradoxes

Abstract

§11.1 argues that the Law of Excluded Middle is not to be blamed for any of the logico-semantic paradoxes.

§11.2 supports this contention by furnishing core proofs for the bits of deductive reasoning involved in the Liar Paradox.

§11.3 explains and defends my proof-theoretic criterion of paradoxicality, according to which the 'proofs' of inconsistency associated with the paradoxes are in principle distinct from those that establish genuine inconsistencies, in that they cannot be brought into normal form. Instead, the reduction sequences initiated by paradox-posing proofs 'of ⊥' do not terminate. This criterion is defended against some recent would-be counterexamples by stressing the need to use Core Logic's parallelized forms of the elimination rules.

§11.4 shows how Russell's famous paradox in set theory is *not* a genuine paradox; for it can be construed as a disproof, in the free logic of sets, of the assumption that the set of all non-self-membered sets exists.

§11.5 shows how (by contrast) the Liar is *still* paradoxical, according to my proof-theoretic criterion of paradoxicality.

§11.6 summarizes, so need not be summarized.

11.1 What role, if any, does the law of excluded middle play in the derivation of the paradoxes?

We say 'no role at all' in answer to this question. Our examination of the results of Eubulides–Tarski, Löb, Montague, and McGee in Tennant [under submission] lend some striking, if inconclusive, support for this stand. But that stand is hardly the considered view among contemporary philosophers writing about the paradox, who reflect on the alleged role of Classical Logic in generating them. The purpose of this section is to examine and rebut some important opposing views to the effect that the law of excluded middle is somehow the culprit in the reasoning that generates the paradoxes.

Consider this quote from Field [2008], at p. 15 (emphasis in original):

> ... we should seriously consider restricting the law of excluded middle (though not in the way intuitionists propose) ... I take excluded middle to be *clearly* suspect only for certain sentences that have a kind of "inherent circularity" because they contain predicates like 'true' ...

But we need to question the wisdom of holding excluded middle in any way responsible for, or essential to, proofs of the semantic paradox. Rather, the following methodological principle is commended:

> *One is dealing with a case of genuine semantic paradox only if it can be derived relevantly, and without using excluded middle (or any of its intuitionistic equivalents).*

Or, to put it another way:

> *If you think you are dealing with a case of genuine semantic paradox, then you had better be able to derive the absurdity involved in* Core Logic.

This might strike the reader as rather surprising. One of the main reasons why its truth is not clearly recognized is that the derivations of absurdity in the case of paradox that are given by so many authors make conspicuous use of the law of excluded middle; or they themselves explicitly diagnose it as *a* source, if not *the* source, of the problem. For example, Priest [1983], even while opposing the idea that the root cause of paradox, in every case, is excluded middle, asserts at p. 161:

> Although some [semantic paradoxes], such as the heterological paradox, go *via* an assertion of the form $\varphi \leftrightarrow \neg\varphi$, *and hence use the law of excluded middle*, the definability paradox, such as Berry's, Richards', and König's do not. (Emphasis added.)

We shall see in §11.2, however, that the italicized part of Priest's claim is in error.

Field, too, is quick to regard excluded middle as the supposed culprit when a paradox is established by deriving an intermediate assertion of the form $\varphi \leftrightarrow \neg\varphi$. Field sets out one way of deriving the explicit contradiction $\varphi \wedge \neg\varphi$ from $\varphi \leftrightarrow \neg\varphi$ (which he calls the Central Argument), and writes at p. 8:

> Without [excluded middle], it isn't immediately obvious why $[\varphi \leftrightarrow \neg\varphi]$ should be contradictory.

To be sure, Field does then go on to observe that Intuitionistic Logic (which lacks excluded middle)

> invalidate[s] the Central Argument for the contradictoriness of $[\varphi \leftrightarrow \neg\varphi]$ *while leaving other arguments intact.* (Emphasis added.)

Field, does not, however, exhibit or source any of these other arguments. (He is, to be sure, correct in saying that they exist: see §11.2.) Instead, he keeps his focus, puzzlingly, on excluded middle, as though it somehow *has* to be the main culprit. He writes at p. 8 (emphasis in original):

> The most obvious route from $[\varphi \leftrightarrow \neg\varphi]$ to $[\varphi \wedge \neg\varphi]$ within in-tuitionism comes from the intuitionist *reductio* rule, which says that if $[\neg\varphi]$ follows from Γ and $[\varphi]$ together then it follows from Γ alone. Although intuitionists have reasons of their own for accepting this rule, the most obvious arguments for accepting it assume excluded middle.[1] For instance: $[\neg\varphi]$ certainly follows from Γ and $[\neg\varphi]$ together, so if it also follows from Γ and $[\varphi]$ together then it must follow from Γ and $[\varphi \vee \neg\varphi]$ together, and hence *assuming excluded middle* it follows from Γ alone.

Priest had gone even further than Field on this score. He claimed ([1983], p. 160) that:

---

[1] Footnote added by the present author: *Au contraire*, surely the most obvious intu-itionistic argument for the *reductio* rule in question is

$$
\underbrace{\Gamma \quad , \quad \overset{\displaystyle \rule{1.5em}{0.4pt}{\scriptstyle(1)}}{\varphi}}
$$

$$
\frac{\dfrac{\overset{\rule{1.2em}{0.4pt}{\scriptstyle(1)}}{\varphi} \qquad \vdots \\ \qquad \neg\varphi}{\underset{\neg\varphi}{\bot}{\scriptstyle(1)}}}{}
$$

This argument involves only the introduction and elimination rules for negation.

> The *reductio* principle $[\varphi \rightarrow \neg\varphi/\neg\varphi]$ is, under very weak assumptions, *equivalent to* the law of excluded middle. (Emphasis added.)

On p. 161, Priest appears to be willing to blame excluded middle for the *set-theoretic* (as opposed to the *semantic*) paradoxes:

> ... although no set theoretic paradox may be provable without the law of excluded middle,[2] the case is different with the semantic paradoxes.

So we see from the record that two leading experts on the paradoxes are at the very least equivocal about the extent to which excluded middle is implicated in precipitating paradox. And if one looks at other writings by philosophical logicians, the theme recurs. The law of excluded middle is frequently invoked in negotiating deductive passage to contradiction, when illustrating for the beginner whatever paradox is at hand.

It is as though, once one has the canons of Classical Logic at hand, one cannot help but use the full arsenal to bring about ruin. But the reason why the genuine semantic paradoxes are such a threat is that *they arise for the constructive relevantist* (the character more pithily called the *Core logician*) whenever they arise for the classicist. Using excluded middle to display the devastation they wreak obscures just how devastating the paradoxes really are.

Pace Priest, excluded middle plays no role at all (or *need not play any role*) in the derivation of contradiction in the case of set-theoretic paradoxes. *Pacete* Priest and Field, excluded middle plays no role at all (or: *need not play any role*) in the derivation of contradiction in the case of the Liar paradox, the heterologicality paradox, the postcard paradox, or any of those semantic paradoxes (involving such notions as *true* or *true of*) in a semantically closed language. Contrary to the impression that can easily arise from a reading of Priest and Field, the law of excluded middle *need not play any role* in the derivation of the explicit contradiction $\varphi \wedge \neg\varphi$ from $\varphi \leftrightarrow \neg\varphi$. In sum, as made clear in Tennant [1978] for Russell's Paradox in set theory, and in Tennant [1982] for the semantic paradoxes:

---

[2] Footnote added by the present author: For an intuitionistic—indeed, core—proof of Russell's Paradox in set theory, see Tennant [1978], at p. 164; and §11.4 of this work. Since that proof can be conducted within free logic (so as to reduce the existential assumption $\exists!\{x|x \notin x\}$ to absurdity), it establishes the non-existence of the Russell set $\{x|x \notin x\}$. (See Tennant [1982] at pp. 276 and 292.) Thus, by the criterion of paradoxicality to be set out below, Russell's so-called paradox (within a free logic of sets) is not really a paradox at all. Rather, it is a negative existential theorem of *constructive* set theory.

> *The set-theoretic paradoxes, and the semantic paradoxes involv-*
> *ing* true *and* true-of, *arise by means of reasoning that is thor-*
> *oughly constructive and relevantist. In no way can the blame for*
> *paradoxicality be laid at the door of the law of excluded middle.*

To blame excluded middle in this way would be to close the barn door after the horse has bolted.

11.2 Excluded middle plays no essential role in the derivation of the paradoxes

The following proofs are all in Core Logic. Note that we are 'lapsing' into a use of the Gentzen–Prawitz proof system, which employs the *serial* forms of the Elimination rules for $\to$ and $\leftrightarrow$. This is because of the dialectical consideration that we wish to show that certain proofs *were to be had*, if only care had been taken to find them. The Gentzen–Prawitz system of natural deduction would have been an obvious early choice for the sorts of proof search that philosophical logicians needed to carry out before discussing the paradoxes and venturing to blame them on excluded middle. (Sometimes a public defender can be rightly convinced that an otherwise scurrilous defendant is not guilty as charged.)

We give complex labels on the left, and the detailed proof on the right.

$$
\begin{array}{ccc}
\varphi \leftrightarrow \neg\varphi & & \cfrac{\varphi \leftrightarrow \neg\varphi \qquad \cfrac{\qquad}{\varphi}^{(1)}}{\neg\varphi} \qquad \cfrac{\qquad}{\varphi}^{(1)}\\[2ex]
\Pi \quad : & & \cfrac{\dfrac{\bot}{\quad}^{(1)}}{\neg\varphi}\\[2ex]
\neg\varphi & &
\end{array}
$$

Field's so-called Central Argument (for the conclusion $\varphi \wedge \neg\varphi$ from the premise $\varphi \leftrightarrow \neg\varphi$) may now be eschewed in favor of the following core proof:

$$
\cfrac{\cfrac{\varphi \leftrightarrow \neg\varphi \qquad \cfrac{\varphi \leftrightarrow \neg\varphi}{\displaystyle \Pi \atop \neg\varphi}}{\varphi} \qquad \cfrac{\varphi \leftrightarrow \neg\varphi}{\displaystyle \Pi \atop \neg\varphi}}{\varphi \wedge \neg\varphi}
$$

Alternatively, if one prefers to establish paradoxicality by means of a proof of $\bot$, one can take the proof

$$
\begin{array}{c}
\varphi \leftrightarrow \neg\varphi \\
\Sigma \\
\varphi
\end{array}
\quad : \quad
\begin{array}{c}
\varphi \leftrightarrow \neg\varphi \\
\Pi \\
\dfrac{\varphi \leftrightarrow \neg\varphi \qquad \neg\varphi}{\varphi}
\end{array}
$$

and embed it twice thus, for a core proof Ω of $\bot$ from the premise $\varphi \leftrightarrow \neg\varphi$:[3]

$$
\begin{array}{c}
\varphi \leftrightarrow \neg\varphi \\
\Omega \\
\bot
\end{array}
\quad : \quad
\dfrac{\dfrac{\varphi \leftrightarrow \neg\varphi \quad \begin{array}{c}\varphi \leftrightarrow \neg\varphi \\ \Sigma \\ \varphi\end{array}}{\neg\varphi} \qquad \begin{array}{c}\varphi \leftrightarrow \neg\varphi \\ \Sigma \\ \varphi\end{array}}{\bot}
$$

So we see that Core Logic furnishes a proof of the conclusion $\varphi \wedge \neg\varphi$ from the premise $\varphi \leftrightarrow \neg\varphi$, as well as a proof of $\bot$ from the same premise. It is simply an error to maintain that the law of excluded middle is in any way responsible for, or essential to, the derivations of absurdity that are involved in any of the proofs of the various paradoxes.

The purpose of this discussion has been to rescue the law of excluded middle from misdirected opprobrium. That dog already has a bad enough name. What is so unsettling about the informal reasoning involved in the logical and semantical paradoxes is that it can be formalized in *Core Logic*. This shows that the source of the paradoxes is to be sought in the (inappropriately formulated?) meaning-conferring rules involved for the central concepts, such as *true* and *true of*. It is not to be found in any of the rules for the usual logical operators. Those rules are just hapless witnesses to a conceptual train wreck occasioned by malfunction of other parts of our cognitive machinery.

How, then, are we to solve the paradoxes? It is not for this study to venture any new suggestions beyond those of Tennant [1982] and Tennant [1995].

---

[3] Note that if we tried instead to form a proof of $\bot$ thus:

$$
\dfrac{\dfrac{\varphi \leftrightarrow \neg\varphi \quad \begin{array}{c}\varphi \leftrightarrow \neg\varphi \\ \Pi \\ \neg\varphi\end{array}}{\varphi} \qquad \begin{array}{c}\varphi \leftrightarrow \neg\varphi \\ \Pi \\ \neg\varphi\end{array}}{\bot}
$$

then we would have a construction that was not in normal form. This is because the major premise $\neg\varphi$ of the final step of $\neg$-Elimination stands, within the proof Π, as the conclusion of a step of $\neg$-Introduction. The foregoing construction does, however, normalize to the proof Ω given in the text.

Those works provided *reductio* proofs, formalized as natural deductions, for all the major paradoxes (including Yablo's paradox—see Yablo [1993]). They showed that all these *reductio* proofs (unlike proofs of straightforward inconsistencies) cannot be converted into normal form. The original proof-theoretic thesis stands:

> Genuine paradoxes are those whose associated *proofs of absurdity*, when formalized as natural deductions, cannot be converted into normal form.

This conjecture provides a proof-theoretic criterion for the identification of genuine paradoxes, as opposed to straightforward inconsistencies. The 'essentially circular' nature of the deductive reasoning involved with any genuine paradox is explicated in terms of its failure to assume normal form.

This is *not* to say that an insistence that proofs (and disproofs) be in normal form will somehow free us from contradictions when we use a semantically closed language.[4] On the contrary, where λ is the Liar sentence, we would still have a normal-form proof Π (say) of $T[\lambda]$, and a normal-form proof Σ (say) of $\neg T[\lambda]$:

$$
\Pi: \quad \cfrac{\cfrac{(1)}{T[\lambda]} \quad \cfrac{\cfrac{(1)\rule{1.5em}{0.4pt}}{T[\lambda]} \ TE}{\cfrac{\lambda}{\neg T[\lambda]} \text{ i.e.}}}{\cfrac{\bot}{\neg T[\lambda]}(1)} \ \neg E
\qquad\qquad
\Sigma: \quad \cfrac{\cfrac{\Pi}{\neg T[\lambda]} \text{ i.e.}}{\cfrac{\lambda}{T[\lambda]} \ TI}
$$

So we would have grounds for both asserting and denying the Liar. But we would *not* have an outright proof, *in normal form*, of *absurdity*. This is because the attempted final step of $\neg$-Elimination (applied to the conclusions of the proofs Π and Σ):

$$
\cfrac{\cfrac{\Pi}{\neg T[\lambda]} \quad \cfrac{\Sigma}{T[\lambda]}}{\bot} \ \neg E
$$

would turn the displayed occurrence of $\neg T[\lambda]$ into a maximal sentence occurrence within the proof, thereby making it abnormal. The occurrence in question of $\neg T[\lambda]$ stands as the conclusion of a final step of $\neg$-Introduction within Σ—so it cannot serve as the major premise for $\neg$-Elimination, on pain of abnormality.

---

[4] This was pointed out by Kevin Scharp.

Our proof-theoretic criterion for paradoxicality still serves, however, to distinguish between genuine contradictions (those proofs of absurdity that are in normal form) and paradoxes (those apparent contradictions whose associated proofs *of absurdity* cannot be brought into normal form). The former are fatal for any theory containing them; the latter are not.

Those proofs of absurdity will use only such logical steps as are available within Core Logic. The non-normalizability of such a proof does not count against this claim (on the grounds that every core *logical* proof must be in normal form). Rather, it underscores the fact that it will be the rules governing the troublesome notions (that is, rules other than those for the logical operators) that are being applied in such a way as to prevent one from converting the proof in question into normal form.

11.3   A proof-theoretic criterion of paradoxicality

11.3.1   Background

Tennant [1982] proposed a proof-theoretic criterion for paradoxicality —that of *non-terminating reduction sequences* initiated by the 'proofs of ⊥' associated with the paradoxes in question (p. 271). In that paper, the subsequent focus was on *looping* reduction sequences. But there are other kinds of non-terminating (sequences of) reduction procedures besides those that enter into loops. One needs to bear this in mind in order to account for the paradox in Yablo [1993]. With Yablo's paradox, as shown in Tennant [1995], the reduction sequence does not so much loop as 'spiral endlessly', ratcheting up a numerical index with each turn.

Tennant [1982] concentrated on logico-semantic paradoxes, but did also examine Russell's Paradox. Prawitz [1965] had shown how certain naïvely formulated introduction and elimination rules in set theory would—despite the fact that they appeared to admit of a reduction procedure—block the proof that all natural deductions can be brought into normal form. Indeed, the blockage was furnished by an obvious formalization of the reasoning in Russell's Paradox. It is not in normal form, and it cannot be brought into normal form by means of the reduction procedure in question. But Prawitz did not examine the possibility of alternative, non-naïve rules for set abstraction (in a *free* logic) that might enable one to obtain a normalization theorem. Tennant [1982] examined this prospect in some detail, using the Introduction and Elimination rules for set abstraction that were framed and proved complete in Tennant [1978]; but found that difficulties still stood in

the way of providing a proof in normal form of Russell's Paradox, recon-
strued now as a proof that the Russell set $\{x|\neg x \in x\}$ does not exist.

11.3.2 The Ekman Problem

Schroeder-Heister and Tranchini [forthcoming] have suggested that the con-
jectural proof-theoretic diagnosis of paradoxicality mistakenly takes looping
reduction sequences (for proofs of $\perp$) as a *sufficient* condition for paradoxicality. They think that the phenomenon of looping reduction sequences
is already manifest in proofs of quite 'ordinary', non-paradoxical cases of
inconsistency. They put forward an example taken from Ekman [1998] in
order to make this clear. The example in question is the proof of inconsis-
tency of $\{A \rightarrow \neg A, \neg A \rightarrow A\}$. Schroeder-Heister and Tranchini examine the
obvious proof of $\perp$ that one would construct in the Gentzen–Prawitz system
of natural deduction, using Modus Ponens as the Elimination rule for $\rightarrow$;
and indeed *that* proof is not in normal form, and its reduction sequence
enters a loop. They conclude that 'Tennant's test is too coarse, as it induces
unmotivated ascriptions of paradoxicality.'

So, have Schroeder-Heister and Tranchini uncovered a good reason to
abandon the criterion, for paradoxicality, of non-terminating reduction sequences? Closer reflection will reveal they have not. Their closing sentence
ventured the suggestion that

> ... non-normalizability is connected to features of implication
> in natural deduction systems. The tenability of this claim is left
> for further inquiries.

It will emerge, from the considerations given below, that the 'Ekman problem' for the paradoxicality criterion is an artefact of the choice of *serial
forms of elimination rules* in natural deduction—in particular, the serial
form of ($\rightarrow$E), or Modus Ponens. The problem *simply does not arise* if one
insists on the use, instead, of the *parallelized* forms of elimination rules, with
MPEs *standing proud*.

When we revisit the problem of finding a normal-form proof of Russell's
Paradox, it turns out that this problem, too, disappears upon the adoption
of suitably parallelized forms of the Elimination rules for set abstraction.

11.3.3 A crucial distinction: serial v. parallelized elimination rules in natural deduction

The Gentzen–Prawitz system allows one to construct natural deductions in
which the conclusion of an application of an introduction rule may stand

as the major premise of an application of the corresponding elimination rule. Such sentence occurrences are called *maximal,* and one can get rid of them by applying the well-known reduction procedures. The normalization theorem is to the effect that any proof (of, say, $\Delta : \varphi$) can be turned into one in normal form, establishing the conclusion φ from some subset of Δ.

Maximal sentence occurrences in proofs represent an undesirable and avoidable prolixity. Another kind of prolixity, which one should be just as anxious to avoid, is that which is evident in the situation that we shall call *subproof compactification.* This is when a proof contains a subproof Σ such that Σ proves $\Delta : \varphi$ (say) *and* Σ contains a *proper* subproof (call it Σ') of some subsequent of $\Delta : \varphi$. Surely, one might think, one should be able to make do with just Σ' in place of Σ, and avoid having the 'extra fluff' that Σ has built up around Σ'? To this end, Ekman had a further reduction procedure, aiming to do just that—at least, in certain straightforward cases. Ekman's reduction procedure is exactly like the one motivated in Tennant [1982] at p. 270 *infra,* and called 'shrinking' in its applications at pp. 286ff.[5] We shall not need to examine the possible scope of these further reduction procedures here. Suffice it to say that the shrinking reduction was used in the analysis of all the paradoxes covered in Tennant [1982], and particularly in the analysis of Russell's Paradox in the non-naïve case discussed above.

Regarding as always eligible for application both the standard Gentzen–Prawitz reduction procedures and the aforementioned additional reduction procedure to get rid of subproof compactification, Schroeder-Heister and Tranchini proceed to examine the reduction sequence that would ensue from the obvious proof in the *Gentzen–Prawitz* system of natural deduction (using the *serial* form of $\rightarrow$-Elimination) that establishes

$$A \rightarrow \neg A, \neg A \rightarrow A : \bot.$$

They demonstrate that the reduction sequence loops. Then, since something like the premise pair of this example is involved in the proof of $\bot$ in the case of the Liar Paradox, they conclude that the proof-theoretic criterion of paradoxicality is looking for paradoxicality in the wrong place.

In order to call into question the cogency of these considerations, the reader is invited to consider not the proof of $A \rightarrow \neg A, \neg A \rightarrow A : \bot$ examined by Schroeder-Heister and Tranchini, but rather the obvious proof of the same result in the natural-deduction system for Core Logic, in which the

---

[5] This was an early precursor of the global restriction on proof formation (the anti-dilution condition) discussed in §5.5.3.

rule of →-Elimination takes the *parallelized* form

$$(\to E)$$

$$
\begin{array}{ccc}
& & \square\!\!-\!\!(i) \\
\Delta & & \Gamma, \psi \\
\vdots & & \vdots \\
\varphi \to \psi \quad \varphi & & \theta \\
\hline
& \theta &
\end{array}\!\!(i)
$$

Recall that the box next to the discharge stroke in (→E) over the 'assumption for discharge' in the major subproof indicates that vacuous discharge is not allowed. That is to say, there really must be an assumption of the indicated form in the subordinate proof in question, available to be discharged upon application of the rule. This makes parallelized (→E) equiform with the sequent rule

$$(\to:) \qquad \frac{\Delta : \varphi \qquad \Gamma, \psi : \theta}{\Delta, \Gamma, \varphi \to \psi : \theta}$$

Let us now see how the parallelized form of (→E) can be used to avoid the Ekman problem.

$$
\text{Let} \quad
\underbrace{A \to \neg A, \neg A \to A}_{\Pi}
\atop A
\quad =_{df} \quad
$$

$$
\begin{array}{c}
\dfrac{}{\neg A}\text{-(1)} \quad \dfrac{}{A}\text{-(2)} \\
(2)\dfrac{}{} \qquad \dfrac{\bot}{}\text{-(1)} \\
\dfrac{A \to \neg A \quad A}{\dfrac{\bot}{\neg A \to A} \qquad \dfrac{\dfrac{\bot}{\neg A}\text{-(2)} \qquad \dfrac{}{A}\text{-(3)}}{A}\text{-(3)}}
\end{array}
$$

Then the following is a proof, in the parallelized system of natural deduction for Core Logic, of the inconsistency of $\{A \to \neg A, \neg A \to A\}$:[6]

$$
\begin{array}{c}
\underbrace{A \to \neg A, \neg A \to A}_{\Pi} \qquad \underbrace{A \to \neg A, \neg A \to A}_{\Pi} \\
\dfrac{A \to \neg A \qquad A \qquad (4)\dfrac{}{\neg A} \qquad \dfrac{\neg A \quad A}{\bot}\text{-(4)}}{\bot}
\end{array}
$$

---

[6] A similar proof, of $\neg((A \to \neg A) \wedge (\neg A \to A))$, using parallelized elimination rules, but using the definition of $\neg \varphi$ as $\varphi \to \bot$, was given by von Plato [2000], at p. 123, by way of solution of Ekman's problem for the serial forms of elimination rules.

This proof, like any core proof, is in normal form. Moreover, it contains no subproof Σ such that Σ both proves Δ : φ (say) and contains a subproof of some subsequent of Δ : φ. That is to say, there is no subproof compactification to worry about. Hence Ekman's so-called paradox is no paradox at all. The inconsistency of {A → ¬A, ¬A → A} has a perfectly straightforward proof in normal form, calling for no reduction whatsoever in order to get rid of prolixities of any kind at all. With the parallelized form of →-Elimination, as we have just seen, *there is no looping* in our proof of Ekman's example. This is because it is already in normal form, so there is no reduction sequence to be embarked upon.

That it might have been thought otherwise (i.e. that Ekman's example would resist any normal-form proof) is an artefact of the mistaken presumption that a system of natural deduction ought to use the *serial* form of →-Elimination (Modus Ponens) rather than the parallelized form used above.

This issue of serial versus parallelized forms of elimination rules is no minor matter. Note that in the original system of natural deduction in Gentzen [1934, 1935], and in the treatment of it in Prawitz [1965], the operators ∧, →, and ∀ had their Elimination rules stated in *serial* form. (This could not be done, of course, for either ∨ or ∃; *their* Elimination rules, accordingly, were in parallelized form.) Moreover, in Gentzen's original system of sequent proof for *Classical* Logic he permitted *multiple succedents*. (The system of Classical Core Logic, by contrast, as we have seen, permits at most one sentence in a succedent.)

These fateful decisions on Gentzen's part led to the vexatiously complicated transformations that were needed in order to convert a natural deduction establishing Δ : φ into a sequent-calculus proof of the same result, and conversely.[7] As soon, however, as one insists that *all* elimination rules in natural deduction be stated in parallelized form, *and* that their MPEs stand proud, with no proof work above them, one ensures that any natural deduction establishing Δ : φ is essentially *isomorphic* to the corresponding (cut-free, thinning-free) sequent proof of the same result. This is subject to the proviso, of course, that in the classical case one insists still on 'single conclusion' sequents and adopts strictly classical rules such as Dilemma or Classical Reductio, as explained in Chapter 6. One can also prove that Cut *can be eschewed*, in the sense that is familiar by now to the reader.

---

[7] In Classical Logic, both of the aforementioned features (serial forms of elimination; multiple conclusions) contribute to this complication. In Intuitionistic Logic, where succedents are at most singletons, only the first feature is at work, but it still causes complications enough.

It is instructive to see the Core sequent proof of Ekman's example, corresponding to the Core natural deduction given above. The following sequent proof will be denoted by $\Pi[A]$.

$$
\cfrac{
\cfrac{
\cfrac{\cfrac{A:A \qquad \neg A, A:}{\cfrac{A \to \neg A, A:}{A \to \neg A : \neg A}} \qquad A:A}{\neg A \to A, A \to \neg A : A}
\qquad
\cfrac{\cfrac{A:A \qquad \cfrac{\cfrac{A:A}{\neg A, A:}}{A \to \neg A, A:}}{\cfrac{A \to \neg A : \neg A}{\neg A \to A, A \to \neg A : A}} \qquad A:A}{\neg A \to A, A \to \neg A, \neg A:}
}{\neg A \to A, A \to \neg A :}
$$

Note that $\Pi[A]$ contains no cuts or thinnings; it uses only the Right and Left rules for $\neg$ and the Left rule for $\to$; and no sequent σ within it is a subsequent of any sequent below σ, on a branch containing σ.

Ekman's example can now be seen to qualify no longer as a 'false positive' embarrassing the proof-theoretic criterion for paradoxicality. Although its proof in a system of natural deduction with *serial* $\to$-Elimination initiates a looping reduction sequence, the same is not true of its proof in the system with *parallelized* $\to$-Elimination. *The form of elimination rule matters*; and we must be careful from now on to frame the criterion for paradoxicality with reference only to the system of parallelized natural deduction.

On reflection, it is only right that there should be a straightforward proof of inconsistency of $\{A \to \neg A, \neg A \to A\}$, for otherwise we would be condemned to regarding as paradoxical the straightforward logical truth that no one shaves all and only those who do not shave themselves. Ekman's sequent has the parametric instance $\{Saa \to \neg Saa, \neg Saa \to Saa\}$, from which the desired result—the sequent $\emptyset : \neg \exists x \forall y (Syx \leftrightarrow \neg Syy)$—would be obtained as follows:

$$
\cfrac{
\cfrac{
\cfrac{
\cfrac{\cfrac{\Pi[Saa]}{Saa \to \neg Saa, \neg Saa \to Saa : \bot}}{Saa \leftrightarrow \neg Saa : \bot}
}{\forall y (Sya \leftrightarrow \neg Syy) : \bot}
}{\exists x \forall y (Syx \leftrightarrow \neg Syy) : \bot}
}{: \neg \exists x \forall y (Syx \leftrightarrow \neg Syy)}
$$

If no core proof like $\Pi[A]$ were available, we would be in a terrible fix. But of course we are not, as we have shown above.

11.4   Revisiting Russell's Paradox

The core proof just given, showing that no one shaves all and only those
who do not shave themselves, can be adapted to show that no set has as
members all and only those sets that do not have themselves as members:

$$\dfrac{\dfrac{\dfrac{\dfrac{\overset{\Pi[a\in a]}{a\in a\to\neg a\in a,\ \neg a\in a\to a\in a:\bot}}{a\in a\leftrightarrow\neg a\in a:\bot}}{\forall y(y\in a\leftrightarrow\neg y\in y):\bot}}{\exists x\forall y(y\in x\leftrightarrow\neg y\in y):\bot}}{:\ \neg\exists x\forall y(y\in x\leftrightarrow\neg y\in y)}$$

The conclusion of this proof exactly captures the thought that there is no set
of all and only those sets that do not contain themselves. This is Russell's
Paradox made unparadoxical (in the sense that occupies us here). The
would-be paradox has been converted into a straightforward (constructive
and relevant) proof of the negative existential theorem

$$\neg\exists x\forall y(y\in x\leftrightarrow\neg y\in y).$$

Understandably, one would expect the same to hold when the proof is of the
different, but theoretically equivalent, conclusion $\neg\exists y\,y=\{x|x\notin x\}$, where
our logically primitive *term-forming operator of set abstraction* is applied to
the open sentence $\neg x\in x$ in order to form the singular term $\{x|\neg x\in x\}$.
One would expect that in the free logic of set terms, with either

(i) suitable natural-deduction rules ({ }I) and ({ }E) for the Introduction
and Elimination of the set-abstraction operator, or

(ii) suitable sequent rules ({ }:) and (:{ }) for the introduction of the set-
abstraction operator on the left or on the right of the colon,

there should be a Core disproof of the *reductio* assumption $\exists y\,y=\{x|x\notin x\}$.

In Tennant [1982] the alternative (i) was investigated. Use was made of
the Introduction and Elimination rules for the set-abstraction operator that
had been formulated, and supplied with a completeness proof, in Tennant
[1978]. The rules in question were:

$$\frac{(i)\underline{\quad} \quad \underline{\quad}(i)}{\underbrace{\Phi^x_a \;,\; \exists!a}}$$

$$\{\,\}\mathrm{I} \qquad \vdots \qquad\qquad \frac{\underline{\quad}(i)}{a\in t}$$

$$\frac{a\in t \qquad \exists!t \qquad \dfrac{}{\Phi^x_a}(i)}{t = \{x|\Phi\}}$$

and, corresponding to the three subordinate proofs for $\{\,\}$I, the three respective Elimination rules

$$\{\,\}\mathrm{E} \qquad \frac{t = \{x|\Phi\} \quad \Phi^x_u \quad \exists!u}{u\in t} \quad \frac{t = \{x|\Phi\}}{\exists!t} \quad \frac{t = \{x|\Phi\} \quad u\in t}{\Phi^x_u} \quad,$$

of which the middle rule is already covered as a special instance of the Rule of Denotation for free logic, namely

$$\frac{A(\ldots,t,\ldots)}{\exists!t} \;, \text{ where } A \text{ is a primitive predicate.}$$

The corresponding sequent form of the Rule of Denotation is the additional rule of initial sequents

$$\frac{}{A(\ldots,t,\ldots) : \exists!t}\;.$$

The investigation in Tennant [1982] of Russell's Paradox on the basis of these rules returned the untoward result that the *reductio* proof for the assumption $\exists y\, y = \{x|x \notin x\}$ initiated a looping reduction sequence—in generating which the reduction that we have called the *shrinking* reduction was always applicable. The conclusion drawn at that stage—a conclusion to which we now *demur*—was the overly pessimistic, because over-hasty, claim that Russell's Paradox had somehow earned its label as a paradox on the proof-theoretic construal of paradox that was formulated in that paper.

That over-hasty conclusion was in error. This has become evident only thanks to the reflections prompted by the interesting challenge posed by Schroeder-Heister and Tranchini [forthcoming]. It turns out that the looping reduction sequence is once again an artefact of the *serial form of the elimination rules for set abstraction*, and in particular the rules that from now on we shall call $\{\,\}\mathrm{E}_1$ and $\{\,\}\mathrm{E}_2$:

$$\{\,\}\mathrm{E}_1 \quad \frac{t = \{x|\Phi\} \quad \Phi^x_u \quad \exists!u}{u\in t} \qquad\qquad \{\,\}\mathrm{E}_2 \quad \frac{t = \{x|\Phi\} \quad u\in t}{\Phi^x_u}$$

These two rules need to be parallelized, in order to enable the construction of a *normal* disproof for Russell's 'Paradox', thereby depriving it of the status of a genuine paradox. We propose the following parallelized versions, which we shall simply call E_1 and E_2:

$$E_1 \qquad \frac{t=\{x|\Phi\} \quad \Phi_u^x \quad \overset{\Box \!\!-\!\!-\!\!-(i)}{\underset{\vdots}{u \in t}} \quad \overset{\vdots}{\psi}}{\psi}(i)$$

$$E_2 \qquad \frac{t=\{x|\Phi\} \quad u \in t \quad \overset{\Box\!\!-\!\!-\!\!-(i)}{\underset{\vdots}{\Phi_u^x}} \quad \overset{\vdots}{\psi}}{\psi}(i)$$

For a Core *reductio* of the assumption $\exists y \; y = \{x|\neg x \in x\}$ we shall avail ourselves of the instances of E_1 and of E_2 where for Φ we take $\neg x \in x$; for ψ we take $\bot$; and for both terms t and u we take the parameter a:

$$E_1 \qquad \frac{a=\{x|\neg x \in x\} \quad \neg a \in a \quad \exists! a \quad \overset{\Box\!\!-\!\!-\!\!-(i)}{\underset{\vdots}{a \in a}} \quad \overset{\vdots}{\bot}}{\bot}(i)$$

$$E_2 \qquad \frac{a=\{x|\neg x \in x\} \quad a \in a \quad \overset{\Box\!\!-\!\!-\!\!-(i)}{\underset{\vdots}{\neg a \in a}} \quad \overset{\vdots}{\bot}}{\bot}(i)$$

Now let $\overset{\overbrace{a=\{x|\neg x \in x\}\,,\;a \in a}}{\underset{\bot}{\Sigma}}$ be the Core natural deduction

$$(E_2)\frac{a=\{x|\neg x \in x\} \quad a \in a \quad (E_1)\frac{a=\{x|\neg x \in x\} \quad \neg a \in a \quad \frac{(2)\;a=\{x|\neg x \in x\} \quad \frac{\overline{}(2)\;\overline{}(1)}{\neg a \in a \quad a \in a}}{\exists! a} \quad \frac{\bot}{}(1)}{\bot}(2)}{\bot}$$

We can now provide a Core *reductio* of the assumption $a = \{x|\neg x \in x\}$:

$$\cfrac{\cfrac{\overbrace{a=\{x|\neg x \in x\}\ ,\ a \in a}^{} \quad \cfrac{}{\Sigma}\ \text{(3)}}{\cfrac{\bot}{a=\{x|\neg x\in x\}}\ \text{(3)}\quad \cfrac{a=\{x|\neg x\in x\}}{\neg a \in a}\quad \cfrac{a=\{x|\neg x\in x\}}{\exists!a}}\quad \cfrac{\overbrace{a=\{x|\neg x\in x\}\ ,\ a \in a}^{}\ \text{(4)}}{\cfrac{\Sigma}{\bot}\ \text{(4)}}}{\bot}\ \text{(E}_1\text{)}$$

A final step of ($\exists$E) then yields the sought Core *reductio* of the existential assumption $\exists y\, y = \{x|\neg x \in x\}$. Inspection will reveal that no shrinking reduction is ever called for.

It is instructive to prove the same result in the sequent calculus for Core Logic, because with sequent proofs it is easier to check that there is no subproof compactification. The two rules E_1 and E_2 become the following left sequent rules:

$$(\{\}:)_1\ \frac{\Delta:\Phi_u^x \quad \Gamma:\exists!u \quad \Theta,u\in t:\psi}{t=\{x|\Phi\},\Delta,\Gamma,\Theta:\psi} \qquad (\{\}:)_2\ \frac{\Delta:u\in t \quad \Gamma,\Phi_u^x:\psi}{t=\{x|\Phi\},\Delta,\Gamma:\psi}$$

The foregoing Core *reductio* is rendered in the Core sequent calculus as follows. Let Σ be the following sequent proof of $a=\{x|\neg x\in x\}, a\in a:\bot$.

$$(\{\}:)_2\ \cfrac{a\in a:a\in a \quad (\{\}:)_1\ \cfrac{\neg a\in a:\neg a\in a \quad a=\{x|\neg x\in x\}:\exists!a \quad \cfrac{a\in a:a\in a}{\neg a\in a,a\in a:\bot}}{a=\{x|\neg x\in x\},\neg a\in a:\bot}}{a=\{x|\neg x\in x\},a\in a:\bot}$$

Then use Σ twice over to construct a Core *reductio* (sequent) proof for Russell's Paradox:

$$(\{\}:)_1\ \cfrac{\cfrac{\overset{\Sigma}{a=\{x|\neg x\in x\},a\in a:\bot}}{a=\{x|\neg x\in x\}:\neg a\in a}\quad a=\{x|\neg x\in x\}:\exists!a \quad \overset{\Sigma}{a=\{x|\neg x\in x\},a\in a:\bot}}{\cfrac{a=\{x|\neg x\in x\}:\bot}{\cfrac{\exists y\, y=\{x|\neg x\in x\}:\bot}{:\neg\exists y\, y=\{x|\neg x\in x\}}}}$$

That completes our discussion here of Russell's Paradox. We have seen that there are perfectly good proofs, in normal form, of both versions:

$$\neg\exists x\forall y(y\in x\leftrightarrow\neg y\in y),\ \text{ and }\ \neg\exists y\, y=\{x|\neg x\in x\}.$$

The proof of the former is in normal form, so there are no Eckmanesque problems with it. The proof of the latter, too, is in normal form, thereby revealing that Russell's Paradox is not a genuine paradox at all, in the sense being explicated by Tennant [1982]. The explicans on offer there was, rather, for genuinely *logico-semantical* paradoxes, of the kind that arise within semantically closed languages, rather than of any of the so-called paradoxes in the foundations of mathematics. The latter paradoxes, once they are resolved by suitable reformulations of first principles, are thereby 'tamed' as straightforward proofs of negative existentials—proofs that *can* be brought into normal form, if they are not already in normal form.

It remains now to ascertain that these insights, and this reclassification of Russell's Paradox, do not impugn the original proof-theoretic test for paradoxicality in terms of non-terminating reduction sequences. To this end we have space only to revisit the Liar Paradox, as the best representative of the class of logico-semantical paradoxes. We shall show that the Liar remains genuinely paradoxical on the amended account on offer, even though that account lays preferential stress on parallelized elimination rules.

11.5 Rules of truth, and id est rules: the Liar is still paradoxical

The following are the rules (*T*I) and (*T*E) for introducing and eliminating the truth predicate, followed by the reduction procedure for *T*; and the rules (λI) and (λE) for introducing and eliminating the Liar sentence λ, followed by the reduction procedure for λ. (λI) and (λE) are the 'id est' rules for the Liar.

In framing (*T*I) and (*T*E) we suppress use of corner quotes, and write '*T*φ' in place of '*T*⌜φ⌝'. Note, however, that 'λ' is the *name* of the sentence '¬*T*λ', so that there would never be any occasion to write '*T*⌜λ⌝'; indeed, it would be sortally incorrect, since the truth predicate can be satisfied only by sentences, not by their names.

$$(TI) \quad \begin{array}{c} \Delta \\ \vdots \\ \varphi \\ \hline T\varphi \end{array} \qquad\qquad (TE) \quad \begin{array}{c} \overbrace{\Delta\,,\ \varphi}^{(i)} \\ \vdots \\ T\varphi \qquad \theta \\ \hline \theta \end{array}{\scriptstyle (i)}$$

Reduction procedure for T:

$$\left[\begin{array}{cc} \begin{array}{c}\Delta \\ \Pi \\ \varphi \\ \hline T\varphi\end{array} & , & \begin{array}{c}\overbrace{\Gamma\,,\ \varphi}^{(i)} \\ \Sigma \\ T\varphi \quad \theta \\ \hline \theta\end{array}{\scriptstyle (i)} \end{array}\right] =_{df} \left[\Pi\,,\Sigma\right]$$

$$(\lambda I) \quad \begin{array}{c} \overbrace{\Delta\,,\ T\lambda}^{(i)} \\ \vdots \\ \bot \\ \hline \lambda \end{array}{\scriptstyle (i)} \qquad\qquad (\lambda E) \quad \begin{array}{c} \overbrace{\Delta\,,\ \neg T\lambda}^{(i)} \\ \vdots \\ \lambda \qquad \theta \\ \hline \theta \end{array}{\scriptstyle (i)}$$

Reduction procedure for λ:

$$\left[\begin{array}{cc} \begin{array}{c}\overbrace{\Delta\,,\ T\lambda}^{(i)} \\ \Pi \\ \bot \\ \hline \lambda\end{array}{\scriptstyle (i)} & , & \begin{array}{c}\overbrace{\Gamma\,,\ \neg T\lambda}^{(j)} \\ \Sigma \\ \lambda \quad \theta \\ \hline \theta\end{array}{\scriptstyle (j)} \end{array}\right] =_{df} \left[\begin{array}{cc} \begin{array}{c}\overbrace{\Delta\,,\ T\lambda}^{(i)} \\ \Pi \\ \bot \\ \hline \neg T\lambda\end{array}{\scriptstyle (i)} & , & \begin{array}{c}\Gamma,\neg T\lambda \\ \Sigma \\ \theta\end{array} \end{array}\right]$$

Using the foregoing natural-deduction rules, one can choose to

(1) both refute λ and prove λ; or

(2) both refute $T\lambda$ and prove $T\lambda$.

(1) offers the shortest way to achieve either of these goals, by constructing the following shortest possible proof of λ and shortest possible refutation of λ.

$$\Omega: \quad \cfrac{\cfrac{\cfrac{(3)\;\cfrac{}{T\lambda}}{(TE)\;\cfrac{\;\;\;}{}}}{(2)\;\cfrac{\lambda}{}} \quad (\lambda E)\cfrac{\;\;\;\cfrac{(1)\cfrac{}{\neg T\lambda}\quad\cfrac{}{T\lambda}(3)}{(\neg E)}\;\;\;}{\cfrac{\bot}{\bot}(1)}\;(2)}{(\lambda I)\cfrac{\bot}{\lambda}(3)}$$

$$\Xi: \quad (\lambda E)\cfrac{\;\cfrac{\lambda}{}\quad\cfrac{\cfrac{(1)\cfrac{}{\neg T\lambda}\quad\cfrac{\lambda}{T\lambda}(TI)}{(\neg E)}}{\bot}(1)\;}{\bot}$$

The sequent rules corresponding to the parallelized natural deduction rules for T and for λ are as follows.

$$(:T)\quad \frac{\Delta:\varphi}{\Delta:T\varphi} \qquad\qquad (T:)\quad \frac{\Delta,\varphi:\theta}{\Delta,T\varphi:\theta}$$

$$(:\lambda)\quad \frac{\Delta,T\lambda:\bot}{\Delta:\lambda} \qquad\qquad (\lambda:)\quad \frac{\Delta,\neg T\lambda:\theta}{\Delta,\lambda:\theta}$$

Using these sequent rules, one can give the following sequent proofs corresponding to Ω and Ξ:

$$\Omega':\quad (T:)\cfrac{(\lambda:)\cfrac{(\neg:)\cfrac{T\lambda:T\lambda}{\neg T\lambda,T\lambda:\bot}}{\lambda,T\lambda:\bot}}{\cfrac{T\lambda:\bot}{:\lambda}(:\lambda)}$$

$$\Xi':\quad (\lambda:)\cfrac{(\neg:)\cfrac{(:T)\cfrac{\lambda:\lambda}{\lambda:T\lambda}}{\neg T\lambda,\lambda:\bot}}{\lambda:\bot}$$

It would be the orthodox expectation that one could now put these two proofs together, 'cutting on λ', so as to produce an 'outright' sequent proof of $: \perp$ (i.e. the empty sequent), thereby completing the embarrassment that is the Liar Paradox. It would be a further expectation on the part of some that the resulting overall 'proof of $\perp$' could be normalized, or made cut-free.

But inspection reveals that the proofs Σ and Ω, despite the reduction procedures just stated for T and λ, cannot be put together to produce a proof, in normal form, of $\perp$. Let us consider the matter from the perspective of the ordinary natural deduction theorist, who, unlike the Core logician, allows major premises for eliminations to be conclusions of non-trivial proof work. Copies of the proof Ω of conclusion λ would have to be grafted onto the undischarged assumption occurrences of λ within the proof Ξ.

$$
\Omega: \qquad
\begin{array}{c}
\cfrac{
\cfrac{
\cfrac{
\cfrac{(3)\rule{1.5em}{0.4pt}}{T\lambda}(TE)
\quad
\cfrac{
\cfrac{(2)\rule{1em}{0.4pt}}{\lambda}
}{\ \ }(\lambda E)
\quad
\cfrac{
\cfrac{(1)\rule{1.5em}{0.4pt}}{\neg T\lambda}
\quad
\cfrac{\rule{1.5em}{0.4pt}(3)}{T\lambda}
}{
\cfrac{\perp}{\perp\ (2)}(\neg E)\,(1)
}
}{\ \ }
}{}
}{\cfrac{}{\lambda}(\lambda I)\,(3)}
\end{array}
$$

$$
\Xi: \qquad
\cfrac{
\cfrac{\lambda}{\ }(\lambda E)
\quad
\cfrac{
\cfrac{(1)\rule{1.5em}{0.4pt}}{\neg T\lambda}
\quad
\cfrac{\lambda}{T\lambda}(TI)
}{
\cfrac{\perp}{\ }(\neg E)\,(1)
}
}{\perp}
$$

Such grafting would make the leftmost occurrence of λ in Ξ, which stands as a major premise for λ-Elimination, stand also as the conclusion (which it is within Ω) of λ-Introduction. So the reduction procedure for λ would be applicable. The reader can check that the 'accumulated' proof

$$
\begin{array}{c}
\Omega \\
(\lambda) \\
\Xi \\
\perp
\end{array}
$$

becomes, upon λ-reduction, a proof calling for an application of $\neg$-reduction; and the result of the latter reduction calls for an application of T-reduction ... whereupon we are back with the input for the initial λ-reduction. It is easy to verify also that the same pathology is evident when for Ω we use the slightly longer proof

$$
\begin{array}{c}
\dfrac{\begin{array}{c}
\dfrac{\quad}{T\lambda}\,(3) \\[2pt]
\dfrac{\begin{array}{c}
\dfrac{\quad}{\lambda}\,(2) \\
\end{array}}{\quad}\,(\lambda E) \quad
\dfrac{\dfrac{(1)\,\dfrac{\quad}{\neg T\lambda} \quad \dfrac{\dfrac{\quad}{\lambda}\,(2)}{T\lambda}\,(TI)}{\bot}\,(\neg E)}{\bot}\,(1) \\
\end{array}}{\dfrac{T\lambda}{\quad}}\,(TE) \\
\dfrac{\bot}{\quad}\,(2) \\
\dfrac{\bot}{\lambda}\,(\lambda I)\,(3)
\end{array}
$$

In pursuit of choice (1) above, we find that putting together any proof of λ with any refutation of λ results in a looping reduction sequence. The same holds with any attempted pursuit of choice (2): putting together any proof of $T\lambda$ with any refutation of $T\lambda$ results in a looping reduction sequence.

The reduction sequences loop, despite the fact that there is no call for subproof compactification. Even with parallelized elimination rules, the Liar Paradox remains genuinely paradoxical according to our proof-theoretic test—unlike Russell's Paradox.

11.5.1 Digression on Cut and transitivity of deduction

If one has two proofs (natural deductions) of the respective forms

$$
\begin{array}{ccc}
\Delta & & \overbrace{\varphi,\Gamma} \\
\Pi & \text{and} & \Sigma \\
\varphi & & \psi
\end{array}\;,
$$

how is one to obtain from them the (usually expected) 'target result'

$$
\Delta,\Gamma:\psi\,?
$$

That is to say, how is one to ensure that one may 'perform the cut' that is invited?

The usual answer from the natural-deduction theorist is that one can simply 'accumulate' the proofs Π and Σ, by placing a copy of Π over every undischarged assumption occurrence of φ within Σ:

$$
\begin{array}{c}
\Delta \\
\Pi \\
\underbrace{(\varphi)\,,\;\Gamma} \\
\Sigma \\
\psi
\end{array}
$$

The usual definition of the notion of proof for the system of natural deduction allows the result of such grafting to count as a proof. In brief: such
accumulations of proofs are always proofs (even if they happen to fail to be
in normal form).

The corresponding usual answer from the sequent-calculus theorist is
that one can apply the (unrestricted) rule of Cut as a rule *of the system*,
thereby obtaining a new proof, in the system, from the two proofs Π and Σ.
In the schematic notation of sequent proofs, this would amount to constructing a sequent proof of the following form:

$$\frac{\begin{array}{cc} \Pi & \Sigma \\ \Delta : \varphi & \Gamma, \varphi : \psi \end{array}}{\Delta, \Gamma : \psi}\text{(CUT)}$$

The usual definition of the notion of proof for the sequent calculus allows
the result of such an application of (unrestricted) Cut to count as a proof.
In brief: such *unrestricted cuts on proofs always produce proofs* (even if they
happen to produce results that fail certain obvious tests for relevance of
premises to conclusion). One of the Holy Grails of proof theory has always
been to prove (for whichever system is under consideration) a Gentzenian
Hauptsatz to the effect that if a given sequent can be proved using Cut,
then it can be proved *without* using Cut. That is, all applications of Cut
are *eliminable* from sequent proofs. It is strange, indeed, that it should be
a central concern to establish the dispensability of a rule on which almost
every logician insists. The peculiarity of this predicament is discussed at
greater length in Tennant [2016].

Let us revisit the question posed above, but express it now wholly in
terms of sequents:

If one has two sequent proofs $\begin{array}{c} \Pi \\ \Delta : \varphi \end{array}$ and $\begin{array}{c} \Sigma \\ \Gamma, \varphi : \psi \end{array}$, how is one to obtain
from them the (usually expected) 'target result'

$$\Delta, \Gamma : \psi \quad ?$$

That is to say, how is one to ensure that one may 'perform the cut' that is
invited?

In Core Logic, as we have seen *ad nauseam*, neither Cut nor Thinning is an
applicable rule of the system of sequent proof. So how is the Core logician
to respond to this very understandable question in a way that can allay
the concern behind it? The concern, of course, is that even with *cut-free*
proofs Π and Σ, one might fail to find a *cut-free* proof of the target result

(let alone: a cut-free, *thinning-free* proof of it). The Core logician takes this
concern very seriously, and has a perfectly adequate answer to the question
posed. As we have seen, there is an effective binary operation on core proofs,
denoted by $\left[\,\Pi_1, \Pi_2\,\right]$, such that

$$\left[\begin{array}{cc} \Pi & \Sigma \\ \Delta : \varphi \,' & \Gamma, \varphi : \psi \end{array}\right]$$

is a core proof of *some subsequent of* the target sequent $\Delta, \Gamma : \psi$. This
provides all the transitivity of deduction that one could possibly need.

11.5.2   The inadmissibility of cut in languages containing paradoxes

The orthodox proof theorist who works with a sequent calculus in which
Cut is an applicable rule will not be able to prove Cut-elimination for that
calculus if the language in question is a semantically closed one in which the
Liar can be formulated. Likewise, the Core logician, who does *not* have Cut
as a structural rule of his sequent calculus, will not be able to prove that
Cut is nevertheless admissible for any such language. To see this, we need
only note that, *with* Cut, we can prove the 'empty sequent' $\emptyset : \bot$, by using
the sequent proofs Ω' and Ξ':

$$
\begin{array}{c}
\dfrac{\dfrac{\dfrac{\dfrac{T\lambda : T\lambda}{\neg T\lambda, T\lambda : \bot}}{\lambda, T\lambda : \bot}}{\dfrac{T\lambda : \bot}{\emptyset : \lambda}} \qquad \dfrac{\dfrac{\dfrac{\lambda : \lambda}{\lambda : T\lambda}}{\neg T\lambda, \lambda : \bot}}{\lambda : \bot}}{\emptyset : \bot}\ (\text{CUT})
\end{array}
$$

But the sequent rules for $\neg$, T, and λ all have at least one sentence on the left
or on the right of their conclusion sequents. So no arrangement of steps in
accordance with those rules could possibly be a proof of the empty sequent.
Hence Cut-Elimination fails for the sequent calculus based on those rules.

This is the fundamental proof-theoretic lesson to be drawn from the
logico-semantical paradoxes. It does not mean, however, that one should
simply throw in the towel and avoid semantically closed languages altogether (as Tarski did). Rather, one should have one's eyes wide open, as it
were, for localized failures of Cut, *because they are symptomatic of paradoxicality.* All would-be proofs 'by transitivity' in the semantically open fragment of a semantically closed language, along with many innocuous such

proofs in the semantically closed part as well, will turn out to be normalizable (equivalently, in the terminology of sequents: they will turn out to have sequent proofs using just the Left and Right rules). The normal-form proofs thus obtained will guarantee transmission of warrants for assertion from their premises to their conclusions. If the proof's conclusion is ⊥, then the proof's premises will have been revealed as jointly incoherent. If, however, a would-be proof of ⊥ 'by transitivity' of normal proofs is not itself normalizable, then no such incoherence will have been established. On the present account, the reasoning associated with paradoxes *does not reveal inconsistency* of the usual suspects. It provides no warrant at all for joint denial. This is because the regimented reasoning cannot be brought into normal form. For further development of this view, with replies to anticipated objections, see Tennant [2015a].

11.6 Summary and conclusion

We have seen that the proof-theoretic criterion of paradoxicality has not been found to rule as paradoxical proofs of absurdity that are not genuinely paradoxical; in particular, Ekman's 'paradox' is no paradox at all. The initial impression to the contrary that Schroeder-Heister and Tranchini [forthcoming] articulated in some detail can be dispelled by observing that there is a crucial difference between serial and parallelized elimination rules. Ekman's proof of ⊥ from $\{A \rightarrow \neg A, \neg A \rightarrow A\}$ appears to initiate a non-terminating (because looping) reduction sequence only because it uses the serial form of Elimination rule for →. We have seen that if the parallelized form is used instead, then the proof can be given in normal form. Moreover, the proof of absurdity using the parallelized elimination rules in the case of the Liar Paradox turns out still to be a starting point for non-terminating (sequences of) reduction procedures, no matter how the reductions are applied. Thus the Liar's intuitive paradoxicality is borne out by the formal proof-theoretic criterion originally proposed, but incorporating now the necessary qualification that all elimination rules should be in parallelized form.

Essentially, this means that we are making our original point about paradoxicality by framing it in terms of cut-free, thinning-free sequent proofs (equivalently: core natural deductions), rather than in terms of natural deductions of Gentzen–Prawitz form. We find also that Russell's Paradox enjoys a proof in normal form, so that it is not genuinely paradoxical. This brings our proof-theoretic criterion of paradoxicality more closely into line with Ramsey's famous (and now, we see, perhaps more deeply principled) distinction between the 'Group A contradictions'—that is, the mathematical paradoxes such as Russell's Paradox—and the 'Group B contradictions'— that is, the logico-semantical paradoxes such as the Liar.[8] The latter para- doxes call for a proof-theoretic clarification of the vicious circles within them—which is what I have sought, and am still seeking, to provide. The former 'paradoxes', however, like Russell's, call for a rethinking of both our abstract ontology and the first principles governing our thinking about it. In the latter case (for example, the Liar Paradox) we cannot have normality of (dis)proof. In the former case (for example, Russell's Paradox) we can; and we thereby obtain important negative existentials.

We cannot, and should not, hope for or expect anything remotely similar to happen in the case of the genuine logico-semantical paradoxes. We simply have to find a way to live with them—such as the way of the anti-realist that is described in Tennant [2015a].

---

[8] See Ramsey [1926], at pp. 352–3.

Chapter 12

Replies to Critics of Core Logic

Abstract

We take on the critics in chronological order of first critique.
§12.1 rebuts JC Beall and Greg Restall.
§12.2 rebuts Dirk Hartmann.
§12.3 defines the notions of perfect validity and entailment.
§12.4 rebuts John Burgess.
§12.5 rebuts Harvey Friedman and Arnon Avron.
The interested reader should not be detained here with any further details.

This book recommends that we adopt Core Logic (or its classicized extension) as the ideal system fulfilling the original Fregean (or even Aristotelian) aspirations concerning a formal logic for the conduct of mathematical and scientific reasoning. The record of reactions to Core Logic thus far shows that certain critics and I are not 'on the same page'. They are either mistaken about certain key features of the Core systems, or over-hasty in their assessment of those features' implications for a choice of logic. This chapter investigates our differences of opinion. In §12.1, §12.2, and §12.4 I reply to published criticisms that have been leveled against Core Logic by JC Beall and Greg Restall, by John Burgess, and by Dirk Hartmann respectively. In §12.5 I reply to criticisms leveled against Core Logic by Harvey Friedman and Arnon Avron in the leading online forum for the foundations of mathematics (fom@cs.nyu.edu).

I shall argue that the criticisms are based on a misunderstanding either of the kind of transitivity of proof that is enjoyed by Core Logic, or of the metalogical reasoning involved in characterizing it. It is clear that we need to address what appears to be a collective lack of recognition of the importance of what I have called 'potential epistemic gain' in those cases where *unrestricted* transitivity fails. Core Logic affords *restricted* transitivity. Nevertheless, what needs to be emphasized is that it is (restricted) *transitivity*. Core Logic furnishes all the transitivity one *needs*, which is to say: *all the transitivity that one may demand*, on pain of irrelevance.

A word or two of methodological caution is in order at the outset. In any process of reaching a reflective equilibrium about descriptive and normative matters, and striking a balance between pre-theoretical intuitions about instances and higher-level hypotheses of greater generality, it is vital to have a properly informed grasp of the true extent of alleged 'losses' or 'gains' for choosing *this* higher-level principle over *that*. One has to avoid over-hasty reactions to considered eschewal of old principles, and really understand what is at issue in making any proposed changes to overarching theory.

12.1   Reply to Beall and Restall

> The imputation of Novelty is a terrible charge amongst those who judge of men's heads, as they do of their perukes, by the fashion, and can allow none to be right but the received doctrines.
> —From the Dedication of Locke's *Essay Concerning Human Understanding*

We turn now to consider the attitudes of two logical *pluralists* to logical systems, such as Core Logic, that moderate the absolutely unrestricted transitivity of orthodox logic. As Read [2006] notes, at p. 198,

> Beall and Restall's logical pluralism tries to be eclectic and all-embracing (up to a point which excludes Aristotle, Smiley and Tennant) ....

Read was referring to Beall and Restall [2000]; but we could add also their subsequent book Beall and Restall [2006]. At p. 476 of the former piece, Beall and Restall state a principle they call (V):

> (V) A conclusion, A, *follows from* premises, Σ, if and only if any case in which each premise in Σ is true is also a case in which A is true.

Their professed pluralist outlook notwithstanding, Beall and Restall write ([2000], p. 489)

> In both cases, with paraconsistent and intuitionist logic, we find a place for these non-classical logics—for both are elucidations

of the pretheoretical notion (V) of logical consequence.

> On the other hand, there might be other logics for which
> we can find no place in our catalogue of True Logics, as much as
> we admire their technical subtlety.

In a footnote, they add:

> Tennant's 'relevant logic' [Tennant [1994a]], which rejects tran-
> sitivity, ... fails to fall under the banner of logical consequence
> given in (V).

Beall and Restall do not, however, explain how it thus 'fails to fall' under
the 'banner' of (V). Indeed, one can argue that it *does* so fall. For every
core-provable sequent $\Sigma : A$, the conclusion A *follows from* the premises Σ
in precisely the sense stated in (V): any case in which each premise in Σ is
true is also a case in which A is true. (Indeed, we have gone one better,
in Chapter 9, by showing that for any model M, any core proof of ψ from
$\varphi_1, \ldots, \varphi_n$ will transform any M-relative truthmakers for $\varphi_1, \ldots, \varphi_n$ into an
M-relative truthmaker for ψ.) That is, Core Logic is *sound*. Moreover, Core
Logic is *complete*. For, let $\Sigma : A$ be any sequent such that any case in which
each premise in Σ is true is also a case in which A is true. Then Core Logic
proves (a subsequent of) $\Sigma : A$. This is so regardless of whether truth is
construed intuitionistically and we take the system of Core Logic; or truth
is construed classically, and we take the correspondingly classicized system,
Classical Core Logic.

In Beall and Restall [2006] the principle (V) is restated as the

> GENERALIZED TARSKI THESIS (GTT): An argument is valid$_X$ if
> and only if, in every case$_X$ in which the premises are true, so is
> the conclusion.

On p. 91 they tell their reader once again that

> Tennant-style 'relevant' logics, which reject transitivity, ... fail
> to fall under the banner of logical consequence given in (GTT)
> ....

This claim—which we have already shown above to be in error—is followed
immediately by the following dogmatic statement of position (emphasis in
original).

> What can we say? We hold the line. The given kinds of non-
> transitive ... systems of 'logical consequence' are logics by cour-
> tesy and by family resemblance, where the courtesy is granted

via analogy with logics *properly* so called. Non-transitive ... systems of 'entailment' may well model interesting phenomena, but they are not accounts of *logical consequence*. One must draw the line somewhere and, pending further argument, we (defeasibly) draw it where we have. We require transitivity ... in logical consequence. We are pluralists. It does not follow that absolutely *anything* goes.

One wonders whether this proposed Maginot line can hold up upon consideration of some historical analogies. Let us try the following, which is not unlike things one can read by philosophers or geometers in the mid to late 1800s who could not comprehend the import of non-Euclidean geometries:

What can we say? We hold the line. The given kinds of non-Euclidean systems of 'distance' are geometries by courtesy and by family resemblance, where the courtesy is granted via analogy with geometry *properly* so called. Non-Euclidean systems of 'distance' may well model interesting phenomena, but they are not accounts of *distance*. One must draw the line somewhere and, pending further argument, we (defeasibly) draw it where we have. We require invariance of geometric figures under Euclidean transformations. We are mathematical pluralists. It does not follow that absolutely *anything* goes.

If this does not persuade the reader that such dogmatism should be avoided in this day and age—at least, on the part of self-proclaimed *pluralists*—let us try the following further variation on the theme, which is not unlike things one can read by those philosophers or logicians in the early decades of the twentieth century who could not comprehend the import of the new intuitionistic or constructive logics:

What can we say? We hold the line. The given kind of non-classical systems of 'logical consequence' are logics by courtesy and by family resemblance, where the courtesy is granted via analogy with logics *properly* so called. Non-classical systems of 'constructive validity' may well model interesting phenomena, but they are not accounts of *logical consequence*. One must draw the line somewhere and, pending further argument, we (defeasibly) draw it where we have. We require that any account of logical consequence should validate Double Negation Elimination. We are pluralists. It does not follow that absolutely *anything* goes.

It would be fair to inquire of Beall and Restall: what exactly *are* 'logics *properly* so called'? They claim to be able to 'find a place' for 'paraconsistent and intuitionistic logic'; so what is so offputting, to them, about the system of Core Logic, which is *both* paraconsistent *and* intuitionistic?—or about the classicized version thereof, which is paraconsistent? What features of a system of deduction need to be highlighted in order for it to be 'granted the courtesy' of being called a logic in their sense (*properly* so called ...)? What exactly *is* this 'transitivity' that they require? How does that requirement follow from their principle (V) (or its restatement as (GTT)), which concerns transmission of truth, rather than transitivity of consequence *per se*? How exactly does Core Logic, in their view, 'fail to fall under the banner of logical consequence' that flutters out from this principle? The rationale for their pronouncements on these matters is opaque to this author.

In Beall and Restall's hands, GTT is a versatile template. They are eager to consider a wide range of *kinds* of cases$_X$. The most familiar and straight-forward are of course the classical cases—models that decide each atomic matter of fact either positively or negatively, but not both. They consider also *incomplete* cases, in order to generate intuitionistic consequence; and *inconsistent* (or *incoherent*) cases, in order to generate paraconsistent logic.

In closing, the final part of the quote deserves to be revisited:

> We are pluralists. It does not follow that absolutely *anything* goes.

Beall and Restall do not explain exactly what it is about Core Logic that banishes it to some untouchable periphery of the domain of their deprecating quantifier 'absolutely *anything*'. What is it about Core Logic that makes it somehow beyond the pale, or impossible to contemplate, or perverse, or manifestly incorrect, or threatening in its implications (or imagined lack of them)? Perhaps the most apt rebuttal is an even terser quantifying phrase: *Some pluralists ...!*

12.2 Reply to Hartmann

Hartmann [2003], at p. 39, writes as follows:

> Tennant, ... while holding on to [Disjunctive Syllogism] as well as to (∨I), is among those few authors questioning the unre-stricted validity of transitivity .... Most people think ... that blaming Cut is not really convincing, as transitivity of valid in-ference seems to be at the very heart of logic.

Hartmann offers no empirical evidence for his claim 'Most people think
...'. And in response to his phrase 'the very heart of logic', we seize this
opportunity to remind the reader of the *OED* entry quoted in the Preface:

> core, *n.* The central or innermost part, the heart of anything.

Our rejoinder is that Core Logic *is* 'the very heart of logic'. It captures
exactly the transitivity of valid inference that squares with uncorrupted
intuition, and that can be shown, theoretically, to be sufficient unto all the
methodological demands we can impose upon a logic. Hartmann goes on to
say, at p. 40,

> Tennant claims that ... transitivity fails in [Core Logic] only
> where it "ought" to fail: namely whenever the premises form an
> inconsistent set.

But, he says, there is the following serious problem:

> ... consistency of sets of formulas in [Core Logic] is not ... de-
> cidable. So, the partial failure of transitivity may sometimes
> leave us in the dark regarding the question of whether we may
> combine a given proof of A from the set of premises Δ with an-
> other given proof of B from A plus Γ to a proof of A from Δ
> and Γ.

Let us address directly Hartmann's worry about the undecidability of consistency. (As with the systems of Minimal Logic **M**, Intuitionistic Logic **I**, and
Classical Logic **C**, it is the *first-order* system that is undecidable, not the
propositional system. Note, though, that Anderson and Belnap's system **R**
is undecidable even at the propositional level.[1]) The undecidability of first-
order Core Logic poses no problem whatsoever for the Core logician. The
Core logician is not 'left in the dark' at all, in the way Hartmann imagines.
What Hartmann appears to be omitting from consideration is that, given
any *core proof* (Π, say) of A from the set of premises Δ, along with a *core
proof* (Σ, say) of B from A plus Γ, the core proof [Π, Σ] *can be constructed*
in a perfectly effective way, so that one can determine from it which subsequent of Δ, Γ : B it establishes. One does not have to decide whether the
combined set Δ ∪ Γ is consistent!—the method of proof conversion that is
provided in our proof of Metatheorem 4 suffices to tell us what (restricted)
transitivity yields in the case at hand.

---

[1] See Urquhart [1984].

This omission of consideration of the cut-elimination operation $[\Pi, \Sigma]$ is evident also at p. 86, where Hartmann writes that the Core logician's requirement that all core proofs be in normal form is to '[lose] transitivity':

> This is because it may very well be the case that there are proofs of normal form establishing $\Delta \vdash A$ and $A, \Delta \vdash B$, but *there's no proof for $\Delta \vdash B$*. (Emphasis added.)

We know now, however, that this complaint is in error. Under the assumptions made, there will indeed be a core proof of (some subsequent of) $\Delta : B$. Moreover, if one is given proofs of the sequents $\Delta : A$ and $A, \Delta : B$, one will be able to determine from them, *effectively*, a core proof of some subsequent of $\Delta : B$. Furthermore, if that subsequent is a *proper* subsequent of $\Delta : B$, one might well have achieved epistemic gain.

The objector who sides with Hartmann might complain here that (in the more general case first discussed above) we are helping ourselves to the assumption that one is *given* a proof Π of A from the set of premises Δ, and that one is *given* a proof Σ of B from A plus Γ. What, though, if this is not the case? What if all we are told is that *there is* a (core) proof of A from the set of premises Δ, and *there is* a (core) proof Σ of B from A plus Γ, but we are *not given* any (core) proofs in witness thereof?

Our reply is that in such a case the Core logician will be no worse off, epistemically, than the ordinary logician who appeals to unrestricted Cut. In the envisaged situation—where all we may assume is that $\Delta \vdash A$ and $A, \Gamma \vdash B$—the ordinary logician will conclude (by unrestricted Cut) that $\Delta, \Gamma \vdash B$. But the Core logician will be able to conclude that, for some $\Theta \subseteq (\Delta \cup \Gamma)$, either $\Theta \vdash B$ or $\Theta \vdash \bot$. Hence, if the Core logician can be assured that $\Delta \cup \Gamma$ is consistent, then he will be able to conclude that $\Theta \vdash B$. And if the Core logician *cannot* be assured that $\Delta \cup \Gamma$ is consistent, then he will at least be no worse off than the ordinary logician, who would have to worry that perhaps the newly combined premises $\Delta \cup \Gamma$ harbor an inconsistency, even if Δ is known to be consistent and Γ is known to be consistent. Even the ordinary logician would have to acknowledge that, in the event that $\Delta \cup \Gamma$ is *in*consistent, he would be disturbed by the thought that he might have derived B from $\Delta \cup \Gamma$ by trading on that very inconsistency via applications of EFQ of which he might not be immediately aware, and which would be brought to light only after normalizing and relevantizing the (non-core) proof in question (of B 'from' $\Delta \cup \Gamma$).

Hartmann observes also that the Replacement Theorem fails (for Core Logic). Adapting Hartmann's example, we have

$$A \wedge \neg A \vdash B \vee (A \wedge \neg A) \quad \text{and} \quad B \vee (A \wedge \neg A) \dashv\vdash B,$$

but

$$A \wedge \neg A \nvdash B.$$

For Hartmann, the failure of the Replacement Theorem, combined with his worry about undecidability of consistency, is

> in my opinion, too big a price to pay to get rid of Lewis' "proof" of EFQ.

Once again, the criticism can be met by attending to the methodological implications of Metatheorem 4. The foregoing failure of *unrestricted* transitivity in Core Logic—which Hartmann considers at least a significant part of the price that is 'too big ... to pay' to avoid Lewis's First Paradox— invites more careful consideration. In what, exactly, does this supposedly distressing (hence detracting) 'failure' consist? Take the obvious core proofs of the two statements of deducibility in Hartmann's example. Then form their reduct:

$$
\left[
\begin{array}{c}
\dfrac{A \wedge \neg A}{B \vee (A \wedge \neg A)} \ , \quad
\dfrac{\dfrac{}{B \vee (A \wedge \neg A)}^{(2)} \quad \dfrac{\dfrac{A \wedge \neg A}{\dfrac{\dfrac{\dfrac{}{\neg A}^{(1)} \quad \dfrac{}{A}^{(1)}}{\bot}}{\bot}^{(1)}}^{(2)}}{B}^{(2)}}{B}
\end{array}
\right] .
$$

Our now-familiar procedure (here, consisting of but a single application of ∨-Reduction) simplifies this to

$$
\dfrac{A \wedge \neg A \quad \dfrac{\dfrac{}{\neg A}^{(1)} \quad \dfrac{}{A}^{(1)}}{\bot}}{\bot}^{(1)} .
$$

So, *there is no proof* of $A \wedge \neg A : B$ covertly lurking within the two core proofs that deliver this reduct! The demand that (unrestricted) transitivity should hold, and *give us* the result $A \wedge \neg A : B$ is—witness this particular instance—simply unwarranted.[2] Hence the most one can demand is a form

---

[2] One wants to ask: how could it have turned out otherwise? This particular example of Hartmann's has the inconsistent premise $A \wedge \neg A$ in the first sequent, and its second sequent—whether proved in Core Logic or in Intuitionistic Logic—*has to have that inconsistency revealed* in order for the proof by cases to go through. So it really is a very delinquent pair of sequents begging the court of logical law for the ill-advised indulgence of saying that together they warrant $A \wedge \neg A : B$. The most they can hope for is that they get off with just the suspension of the arbitrary sentence B.

of transitivity that is suitably restricted in some principled way. And that is what is vouchsafed by Metatheorem 4.

Since we have already disposed of the worry about undecidability of consistency, it remains only to rebut the criticism that the Replacement Theorem *ought* to hold (tacitly: in any decent logical system), but *fails* to hold for Core Logic. We shall challenge the first, deontic, conjunct.

To the critic, like Hartmann, who reacts to what they might regard as unwarranted irreverence towards supposedly sacred meta-principles in logic, it is time to put some disarming questions:

> Of what *use* is this vaunted Replacement Principle? Of what *use* is that old saw called the Deduction Theorem, in the erstwhile direction that, we now see, fails? Do you see any mathematicians or scientists wringing their hands at supposedly having them thus tied?

The answer is resoundingly negative. No foundation for mathematics, no program for the automation of mathematical proof verification or proof discovery, no proof of any mathematical theorem, no disproof of any scientific theory on the basis of observable evidence, has any *need* for either of these two meta-principles. The deductive reasoning involved in these intellectual endeavors is to be regimented *by means of formal proofs*; the mathematicians and scientists themselves do not reason *about* proofs.[3] The two meta-principles are simply things that find their way into introductory logic texts, and to which obeisance (or unquestioning observance) has been taught. But that has allowed critical acumen, even if not language, to go on holiday.

12.3   Interlude on perfect validity and entailment

The set-based sequent conception is useful for the purpose of talking of *subsequents*. In general, if $\Delta' \subseteq \Delta$ and $\Gamma' \subseteq \Gamma$, then we say that $\Delta' : \Gamma'$ is a *subsequent* of $\Delta : \Gamma$. If either of these inclusions is proper, then we speak of a *proper* subsequent. For example, we have both

$$\varphi_1, \ldots, \varphi_n : \bot \text{ is a proper subsequent of } \varphi_1, \ldots, \varphi_n : \psi$$

and

$$\varphi_1, \ldots, \varphi_{i-1}, \varphi_{i+1}, \varphi_n : \psi \text{ is a proper subsequent of } \varphi_1, \ldots, \varphi_n : \psi.$$

---

[3] Unless, of course, they are doing metalogic or metamathematics—in which case the point being made here is to be directed, not at the metalevel, but at the metametalevel.

By 'dropping a sentence' on the left (a premise) or on the right (the conclusion), one can *logically strengthen* a sequent, or the argument that it
represents. This is particularly so if one drops a premise that is not logically
implied by the remaining premises, or if one replaces a sentential conclusion
that is not itself inconsistent by the absurdity sign ⊥.

   With an eye to the fact that one can strengthen arguments (or sequents)
in this way, we can motivate the following definition of a more exigent notion
of sequent validity. We supply now some more detailed definitions to develop
ideas touched upon in §4.5.

Definition 37. *A sequent is* perfectly *valid if and only if it is valid and has
no valid proper subsequents.*

   Armed with the notion of perfect validity, we can define the notion of
an *entailment* as follows.

Definition 38. *A sequent is an* entailment *if and only if it is a substitution
instance of a perfectly valid sequent.*

   With a perfectly valid sequent $\varphi_1, \ldots, \varphi_n : \psi$ (where $n \geq 1$ and $\psi \neq \perp$),
one has the assurance that

   (i) every premise $\varphi_1, \ldots, \varphi_n$ is 'doing some work' in order to secure the
       conclusion;

  (ii) the premises $\varphi_1, \ldots, \varphi_n$ are jointly satisfiable; and

 (iii) the conclusion ψ is not logically true.

Moreover, the following claims about perfect validity are easily proved as
metatheorems.

   1. Every valid sequent has a perfectly valid subsequent.

   2. If φ is logically true, then the sequent $\emptyset : \varphi$ is perfectly valid.

   3. If Δ is unsatisfiable but every proper subset of Δ is satisfiable, then
      the sequent $\Delta : \perp$ is perfectly valid.

   4. If $\Delta : \psi$ is perfectly valid, then some atomic sentence occurs within
      some member of Δ and within ψ.

12.4   Reply to Burgess

In §3 of Burgess [2005] a number of comments and criticisms are made about what Burgess calls 'perfectionism' as a brand of relevantism. The present author is taken as representative of this brand. In §5.3, Perfectionist Logic, of Burgess [2009] these comments and criticisms are repeated. Our first order of business will be to correct certain misrepresentations of the nature of Core Logic, *qua* 'perfectionist' system. Our second order of business will be to rebut the arguments that Burgess offers against what one might call 'Core-logical reformism', or 'the Core logician's criticism of Classical Logic'. We shall refer to the '2005 version' and the '2009 version' of Burgess's critical comments when quoting and rebutting.

12.4.1   The distinction between prescriptive and descriptive criticism of Classical Logic

At the very outset let us present an extremely useful distinction that Burgess introduces into the discussion in §1 of his 2005 paper—that between 'prescriptive criticism' and 'descriptive criticism' of Classical Logic. The *prescriptive* critic alleges (p. 729)

> that classical logic correctly describes unacceptable practices in the community of orthodox mathematicians,

while the *descriptive* critic alleges

> that classical logic endorses incorrect principles that are *not* operative in the practice of that community.

An example of a *prescriptive* critic is the intuitionist. The intuitionist maintains that, although the use of strictly classical principles (such as classical *Reductio ad Absurdum*, or Double Negation Elimination) is indeed part of the practice within 'the community of orthodox mathematicians', this is nevertheless *unacceptable*, as one would come to realize after conducting an appropriate philosophical critique. Eschewing the offending classical principles is necessary in order for the practice to be made *respectable*.

An example of a *descriptive* critic is the relevantist. The relevantist maintains that, although Classical Logic enshrines certain inferences such as the First Lewis Paradox $A, \neg A : B$, nevertheless such a principle, on closer inspection, is *not* 'operative in the practice of [the mathematical] community'.[4] Eschewing the offending classical principles is necessary in order for the practice to be *respected*.

The intuitionist's prescriptive criticism is of course much more venturesome than the relevantist's descriptive criticism. As it happens, our concern here need only be with the latter. For the points to be disputed with Burgess regarding our favored brand of relevantism concern only the differences between Classical Logic, on one hand, and the *classicized* version of Core Logic, on the other. It is also worth noting, in apposition to this point, that, insofar as the Core logician offers criticism also of Intuitionistic Logic (on grounds of relevance), this too is *descriptive* criticism in Burgess's sense. The Core logician maintains that, although Intuitionistic Logic enshrines certain inferences such as the First Lewis Paradox $A, \neg A : B$, nevertheless such a principle, on closer inspection, is *not* operative in the practice of the community of *intuitionistic* (or *constructive*) mathematicians.[5] Eschewing the offending principles of Intuitionistic Logic is necessary in order for the practice to be *respected*. The claim, in essence, concerns the question of which formal logical canons are necessary and sufficient for the fully formal codification of the body of (inerrant, expert) practice in question—practice, we might add, that has not itself been 'corrupted' by habitual conformity of certain of its practitioners to rules of an ill-chosen formal logical system.

12.4.2   Correcting some misdescriptions of our brand of perfectionism

On whether one ought to be able to rely on transitivity

Burgess describes a 'revisionist version of perfectionism' (p. 738, emphasis in original)

> according to which, though mathematicians in practice *do* give "proofs" relying on transitivity, they in principle *ought not* to do so ... ,

and he says that this position

> ... might be thought to have certain attractions in view of some formal results of Tennant's.

---

[4] Friedman and Avron do not accept this claim by the relevantist. The matter will be argued further in §12.5.

[5] See Tennant [1994b].

We hasten, however, to distance ourselves from this 'revisionist version' of perfectionism. What we are at pains to point out, from the vantage point of Core Logic, is that the mathematician (classical or intuitionist) does not *need* to apply the Cut rule as a step within a formal proof.

This amounts to a straightforward denial of the following extended claim from Burgess [2009], at p. 106:

> The cut rule directly expresses the classical doctrine that entailment is transitive. Moreover, its use is indispensable if one is to get a sequent calculus demonstration that anywhere near directly formalizes mathematicians' proofs.

We therefore need to make it clear why Burgess is mistaken on this score. What exactly *is* involved, then, in formalizing mathematicians' proofs? Burgess appears not to be willing to grant that it is enough just to have the *cut-free* bits of proof in hand.

Indeed, it would be enough to have just the cut-free, *thinning-free* bits of proof in hand, if the mathematicians have been formally careful enough to indicate any actual applications of thinning—which in fact they never do. Rather, they make informal inferential moves that are directly regimentable by the core rules *without* any thinnings. Lower on the same page as the passage just quoted, Burgess shows that he is aware of this possibility of effectively transforming any cut-free proof of Gentzen's into a cut-free, thinning-free proof.

This is what I have called the Extraction Theorem.[6] Burgess writes of a 'metatheorem [that] tells us' that

> from a classical sequent demonstration without cut, one can extract a perfectionistically acceptable demonstration of the same sequent or a subsequent. The extraction procedure moreover *does not much increase the lengths of demonstration.* (Emphasis added.)

Burgess is right about the metatheorem, but the gloss he has supplied is too weak. The extraction algorithm (call it ϵ), which is a linear-time one, actually yields as output—for every cut-free proof Π (of a sequent $\Delta : \varphi$) as input—a cut-free, thinning-free proof $\epsilon(\Pi)$ (of either $\Gamma : \varphi$ or $\Gamma : \bot$, for

---

[6] See §6.17. The Extraction Theorem can be understood as saying that any cut-free intuitionistic proof can be turned into a core proof of a result at least as strong. The result easily generalizes to the classical case, upon classicizing with the rule of Dilemma as in Tennant [2015b]: any cut-free *classical* proof can be turned into a *classical* core proof of a result at least as strong.

some $\Gamma \subseteq \Delta$) that is *no longer than* Π, and, in most of the interesting cases, considerably *shorter*. A signal example of this is

$$
\cfrac{
  \cfrac{A:A}{A:A\vee B} \quad
  \cfrac{\cfrac{\cfrac{A:A}{\neg A, A:\emptyset}\text{(Thin)}}{\neg A, A:B} \quad B:B}{A\vee B, \neg A:B}\text{(Cut)}
}{\neg A, A:B}
\quad
\underset{\substack{\text{Cut-}\\\text{elimination}}}{\longmapsto}
\quad
\cfrac{\cfrac{A:A}{\neg A, A:\emptyset}\text{(Thin)}}{\neg A, A:B}
\quad
\underset{\substack{\text{(i.e. Thin-}\\\text{elimination)}}}{\overset{\substack{\text{Extraction}\\\epsilon}}{\longmapsto}}
\quad
\cfrac{A:A}{\neg A, A:\emptyset}
$$

where, note, $\neg A, A : \emptyset$ is a (proper) subsequent of $\neg A, A : B$, and the final proof is shorter than the cut-free one before it.

Our Metatheorem 4 can be regarded as *providing a pragmatist explanation* of mathematicians' practice of being ever-ready to 'apply Cut'. They want to *have their cake and eat it*. The cake is the imagined finished formal proof from their mathematical axioms, say, to their sought mathematical theorem. Eating it is a matter of simply juxtaposing the bits of *core* proof that they have created, and claiming that the result *counts as a formal proof*. The points of such juxtaposition are precisely the cuts that will be invoked.

From our point of view, the mathematician is *entitled* to make these cuts, provided only that he does not lose sight of what has thereby *really* been proved. If the resulting overall sequent thus 'proved' (by aggregation of bits of cut-free proofs) is $\Delta : \varphi$, then all that the mathematician is entitled to claim, on the basis of the proof containing the cuts to which he might wish to help himself,[7] is that Δ logically implies φ (in the standard Tarskian sense, which is indifferent to issues of relevance). But he will *not* know, on the basis of such a proof, whether *all of Δ is needed* in order to establish φ— even though those very proof fragments already in hand might contain the surprising and welcome answer that *not* all of Δ is needed. Moreover, if Δ contains hypotheses not known to be consistent with the mathematical axioms used in the proof, then the mathematician will not necessarily know, either, on the basis of his proof, whether Δ 'succeeds' in implying φ only by dint of some irrelevantist 'funny business'. And there is yet another possibility, regarding the epistemic predicament of the happy-go-lucky cutter: he

---

[7] We need to bear in mind that the vast majority of mathematicians simply proceed intuitively with their deductions, and do not consciously take themselves to be applying rules such as Cut while constructing their proofs. It is only those mathematicians who (within living memory) have studied and 'internalized' the inferential moves within a particular traditional formal system (such as **C** or, more recently, **I**) who might take themselves to be 'applying Cut' when proving a lemma and then appealing to it. Such mathematicians form a decided minority within the whole community of mathematicians. So one can still hope that at least the other members of that community might eventually be persuaded that the best regimentation to be supplied for *their* expert but informal deductive reasoning is, in fact, Core Logic, or Classical Core Logic, as the case may be.

will not necessarily know, on the basis of his proof, whether he needs *any* premises in Δ in order to establish φ; for φ *might*, unbeknownst to him, be a *logical* theorem.

Those who lay ready claim to the fruits of transitivity (courtesy of unrestricted Cut) and who would protest at having their deductive windfalls denied them, need to pause to reflect on the realities of their epistemic situation. As cuts accumulate, so grows the danger of inconsistency in the expanding set of premises on which the overall conclusion is being made to rest. A stark if degenerate example is that of Lewis's proof of his First Paradox. The sequent $A : A \vee B$ is perfectly valid (hence also an entailment). So is the sequent $A \vee B, \neg A : B$. But the sequent resulting from applying Cut, namely $A, \neg A : B$ is *not* perfectly valid. Indeed, it is not even an entailment. This is because its set of premises is inconsistent.

Now of course in the case of the sequent $A, \neg A : B$ one can just *see* that its set of premises is inconsistent. But this is only because it is a degenerate example of the wider phenomenon about which we are cautioning the reader. In general, when Cut is applied to two proofs

$$
\begin{array}{cc}
\Delta & \varphi, \Gamma \\
\Pi & \Sigma \\
\varphi & \psi
\end{array}
\qquad \text{(where } \varphi \notin \Gamma \text{ and both } \Delta \text{ and } \Gamma \text{ are non-empty)}
$$

to obtain, *directly, in one step*, a proof of the overall sequent $\Delta, \Gamma : \psi$, we have no *immediate* way of telling whether the collection Δ, Γ of premises is consistent—or whether, even if it should turn out to be inconsistent, the overall proof does not actually exploit the inconsistency by conducting any 'funny business'.

This is where, from the perspective of Core Logic, the recommended eschewal of Cut can render important epistemic service. The things that we have said the irrelevantist logician would not know on the basis of his proof acquired by Cut are things concerning which, by contrast, the proponent of Core Logic might very well have something informative to offer, upon determining the core proof $[\Pi, \Sigma]$ whose existence, and effective determinability, are guaranteed by Metatheorem 4. For $[\Pi, \Sigma]$ will be a *core* proof of *some subsequent of* $\Delta, \Gamma : \varphi$. If that subsequent is a proper one, then we shall have epistemically gainful insights that would be completely missed by the happy-go-lucky cutter. We are, as it were, offering him a sensitive and subtle tool, to use to his advantage, if he is interested in doing so. But we are *certainly* not telling the practicing mathematician that he is not permitted to apply Cut within his proofs if he so wishes. We are saying only that every

time Cut is applied, and the matter regarded as thereby closed, an important opportunity is being missed—an opportunity to potentially strengthen
the overall logical results established, *by the very means implicit in the proof
fragments* Π *and* Σ *created thus far.*

We must make one final point about these growing collections Δ, Γ of
premises that result from applications of Cut. In the mathematical case,
when both Δ and Γ are subsets of the set $\mathcal{A}$ of axioms with which the mathematician is working, he can have recourse to the overriding presumption
that $\mathcal{A}$ is consistent. This goes a long way to explaining why it is that mathematicians do not, generally, need to be much concerned about applying Cut.
For, if $\mathcal{A}$ is indeed consistent, then at least it will be Σ's conclusion ψ, and
not $\perp$, that features as the conclusion of the core proof $[\Pi, \Sigma]$ (should the
mathematician heed the Core logician's advice to determine the latter). And
it may even be that the mathematician just does not value highly enough
the epistemic gain to be had by discovering that some *proper* subset Θ of
$\Delta \cup \Gamma$ features in the concluding sequent $\Theta : \psi$ of $[\Pi, \Sigma]$. The thought may
be: 'So long as we are still working within the system $\mathcal{A}$ of axioms, it really
does not matter *which* of these axioms we end up appealing to in order to
establish ψ as a theorem of the theory concerned.' If that is indeed the case,
then it speaks of a certain lack of logical refinement (or lack of logical curios-
ity) on the part of such a mathematician. One is generally in a much better
epistemic situation, the more keenly one can discern what is needed, logically, for the proofs of certain results. Certainly, the whole field of Reverse
Mathematics founded by Harvey Friedman would never have been founded
and developed, had this sort of indifference to matters of logical intricacy
completely prevailed.[8] (This lends a certain irony to the debate to be taken
up in §12.5.)

12.4.3   Rebuttal of Burgess on Core-logical reformism

We turn now to address the two most important and challenging criticisms
raised against the position of Core Logic regarding transitivity. They are
to be found in the last two paragraphs of Burgess [2005], at pp. 739–40,
respectively. We shall take them in turn.

The issue of feasibility

Referring to mathematicians' informal proofs, Burgess writes (emphasis in
original):

---

[8] See Simpson [1999].

When these proofs are formalized in sequent calculus, instances of transitivity become instances of Cut. Obtaining perfectionist replacements for the proofs would require first eliminating Cut. But in general, Cut can be eliminated only at the cost of making proofs infeasibly long.[fn] So even though *in principle* almost anything classically provable will be perfectionistically provable or else something even better will be, *in practice* the proofs mathematicians actually give not only *do* fail to adhere to the restrictions of perfectionism, but *must* do so if they are to be kept to a humanly comprehensible length.

Let us suppose that all the appeals to transitivity (which then require formalization as applications of Cut) have been identified. It follows that all the bits of proof in between them can be formalized as *core proofs*. Indeed, it is a mark of good intuitive mathematical practice that the passages of reasoning identified as 'proofs' in mathematical publications are the ones beginning and ending with the interpolants that will be the cut sentences upon eventual formalization as a sequent proof. Not that practicing mathematicians ever themselves consciously think of this as a mark of good practice—but the logician attending to the rigorously regimentable structure of mathematical texts will find that interpolants are almost always stated separately as lemmas, and that the reasoning that comes under the heading 'Proof' is bounded by such interpolants (and of course axioms), but seldom contains any of them as an 'interior' point. Whenever a bit of proof is thus punctuated by an interpolant as an interior point, one can always split the proof into two different and shorter proofs—one to the interpolant in question, and the other away from it—that themselves do not contain any interpolants as interior points.

Let us now raise a first cautionary point: logicians who have never actually tried to give a fully regimented formal proof of a non-trivial passage of mathematical reasoning tend to underestimate just how long such a formal proof tends to be. This is the case even when the bit of informal reasoning that is being formalized (as a sequent proof, say) strikes the mathematician as relatively easy and 'short'. 'Long' and 'short' are *very* relative terms when it comes to the comparison of informal with formal proofs. So a first issue that Burgess needs to address is this: just how confident can he be that even the *cut-free* bits of mathematical reasoning (in a mathematical text) that he imagines can be formalized as sequent proofs are not, themselves, going to have to be supplied with formalizations that are *already* infeasibly long? His warning about infeasibility being induced by Cut-elimination might be a case of closing the barn door after the horse has bolted.

Every bit of (informal) cut-free reasoning in a mathematical text is in principle formalizable as a sequent proof in Core Logic. Suppose now that one concedes the unlikely point that every bit of (informal) cut-free reasoning in a mathematical text is *feasibly* formalizable as a sequent proof in Core Logic. On this supposition, let us address Burgess's complaint that eliminating the cuts that are required in order to stitch together a sequence of several such proofs will sometimes be a prohibitive task—the resulting cut-free proof, he warns, could turn out to be infeasibly long.

The same complaint could well be raised against the intuitionist who distinguishes between *direct* and *indirect* warrants for mathematical assertions.[9] The direct warrants, often called canonical proofs, are the intuitionist's 'ultimate' justifications for assertions. According to the intuitionist logician, intuitionistic mathematicians can also make assertions on the basis of indirect warrants. These are constructions encoding an effective method that would, in principle, determine for the assertion in question a *direct* warrant. It is the intuitionist mathematician's standing habit, or practice, never actually to carry out such determinations—for they are pointless. With the background guarantee that they *could* in principle be carried out, he forges ahead, providing more and more theorems with no better than indirect warrants—dignified with the label 'Proof', mind you, in textbooks.

This situation is completely analogous to the one in which the perfectionist finds himself, with regard to the issue of cuts. Certainly, they need never be applied—such is the lesson, in principle, of Metatheorem 4. But, *if* they *were* to be applied, there might be a bountiful harvest of logical strengthenings to be had. The obvious trade-off to be struck is between the time (and resources) available for further research, and the quality (or logical captiousness) of the results already 'secured'. Most mathematicians want to conquer new territories, rather than spend any of their time straightening the streets of already occupied towns.

Burgess's classical mathematician who believes in the regimentability of his informal proofs as proofs in the classical sequent calculus is neither terribly concerned with working at the level of micro-detail called for in sequent proofs, nor terribly concerned with the question whether the formal proofs should involve cuts. But from the way the mathematician forges ahead, from axioms to lemmas, then from lemmas plus other axioms to theorems, and thence to corollaries, speaks of a readiness to have Cut applied liberally on his behalf, once the work of formalization is undertaken in earnest. What,

---

[9] See, for example, Dummett [1977].

then, really ends up being formalized? Answer: just the cut-free bits of reasoning *in between* all those controversial cuts. *Those* are the nitty-gritty passages of reasoning that would end up being regimented as sequent proofs.

The perfectionist (a.k.a. the Classical Core logician) and the Burgess-style classical formalizer will end up contemplating the same set of core proofs that result from the project of regimenting mathematical reasoning down to the last micro-details of logical inference. The question is only what they would then proceed, respectively, to do. The latter will happily apply Cut, and declare his work done. The former will acquiesce in all the 'double turnstile' claims that result from what his opponent, the cutter, has done. For he has Metatheorem 4 in his pocket. But he *also* has—if the spirit moves him—an effective method for determining potentially stronger logical results than the mathematician himself might have realized could be obtained from all the 'raw informal' reasoning that he had pioneered. The perfectionist offers the prospect of an interesting 'mop-up' operation, revealing logical tightenings, or strengthenings, that are *implicit in* the very raw materials that he is being invited to regiment. That is an intellectual undertaking to which the logical quietism of the Burgess-style classical formalizer will tend to make him indifferent, if not blind. The latter will tend to wait until a new (informal) mathematical proof might be offered, that proves a proper subsequent of some previous result (if such be available). The Core logician, however, stands to discover such an improvement by carefully processing the raw informal, and also formal, materials already at his disposal from his analysis of the mathematical reasoning already on record.

The issue of necessary use of principles whose eschewal is being recommended

We turn now to the second important criticism in Burgess [2005], on p. 740. He claims that 'Tennant's result' (for which let us read: Metatheorem 4) has been given 'only a classically and not a perfectionistically acceptable "proof" ...'. (Actually, the proof of the metatheorem is constructive, even for the classical case.) Burgess then asks, rhetorically,

> How far can a logician who *professes* to hold that perfectionism is the correct criterion of valid argument, but who freely accepts and offers standard mathematical proofs, in particular for theorems about perfectionist logic itself, be regarded as *sincere* or *serious* in objecting to classical logic?

A similar objection is raised in Burgess [2009], at p. 107:

> [I]f one is convinced that one's axioms are true, and therefore
> do not entail a contradiction, even without carrying through the
> procedure [to extract a core proof from a proof that contains
> cuts—NT] a perfectionist can conclude that the theorem must
> be true, since it is entailed by a subset of the axioms.
>
> Or, rather, the perfectionist could conclude this *provided
> the metatheorem has been proved in a perfectionistically accept-
> able manner.* But this it has not been, and it is doubtful if
> it ever will be. Thus in practice one who is really, truly, and
> sincerely a perfectionist cannot accept mathematicians' proofs.
> In mathematical practice, *transitivity of entailment is indispens-
> able.* [Emphases in original.]

I readily admit that it is always possible that one underestimates one's
own capacity for self-deception. One might feel both sincere and serious, yet
not really be so. But that possibility is orthogonal to the real issue, since
one can be *genuinely* sincere and serious and yet still be *seriously mistaken*
about the grounds that one believes one has for one's beliefs. It is the danger
of a mistake of the latter kind against which we have to protect ourselves.

The constructive proof offered above of Metatheorem 4 can, I submit, be
feasibly formalized as a core proof in a suitable theory of iterated inductive
definitions. It is, after all, almost completely formalized already, with no
unexamined cases, no passages left to the reader as 'obvious', and no cases
crisply disposed of as 'similar to such-and-such case'.[10]

The (admittedly *informal*) metamathematical proof above of Metatheorem 4 should suffice, as such proofs always do among practicing logicians and
mathematicians, as a basis upon which to proceed with further philosophical
discussion (even though the various cases in the inductive step were left to
the reader as straightforward by inspection). For, *qua* classical mathematician, Burgess would concede that Metatheorem 4 has indeed been proved.
And *this* author, even if he *could* be shown where only expensively eschew-
able cuts might have to be called upon in order to formalize his existing
proof of Metatheorem 4 (which he doubts—since, on careful inspection, he
does not find any), would be entitled (in Burgess's view) to the proven *truth*
of the claim that, upon 'eliminating' such cuts (in the manner afforded by
Metatheorem 4 itself!) he (the present author) would thereby be able, in
principle, to find a core proof of Metatheorem 4.

If that line of rebuttal still fails to move Burgess, then there is a 'burden
of proof' reply to be entered. This points out that the burden of proof

---

[10] I am of course referring here to the proof as laid out in Tennant [2015b].

is not a burden of core proof fairly placed upon the present author. For, consider the dialectical situation: I need to convince Burgess that Core Logic suffices for all serious scientific purposes. I propose to do that by appeal to Metatheorem 4. That result has been established adequately by *Burgess's* lights. As far as *Burgess* is concerned, the result is *true*. It follows, then, that the lessons one can draw from that result, concerning the methodological adequacy of Core Logic, ought to be granted by Burgess himself. He has used a ladder to reach a certain point of enlightenment—a point from which he can now appreciate that a certain part of the ladder can be kicked away.

Early in this study, in §1.4, we took issue with a quote from the opening pages of Shapiro [2014]. I shall end this reply to Burgess by concurring with a later claim of Shapiro ([2014], at pp. 203–4), even though it challenges my 'burden of proof' point against Burgess just entered above. Referring to this dispute, Shapiro writes the following, and has confirmed[11] that my clarifying regimentation of its conditional form, with the fourfold conjunction as consequent, is in order:

> It is crucial to Tennant's philosophical agenda that he win this dispute. ... [I]f Burgess's charge is correct, then Tennant is not entitled to claim that there is epistemic gain in using his system [and] [t]he most he could claim would be that, from the (incorrect) perspective of the classical mathematician, we can see that core logic can do as well, or better, in deducing arguments [and] by his own lights, the perspective from which one can say this is itself flawed, invoking what Tennant takes to be invalid argument forms [and] the ladder cannot be kicked away, at least not while we are on it and taking advantage of what we can see from up there.

I am happy to concede the truth of this conditional. This is because its antecedent is false. Core Logic to the rescue again . . . .

12.5   Reply to Friedman and Avron

Core Logic $\mathbb{C}$, and its classical extension Classical Core Logic $\mathbb{C}^+$, *neither* of which contains the rule EFQ, are nevertheless complete with respect to their respective logical consequence relations (intuitionistic and classical, respectively). That much is established by appropriate metatheorems expounded

---

[11] Personal communication.

above. So we know, in principle, that Core Logic is adequate for the formalization of constructive mathematics, and Classical Core Logic likewise for classical mathematics. Still, this suggestion has provoked some skepticism on the part of certain logicians who remain wedded to tradition.

12.5.1   On the empty set being a subset of every set

In late August and early September 2015, on the moderated email list fom@cs.nyu.edu, there was a discussion of whether, why or how the rule EFQ is actually used, or needs to be used, in the formalization of mathematical reasoning. The discussion was puzzling, to say the least, because those 'on the other side' could only have been proceeding in full innocence of the metatheorems just mentioned. Both Harvey Friedman and Arnon Avron argued that one makes *indispensable* use of EFQ in standard 'set pieces' of mathematical reasoning. Because this commitment to EFQ as part of the modern orthodoxy in logic manifested itself to such a remarkable degree 'on the other side', it is worth taking pains here to dispose, in the most illuminating way possible, of the main example that Avron and Friedman advanced in support of this indispensability claim.

Here is how Friedman insisted on the allegedly *indispensable* use of EFQ (with typos corrected):[12]

> THEOREM. For any set A, emptyset is a subset of A.
>
> In the usual formalization of mathematics, the proof goes like this. First of all one either formalizes this as
>
> THEOREM 1. {x: absurdity} containedin A.
>
> with absurdity sign. This would be common in some automated proof systems. Or there would be a constant symbol introduced, emptyset, by instantiation, with the condition
>
> not(x in emptyset).
>
> And then the formalization would read
>
> THEOREM 2. emptyset containedin A.
>
> Proof of Theorem 1: Let A be a given set. It suffices to show that y in {x: absurdity} implies y in A. Suppose y in {x: absurdity}. Then absurdity. Hence y in A. QED
>
> Proof of Theorem 2: Let A be a given set. It suffices to show that y in emptyset implies y in A. Suppose y in empt[yse]t. Since not(y in empt[yse]t), we have y in A. QED

---

[12] http://www.cs.nyu.edu/pipermail/fom/2015-September/018977.html

> Now these proofs are still a[t] the semiformal level, as the use of
> mathematical definitions, and definitions by instantiation, etc.,
> are not yet formal. But the basic structure is clear.
>
> So when we get to the ACTUAL mathematics, we see that EFQ
> is used, and used without any qualms here.

The reader will not begrudge a use of LATEX symbols in math mode to make this easier to follow. Here is a faithful rendering of the foregoing quote from Friedman.

Theorem. For any set A, $\emptyset \subseteq A$.

In the usual formalization of mathematics, the proof goes like this. First of all one either formalizes this as

Theorem 1. $\{x|\bot\} \subseteq A$.

with absurdity sign $\bot$. This would be common in some automated proof systems.

Proof. Let A be a given set. It suffices to show that $y \in \{x|\bot\}$ implies $y \in A$. Suppose $y \in \{x|\bot\}$. Then $\bot$. Hence $y \in A$. □

[Alternatively,] there would be a constant symbol introduced, $\emptyset$, by instantiation, with the condition

$$\neg(x \in \emptyset).$$

And then the formalization would read

Theorem 2. $\emptyset \subseteq A$.

Proof. Let A be a given set. It suffices to show that $y \in \emptyset$ implies $y \in A$. Suppose $y \in \emptyset$. Since $\neg(y \in \emptyset)$, we have $y \in A$. □

> Now these proofs are still at the semiformal level, as the use of mathematical definitions, and definitions by instantiation, etc. are not yet formal. But the basic structure is clear.
>
> So when we get to the *actual* mathematics, we see that EFQ is used, and used without any qualms here.

The Core logician begs to differ. As Friedman remarks, these proofs are still at 'the semiformal level'. This means that they not only carry heavy traces of 'logician's English'; they are *also* informed by a longstanding tradition (now of several decades' duration) of saying things in certain ways *in which the more 'formal-logically minded' have been trained to say them*. Classical Logic has been *interacting* with ordinary mathematical practice for well over a century now. But the role of formal logic is realized only when we try to attain an *ultimate* degree of formalization, dotting every 'i' and crossing every 't'. That is, we have to convert informal mathematical proofs into strictly formal proofs involving only sentences of the formal object language. Such proofs must proceed by means of primitive rules of inference. Let us do this now for the preceding discourse containing just the statements of theorems and their proofs.

First we observe that $\bot$, as we intend to use it, is not a sentence of the formal object language; it is only ever used as a punctuation marker in proofs, to indicate that an absurdity or contradiction has been reached. Thus the would-be singular set-abstractive term '$\{x|\bot\}$' is actually *ill-formed*. It needs to be replaced with '$\{x|\neg x=x\}$'. Theorem 1 now becomes Theorem 3.

Theorem 3. $\{x|\neg x=x\} \subseteq A$

Proof. Let A be a given set. It suffices to show that

$$y \in \{x|\neg x=x\} \text{ implies } y \in A$$

Suppose $y \in \{x|\neg x=x\}$. Then $\neg y=y$. But $y=y$. Contradiction. Hence $y \in A$. $\square$

Now let us look at Friedman's other alternative, in which a primitive *name* $\emptyset$ is introduced into the language and made subject to the condition

$$\neg(x \in \emptyset).$$

This could also be expressed by the rule of inference

$$\frac{x \in \emptyset}{\bot} .$$

Note that this rule of inference, on its own, does not suffice to secure for the name $\emptyset$ the empty set as its denotation. For it could well be the case that there are *urelements*, none of them containing any members. For any urelement e, the rule of inference

$$\frac{x \in e}{\bot}$$

would hold good.

Still, let us grant Friedman the use of his rule of inference (involving the name $\emptyset$) as capturing at least part, even if not all, of what it is for $\emptyset$ to be the empty set. Theorem 2 now becomes Theorem 4.

Theorem 4. $\emptyset \subseteq A$.

Proof. Let A be a given set. It suffices to show that

$$y \in \emptyset \text{ implies } y \in A$$

Suppose $y \in \emptyset$. Contradiction. So we have $y \in A$. □

Let us now formalize a little further, once again incrementing the numbers of the two theorems by 1.

Theorem 5. $\{x | \neg x = x\} \subseteq A$.

Proof. Let A be a given set. It suffices to show that

$$y \in \{x | \neg x = x\} \to y \in A.$$

Suppose $y \in \{x | \neg x = x\}$. Then $\neg y = y$. But $y = y$. Contradiction. Hence $y \in \{x | \neg x = x\} \to y \in A$. □

Theorem 6. $\emptyset \subseteq A$.

Proof. Let A be a given set. It suffices to show that

$$y \in \emptyset \to y \in A.$$

Suppose $y \in \emptyset$. Contradiction. So we have $y \in \emptyset \to y \in A$. □

Of course, it has not yet been revealed what $\subseteq$ really *means*. This relation symbol (although so familiar) has not yet been formally defined, or made subject to any meaning-conferring axioms or rules of inference. So let us make good that lack in the conventional way, treating $x \subseteq y$ as an *abbreviation* of a longer formula with just the variables x and y free.

Definition 39. $x \subseteq y \equiv_{df} \forall z(z \in x \to z \in y)$.

Now we can revisit the two theorems and their proofs, making proper use of Definition 39. The theorem numbers ratchet up by 1 yet again.

Theorem 7. $\emptyset \subseteq A$; *that is, by Definition 39,*

$$\forall y(y \in \{x | \neg x = x\} \to y \in A).$$

Proof. Let y be arbitrary. It suffices to show that

$$y \in \{x | \neg x = x\} \to y \in A.$$

Suppose $y \in \{x | \neg x = x\}$. Then $\neg y = y$. But $y = y$. Contradiction. Hence $y \in \{x | \neg x = x\} \to y \in A$. □

Theorem 8. $\emptyset \subseteq A$; that is, by Definition 39,

$$\forall y(y \in \emptyset \to y \in A).$$

Proof. Let y be arbitrary. It suffices to show that

$$y \in \emptyset \to y \in A.$$

Suppose $y \in \emptyset$. Contradiction. So we have $y \in \emptyset \to y \in A$. □

We are *still* at a 'semiformal' level. It is time now to formalize fully, using a system of natural deduction in which parameters $a, b, c, \ldots$ are used for applications of $\forall$I and $\exists$E.

Theorem 9. $\forall y(y \in \{x | \neg x = x\} \to y \in A)$.

Proof.

$$\cfrac{\cfrac{\cfrac{\cfrac{}{a \in \{x | \neg x = x\}}^{(1)}}{\neg a = a} \qquad \cfrac{}{a = a}}{\perp}}{\cfrac{a \in \{x | \neg x = x\} \to a \in A}{\forall y(y \in \{x | \neg x = x\} \to y \in A)}}^{(1)}$$

□

Theorem 10. $\forall y(y \in \emptyset \to y \in A)$.

Proof.

$$\cfrac{\cfrac{\cfrac{}{a \in \emptyset}^{(1)}}{\perp}}{\cfrac{a \in \emptyset \to a \in A}{\forall y(y \in \emptyset \to y \in A)}}^{(1)}$$

□

Pace Friedman, and adapting his own words, we can summarize our results as follows.

> The basic structure is clear. When we get to the *actual formalization* of (this bit of) mathematics, we see that EFQ *need not* be used, and may be avoided altogether, without any qualms here.

All one needs is that the rule of →I include a part to the effect that when one has refuted φ, one may infer the conditional $\varphi \to \psi$, thereby discharging all assumption occurrences of φ:

The box indicates that the assumption φ must actually have been used in order for (this part of) the rule →I to be applicable. This is faithful to the truth table for →, which tells us that the *falsity of the antecedent* suffices for the truth of the conditional. That is dual to the other insight yielded by the truth table, to the effect that the *truth of the consequent* suffices for the truth of the conditional; and this is provided for by the part of the rule →I that allows one to infer the conditional from its consequent, without having made any use of the antecedent as an assumption:

$$\frac{\psi}{\varphi \to \psi}$$

And of course one also has, in Core Logic, the remaining part of →I that allows one to infer a conditional $\varphi \to \psi$ upon deducing its consequent ψ from its antecedent φ used as an assumption (and, in doing so, discharging those assumption occurrences of φ):

$$\frac{\overline{\quad}^{(i)}}{\varphi} \\ \vdots \\ \frac{\psi}{\varphi \to \psi}^{(i)}$$

The question is not whether EFQ is actually used in mathematics; it is rather whether it *has* to be used, and whether, if one avoids using it, there is any non-negligible cost in doing so. From the foregoing, the answer thus far is clear: EFQ does *not* have to be used; and there is *no cost at all* in avoiding its use. There is certainly no increase in length of proof. In fact, there is a *decrease*, since one does not need to interpolate an application of EFQ immediately before the application of →I.

One can give up the rule EFQ:

$$\frac{\bot}{\varphi}$$

and refuse to acknowledge it as holding across the board. Core Logic is
the logic for the constructivist who wishes to be free of such freighted
commitment. Classical Core Logic is the logic for the classicist (the non-
constructivist) who shares that wish.

12.5.2 Core Logic is *not* merely 'partially' complete

Finally, I need to deal with another, this time rather different, failure to
grasp the subtle way in which the Core logician takes the deductive system
of Classical Core Logic to be *complete* with respect to the relation $\models$ of
classical logical consequence (despite the fact that the latter relation contains
irrelevancies like the Lewis Paradox). In one of my contributions to the
discussion on fom@cs.nyu.edu (Monday, September 7, 2015 at 10:37 p.m.),
I had written

> ...we know that Classical Core Logic is complete:
>
> $X \models A \Rightarrow$ there is a classical core proof of A or of $\bot$ from (members in) X

to which Friedman replied

> This is of co[u]rse different than
>
> $X \models A \Rightarrow$ there is a Classical proof of A from (members in) X.
>
> So *you don't have literal completeness, but a partial result.* (Emphasis added.)

The 'partialness' that Friedman is alleging here is clearly meant to be understood as some kind of *defect*, or *falling short*, on the part of the proof system
whose completeness I am claiming. But this is both confused and mislead-
ing. Friedman would clearly not complain in like fashion about a proof
system furnishing proofs of A, given an *infinite* set X of premises, whose
premises form only *finite*—hence *proper*—subsets of X. (Indeed, how could
he?—proofs, by their very nature, have to be *finite*!) To drive the point
home: let X be the set of axioms of any axiomatizable but not finitely
axiomatizable theory, such as Peano Arithmetic, or Zermelo–Fraenkel set
theory. It would be clearly absurd to complain about (the allegedly only)
partial completeness of first-order logic on the grounds that every proof it
furnishes for a theorem of PA or of ZFC uses—*tsk, tsk!*—only *finitely* many

of the axioms available (hence only: a *proper* subset of the set of available axioms). Standard completeness, for Friedman, is not *partial* completeness; it is *complete* completeness.

So too, by the same token—so I maintain—is completeness that takes the (potentially epistemically gainful) form

$$X \models A \Rightarrow \text{there is a classical core proof of } A \text{ or of } \bot \text{ from (members in) } X.$$

We have already explained this point of view in §1.2. For nothing is lost if we allow that the succedent of the provable sequent may be *some subset of* the succedent $\{A\}$ of the given valid sequent. The considerations, as between antecedent and succedent of any given sequent, are absolutely symmetric— and should be appreciated, especially by any Classical logician, as so being. For the Classical logician is one to whom certain symmetries should hold a certain appeal.

Friedman is a Classical logician. Yet he seems unprepared to allow that the *right*-hand side Y of a sequent

$$X : Y$$

is entitled to the same consideration as is usually given to the *left*-hand side X. For a language $\mathcal{L}$ we can define the classical consequence relation $X \models Y$ in the following illuminating way:

> No interpretation (model) M of the non-logical expressions of $\mathcal{L}$
> partitions $\mathcal{L}$ thus:

True in M	*False in M*
X	Y

Bear in mind that this definition covers cases where X or Y (or both) may be infinite. The *completeness* claim for a system $\mathcal{S}$ of sequent proof is to the following effect:

> For all sets X, Y of sentences in $\mathcal{L}$, if $X \models Y$ then there is a proof, in the system $\mathcal{S}$, of some sequent $X' : Y'$, where X' is a finite subset of X and Y' is a finite subset of Y.

It is therefore perfectly natural to anticipate the prospect of *subsetting down*, both on the left *and* on the right, as we pass from the sets X, Y mentioned

in the logical-consequence claim to the sets X', Y' that can be read off from
a proof. Such subsetting down is in general associated with *epistemic gain*.

So, if one stresses the fact that one can subset down on the right *as well
as* on the left, why is this to be dismissed as mere *partial* completeness?
The situation here is utterly symmetric, as between X and Y.

Moreover, it *remains* symmetric even if one decides to limit Y to be at
most a singleton. One could still subset down on the right, to the *empty*
set, and achieve completeness. For, if one's proof system yields a proof of
the sequent $X : \emptyset$, this takes care of all classical consequences of the form
$X : Y$, whatever (singleton) Y may be. And note that this does *not* commit
one to saying that the rule EFQ holds across the board.

It is a common criticism of an attempt to prove a sequent $X : A$ by
using certain members of X to derive A that *one did not in fact need to.*
Consider, for example, the sequent

$$B : A \to A .$$

If a beginning student were to actually try to 'use' B as a premise they would
be criticized by their instructor, and for good reason: it would be evidence
that they had failed to grasp that $A \to A$ is a logical truth, provable from the
empty set of premises. Moreover, if the student were to persevere doggedly,
and then triumphantly offer up the would-be proof

$$\frac{\dfrac{B \quad \overline{A}^{(1)}}{B \wedge A}}{\dfrac{A}{A \to A}^{(1)}}$$

the instructor would drily point out that this proof is not in normal form: it
has a maximal occurrence of $B \wedge A$, inviting an application of the reduction
procedure for $\wedge$ to turn it into a proof in normal form. And the latter proof
would be none other than

$$\dfrac{\overline{A}^{(1)}}{A \to A}^{(1)}$$

This drives home the point that the premise B is utterly irrelevant to the
task of establishing what $A \to A$ *really* follows from. The alien operator $\wedge$
being intruded into the proof was a dead give-away that something rum was
being done.

In the single-premise, single-conclusion case the situation is absolutely symmetric. Consider now the sequent

$$A \wedge \neg A : B \, .$$

If a beginning student were to actually try to 'deduce' B as a conclusion from the premise given, they *ought* be criticized by their instructor, for the following good reason: it would be evidence that they had failed to grasp that $A \wedge \neg A$ is a logical falsehood, implying the *empty* set of conclusions, which logicians call $\bot$. Moreover, if the student were to persevere doggedly, and then triumphantly offer up the would-be proof

$$
\cfrac{\cfrac{\cfrac{A \wedge \neg A}{A}}{A \vee B} \quad \cfrac{\cfrac{A \wedge \neg A}{\neg A} \quad \cfrac{A \wedge \neg A}{A}{}^{(1)}}{\bot} \quad \cfrac{}{B}{}^{(1)}}{B}{}^{(1)}
$$

the instructor would drily point out that this proof is not in normal form: it has a maximal occurrence of $A \vee B$, inviting an application of the reduction procedure for $\vee$ to turn it into a proof in normal form. And the latter proof would be

$$
\cfrac{\cfrac{A \wedge \neg A}{\neg A} \quad \cfrac{A \wedge \neg A}{A}}{\bot}
$$

This drives home the point that the would-be conclusion B is utterly irrelevant to the task of establishing what *really* follows from $A \wedge \neg A$. The alien operator $\vee$ being intruded into the proof was a dead give-away that something rum was being done.

Does eschewing, in each of these two examples, the former abnormal proof (as the Core logician does) and recognizing only the latter proof that is in normal form (as far as these problems are concerned) deserve to incur an accusation of furnishing only a *partially* complete proof system, as Friedman maintains? Surely not! As was pointed out in §1.2, the (total!) completeness of a proof system $\mathcal{S}$ with respect to a logical consequence relation $\models$ amounts to no more than this:

> *Every sequent in the extension of* $\models$ *has a subsequent in the extension of* $\vdash_{\mathcal{S}}$.

And for $\mathcal{S}$ we can take $\mathbb{C}^{+}$.

Bibliography

Alan Ross Anderson and Nuel D. Belnap. *Entailment: The Logic of Relevance and Necessity. Vol. I.* Princeton University Press, Princeton, NJ, 1975.

D. M. Armstrong. *Truth and Truthmakers.* Cambridge University Press, Cambridge, 2004.

Jon Barwise and John Etchemendy. Heterogeneous Logic. In Gerard Allwein and Jon Barwise, editors, *Logical Reasoning with Diagrams*, pages 179–200. Oxford University Press, New York, 1996.

JC Beall and G. Restall. Logical Pluralism. *Australasian Journal of Philosophy*, 78: 475–93, 2000.

JC Beall and Greg Restall. *Logical Pluralism.* Clarendon Press, Oxford, 2006.

Douglas Bridges. Constructive Mathematics and Unbounded Operators—A Reply to Hellman. *Journal of Philosophical Logic*, 24: 549–61, 1995.

Douglas Bridges. Can Constructive Mathematics be Applied in Physics? *Journal of Philosophical Logic*, 28(5): 439–53, 1999.

Douglas Bridges and Fred Richman. A Constructive Proof of Gleason's Theorem. *Journal of Functional Analysis*, 162: 287–312, 1999.

L. E. J. Brouwer. Über Definitionsbereiche von Funktionen. In A. Heyting, editor, *Collected Works. Volume I. 1975*, pages 390–405. North-Holland/American Elsevier, Amsterdam, New York, 1927.

John Burgess. No Requirement of Relevance. In Stewart Shapiro, editor, *The Oxford Handbook of Philosophy of Mathematics and Logic*, pages 727–50. Oxford University Press, New York, 2005.

John P. Burgess. *Philosophical Logic*. Princeton University Press, Princeton and Oxford, 2009.

Alexander Chagrov and Michael Zakharyashchev. The Disjunction Property of Intermediate Propositional Logics. *Studia Logica*, L(2): 189–216, 1991.

Alonzo Church. The Weak Theory of Implication. In Wilhelm Britzelmayr, Albert Menne, and Alexander Wilhelmy, editors, *Kontrolliertes Denken*, pages 22–37. Kommissions Verlag Karl Alber, Munich, 1951.

Pablo Cobreros, Paul Egré, David Ripley, and Robert van Rooij. Tolerant, Classical, Strict. *Journal of Philosophical Logic*, 41: 347–85, 2012.

Roy Cook. Should Anti-Realists be Anti-Realists About Anti-Realism? *Erkenntnis*, 79: 233–58, 2014.

W. Craig. Linear Reasoning. A New Form of the Herbrand-Gentzen Theorem. *Journal of Symbolic Logic*, 22(3): 250–68, 1957.

Michael Dummett. *Elements of Intuitionism*. Clarendon Press, Oxford, 1977.

Jan Ekman. Propositions in Propositional Logic Provable Only by Indirect Proofs. *Mathematical Logic Quarterly*, 44(1): 69–91, 1998.

John Etchemendy. *The Concept of Logical Consequence*. Harvard University Press, Cambridge, MA, 1990.

Hartry Field. *Saving Truth from Paradox*. Oxford University Press, New York, 2008.

Gottlob Frege. *Die Grundlagen der Arithmetik: eine logisch mathematische Untersuchung über den Begriff der Zahl*. Georg Olms Verlagsbuchhandlung, Hildesheim, 1884; reprinted 1961.

Harvey Friedman. The Consistency of Classical Set Theory Relative to a Set Theory with Intuitionistic Logic. *Journal of Symbolic Logic*, 38: 315–19, 1973.

Harvey Friedman and Robert K. Meyer. Whither Relevant Arithmetic? *Journal of Symbolic Logic*, 7: 824–31, 1992.

Gerhard Gentzen. Über die Existenz unabhängiger Axiomensysteme zu unendlichen Satzsystemen. *Mathematische Annalen*, 107: 329–350, 1932.

Gerhard Gentzen. Untersuchungen über das logische Schliessen. *Mathematische Zeitschrift*, I, II: 176–210, 405–31, 1934, 1935. Translated as 'Investigations into Logical Deduction', in *The Collected Papers of Gerhard Gentzen*, edited by M. E. Szabo, North-Holland, Amsterdam, 1969, pp. 68–131.

Kurt Gödel. Zum intuitionistischen Aussagenkalkül. *Anzeiger der Akademie der Wissenschaften in Wien*, 69: 65–66, 1932.

Susan Haack. *Philosophy of Logics*. Cambridge University Press, Cambridge, 1978.

Dirk Hartmann. *On Inferring: An Enquiry into Relevance and Validity*. Mentis Verlag, Paderborn, 2003.

Geoffrey Hellman. Gleason's Theorem is Not Constructively Provable. *Journal of Philosophical Logic*, 22: 193–203, 1993a.

Geoffrey Hellman. Constructive Mathematics and Quantum Mechanics: Unbounded Operators and the Spectral Theorem. *Journal of Philosophical Logic*, 22: 221–48, 1993b.

Geoffey Hellman. Quantum Mechanical Unbounded Operators and Constructive Mathematics—a Rejoinder to Bridges. *Journal of Philosophical Logic*, 26: 121–7, 1997.

Henry Hiż. Extendible Sentential Calculus. *Journal of Symbolic Logic*, 24: 193–202, 1959.

Jaakko Hintikka. *Logic, Language-Games and Information: Kantian Themes in the Philosophy of Logic*. Clarendon Press, Oxford, 1973.

I. Johansson. Der Minimalkalkül, ein reduzierter intuitionistischer Formalismus. *Compositio Mathematica*, 4: 119–36, 1936.

Aleksandar Kron. Deduction Theorems for Relevant Logics. *Zeitschrift für Mathematische Logik und Grundlagen der Mathematik*, 19(1): 85–92, 1973.

Roger C. Lyndon. An Interpolation Theorem in the Predicate Calculus. *Pacific Journal of Mathematics*, 9(1): 129–42, 1959.

D. C. McCarty. Intuitionism in mathematics. In Stewart Shapiro, editor, *The Oxford Handbook of Philosophy of Mathematics and Logic*, pages 356–86. Oxford University Press, New York, 2005.

J. C. C. McKinsey and A. Tarski. On Closed Elements in Closure Algebras. *Annals of Mathematics*, 47(1): 122–62, 1946.

J. C. C. McKinsey and A. Tarski. Some Theorems about the Sentential Calculi of Lewis and Heyting. *The Journal of Symbolic Logic*, 13(41): 1–15, 1948.

L. L. Maksimova. Craig's Theorem in Superintuitionistic Logics and Amalgamable Varieties of Pseudo-Boolean Algebras. *Algébra i Logika*, 16(6): 643–81, 1977.

Edwin Mares. The Logic R: Supplement to Relevance Logic. In Edward N. Zalta, editor, *The Stanford Encyclopedia of Philosophy (Summer 2012 Edition)*. URL = <plato.stanford.edu/entries/logic-relevance/logicr>, 2012.

Karl Popper. A Realist View of Logic, Physics and History. In A. D. Breck and W. Yourgrau, editors, *Physics, Logic and History*, pages 1–30. Plenum Press, New York, London, 1970.

Dag Prawitz. *Natural Deduction: A Proof-Theoretical Study*. Almqvist & Wiksell, Stockholm, 1965.

Dag Prawitz. On the Idea of a General Proof Theory. *Synthese*, 27: 63–77, 1974.

Graham Priest. The Logical Paradoxes and the Law of Excluded Middle. *Philosophical Quarterly*, 33(131): 160–5, 1983.

Graham Priest. Paraconsistent Logic. In D. Gabbay and F. Guenthner, editors, *Handbook of Philosophical Logic, Volume 6*, pages 287–393. Kluwer Academic Publishers, Dordrecht, 2002.

F. P. Ramsey. The Foundations of Mathematics. *Proceedings of the London Mathematical Society*, 25: 338–84, 1926.

Stephen Read. Monism: The One True Logic. In D. Devidi and T. Kenyon, editors, *A Logical Approach to Philosophy*, pages 193–209. Springer, Dordrecht, 2006.

David Ripley. Revising Up: Strengthening Classical Logic in the Face of Paradox. *Philosophers' Imprint*, 13(5): 1–13, 2013.

Tor Sandqvist. Base-Extension Semantics for Intuitionistic Sentential Logic. *The Logic Journal of the IGPL*, published online July 13, 2015 doi:10.1093/jigpal/jzv021, 2015.

Peter Schroeder-Heister. A Natural Extension of Natural Deduction. *Journal of Symbolic Logic*, 49: 1284–1300, 1984.

Peter Schroeder-Heister and Luca Tranchini. Ekman's paradox. *Notre Dame Journal of Formal Logic*, forthcoming.

Stewart Shapiro. *Varieties of Logic*. Oxford University Press, New York, 2014.

Moh Shaw-Kwei. The Deduction Theorems and Two New Logical Systems. *Methodos*, 2: 56–75, 1950.

D. J. Shoesmith and T. J. Smiley. Deducibility and Many-Valuedness. *The Journal of Symbolic Logic*, 36(4): 610–22, 1971.

Stephen G. Simpson. *Subsystems of Second Order Arithmetic*. Springer, Berlin, Heidelberg, New York, 1999.

T. J. Smiley. Entailment and Deducibility. *Proceedings of the Aristotelian Society, n.s.*, 59: 233–254, 1959.

Timothy Smiley. The Independence of Connectives. *The Journal of Symbolic Logic*, 27: 426–36, 1962.

Alfred Tarski. The Concept of Truth in Formalized Languages. In J. H. Woodger, editor, *Logic, Semantics, Metamathematics*, pages 152–278. Clarendon Press, Oxford, 1956.

Alfred Tarski. On the Concept of Logical Consequence. In J. H. Woodger, editor, *Logic, Semantics, Metamathematics*, pages 409–20. Clarendon Press, Oxford, 1956; first published in 1936.

Neil Tennant. *Natural Logic*. Edinburgh University Press, 1978.

Neil Tennant. Entailment and Proofs. *Proceedings of the Aristotelian Society*, LXXIX: 167–89, 1979.

Neil Tennant. A Proof-theoretic Approach to Entailment. *Journal of Philosophical Logic*, 9: 185–209, 1980.

Neil Tennant. Proof and Paradox. *Dialectica*, 36: 265–96, 1982.

Neil Tennant. Perfect Validity, Entailment and Paraconsistency. *Studia Logica*, XLIII: 179–98, 1984.

Neil Tennant. Minimal Logic is Adequate for Popperian Science. *British Journal for Philosophy of Science*, 36: 325–9, 1985.

Neil Tennant. Natural Deduction and Sequent Calculus for Intuitionistic Relevant Logic. *Journal of Symbolic Logic*, 52: 665–80, 1987a.

Neil Tennant. *Anti-Realism and Logic: Truth as Eternal*. Clarendon Library of Logic and Philosophy, Oxford University Press, 1987b.

Neil Tennant. Truth Table Logic, with a Survey of Embeddability Results. *Notre Dame Journal of Formal Logic*, 30: 459–84, 1989.

Neil Tennant. *Autologic*. Edinburgh University Press, Edinburgh, 1992.

Neil Tennant. Transmission of Truth and Transitivity of Proof. In Dov Gabbay, editor, *What is a Logical System?*, pages 161–77. Clarendon Press, Oxford, 1994a.

Neil Tennant. Intuitionistic Mathematics Does Not Need *Ex Falso Quodlibet*. *Topoi*, 13(2): 127–33, 1994b.

Neil Tennant. On Paradox without Self-Reference. *Analysis*, 55: 199–207, 1995.

Neil Tennant. Delicate Proof Theory. In Jack Copeland, editor, *Logic and Reality: Essays on the Legacy of Arthur Prior*, pages 351–85. Clarendon Press, Oxford, 1996a.

Neil Tennant. The Law of Excluded Middle is Synthetic A Priori, if Valid. *Philosophical Topics*, 24: 205–229, 1996b.

Neil Tennant. *The Taming of The True*. Oxford University Press, 1997.

Neil Tennant. Negation, Absurdity and Contrariety. In Dov Gabbay and Heinrich Wansing, editors, *What is Negation?*, pages 199–222. Kluwer, Dordrecht, 1999.

Neil Tennant. Ultimate Normal Forms for Parallelized Natural Deductions, with Applications to Relevance and the Deep Isomorphism between Natural Deductions and Sequent Proofs. *Logic Journal of the IGPL*, 10(3): 1–39, 2002.

Neil Tennant. Relevance in Reasoning. In Stewart Shapiro, editor, *The Oxford Handbook of Philosophy of Mathematics and Logic*, pages 696–726. Oxford University Press, New York, 2005.

Neil Tennant. Natural Logicism via the Logic of Orderly Pairing. In Sten Lindström, Erik Palmgren, Krister Segerberg, and Viggo Stoltenberg-Hansen, editors, *Logicism, Intuitionism, and Formalism: What has become of them?*, pages 91–125. Synthese Library, Springer Verlag, Dordrecht, 2009.

Neil Tennant. Inferential Semantics. In Jonathan Lear and Alex Oliver, editors, *The Force of Argument: Essays in Honor of Timothy Smiley*, pages 223–57. Routledge, London, 2010.

Neil Tennant. Cut for Core Logic. *Review of Symbolic Logic*, 5(3): 450–479, 2012a.

Neil Tennant. *Changes in Mind: An Essay on Rational Belief Revision*. Oxford University Press, Oxford, 2012b.

Neil Tennant. Aristotle's Syllogistic and Core Logic. *History and Philosophy of Logic*, 35(2): 120–47, 2014a.

Neil Tennant. Logic, Mathematics, and the *A Priori*, Part I: A Problem for Realism. *Philosophia Mathematica*, 22: 308–20, 2014b.

Neil Tennant. Logic, Mathematics, and the *A Priori*, Part II: Core Logic as Analytic, and as the Basis for Natural Logicism. *Philosophia Mathematica*, 22: 321–44, 2014c.

Neil Tennant. A New Unified Account of Truth and Paradox. *Mind*, 124 (494): 571–605, 2015a.

Neil Tennant. Cut for Classical Core Logic. *Review of Symbolic Logic*, 8(4): 236–56, 2015b.

Neil Tennant. On Gentzen's Structural Completeness Proof. In Heinrich Wansing, editor, *Dag Prawitz on Proofs and Meaning,* in the *Studia Logica* series *Outstanding Contributions to Logic*, pages 385–414. Synthese Library, Springer Verlag, Cham, Switzerland, 2015c.

Neil Tennant. The Relevance of Premises to Conclusions of Core Proofs. *Review of Symbolic Logic*, 8(4): 743–84, 2015d.

Neil Tennant. Rule-Irredundancy and the Sequent Calculus for Core Logic. *Notre Dame Journal of Formal Logic*, 57(1): 105–25, 2016.

Neil Tennant. Normalizability, Cut Eliminability and Paradox. *Synthese*, forthcoming a.

Neil Tennant. A Logical Theory of Truthmakers and Falsitymakers. In Michael Glanzberg, editor, *The Oxford Handbook of Truth*. Oxford University Press, New York, forthcoming b.

Neil Tennant. Truth, Provability and Paradox: On some theorems of Löb, Eubulides–Tarski, Montague and McGee; and a Conjecture about Constructivizability. under submission.

Alasdair Urquhart. The Undecidability of Entailment and Relevant Implication. *Journal of Symbolic Logic*, 49: 1059–73, 1984.

Jan von Plato. A Problem of Normal Form in Natural Deduction. *Mathematical Logic Quarterly*, 46(1): 121–4, 2000.

S. Yablo. Paradox without Self-Reference. *Analysis*, 53: 251–2, 1993.

Elia Zardini. A Model of Tolerance. *Studia Logica*, 90: 337–68, 2008.

Index

346